# Spectral Lines: The Theory of Line Shape in Astrophysics

# Spectral Lines: The Theory of Line Shape in Astrophysics

### Edited by
### Lisa Fisher

www.willfordpress.com

Published by Willford Press,
118-35 Queens Blvd., Suite 400,
Forest Hills, NY 11375, USA

ISBN: 978-1-64728-536-4

**Cataloging-in-Publication Data**

Spectral lines : the theory of line shape in astrophysics / edited by Lisa Fisher.
    p. cm.
Includes bibliographical references and index.
ISBN 978-1-64728-536-4
1. Spectrum analysis. 2. Astrophysics. 3. Spectral line broadening. I. Fisher, Lisa.
QC451 .S64 2023
535.84--dc23

For information on all Willford Press publications
visit our website at www.willfordpress.com

# Contents

# Preface

This book was inspired by the evolution of our times; to answer the curiosity of inquisitive minds. Many developments have occurred across the globe in the recent past which has transformed the progress in the field.

Astrophysics is a branch of space science, wherein the laws of physics and chemistry are applied in order to understand the universe and its components. Spectral line shape refers to the shape of a feature observed in spectroscopy that represents an energy change in an ion, atom or molecule. The shapes and widths of spectral lines are effective instruments for absorbing and emitting gas diagnostics in various astrophysical objects. The absorption and emission lines of astrophysical objects are formed in a variety of kinematical and physical circumstances and at a distance from the observer. Data on spectral lines and their profiles is crucial for modeling, diagnostics and analysis of laser development and design, fusion plasma, and plasma produced through laser. It is also important for various plasmas in technologies such as the piercing and welding of metals by laser-produced plasma and plasma based light sources. This book provides comprehensive insights into the theory of spectral line shapes in astrophysics. Also included herein is a detailed explanation of the various roles and applications of spectral line shapes in astrophysics. This book is a vital tool for all researching and studying this area of astrophysics.

This book was developed from a mere concept to drafts to chapters and finally compiled together as a complete text to benefit the readers across all nations. To ensure the quality of the content we instilled two significant steps in our procedure. The first was to appoint an editorial team that would verify the data and statistics provided in the book and also select the most appropriate and valuable contributions from the plentiful contributions we received from authors worldwide. The next step was to appoint an expert of the topic as the Editor-in-Chief, who would head the project and finally make the necessary amendments and modifications to make the text reader-friendly. I was then commissioned to examine all the material to present the topics in the most comprehensible and productive format.

I would like to take this opportunity to thank all the contributing authors who were supportive enough to contribute their time and knowledge to this project. I also wish to convey my regards to my family who have been extremely supportive during the entire project.

**Editor**

**1**

# Stark Broadening of Co II Lines in Stellar Atmospheres


Zlatko Majlinger [1], Milan S. Dimitrijević [1,2,*] and Vladimir A. Srećković [3]

[1] Astronomical Observatory, Volgina 7, 11060 Belgrade, Serbia; zlatko.majlinger@gmail.com
[2] Sorbonne Université, Observatoire de Paris, Université PSL, CNRS, F-92195 Meudon, France
[3] Institute of Physics Belgrade, University of Belgrade, 11001 Belgrade, Serbia; vlada@ipb.ac.rs
[*] Correspondence: mdimitrijevic@aob.bg.ac.rs

**Abstract:** Data for Stark full widths at half maximum for 46 Co II multiplets were calculated using a modified semiempirical method. In order to show the applicability and usefulness of this set of data for research into white dwarf and A type star atmospheres, the obtained results were used to investigate the significance of the Stark broadening mechanism for Co II lines in the atmospheres of these objects. We examined the influence of surface gravity (log g), effective temperature and the wavelength of the spectral line on the importance of the inclusion of Stark broadening contribution in the profiles of the considered Co II spectral lines, for plasma conditions in atmospheric layers corresponding to different optical depths.

**Keywords:** atomic data; stark broadening data; line profiles; Co II; white dwarfs; A-type stars

## 1. Introduction

The importance of Co II spectral lines, weak or strong (weak lines of Co II could help in the better adjustment of cobalt abundance measured on the basis of existing strong Co I lines), for the cobalt abundance determination in the spectra of A to F type stars, has been discussed elsewhere [1]. For this reason, Stark full widths at half maximum for 46 Co II multiplets have been calculated [2,3] to be helpful for astrophysical purposes. Calculation for all 46 multiplets were done using the modified semiempirical method (MSE) [4]. Stark broadening of spectral lines is the dominant broadening mechanism in the cases of high-temperature and dense plasma which can be found in hot star atmospheres. It is noticed that disregarding the Stark broadening effect in the process of spectral line synthesis can produce a worse fit of synthetic with observed spectral lines (see, for example, [5]), or can cause errors in abundance determination, especially for A-type stars ([6], for example).

In this paper, the applicability and usefulness of an electron-impact broadening dataset for Co II lines for investigations of white dwarf and A-type star atmospheres is analyzed. Stark broadening of the lines in the spectra of hot and dense celestial objects such as white dwarfs (WD), because of specific conditions of high electron density and high temperature in their atmospheres, usually dominates on Doppler broadening. Consequently, particular attention has been payed to hydrogen-rich (DA) and helium-rich (DB) types of WD, trying to figure out if a change in the physical conditions in their atmospheres, such as effective temperature or surface gravity, affects the relationship between thermal Doppler and electron-impact broadening for particular spectral lines.

## 2. Dataset and Methods of Research

Stark broadening theory has its application both in laboratory research as well as in astrophysical plasma [4,7–12]. For example, from the astrophysical point of view, Stark broadening data are always of interest when the Stark broadening contribution to the considered line profile is not negligible, such as in the cases of the interpretation, synthesis and analysis of stellar spectral lines, the determination of chemical abundances of elements from equivalent widths of absorption lines, the calculation of radiative transfer through stellar atmospheres and subphotospheric layers, opacity calculations, radiative acceleration considerations, nucleosynthesis research and other astrophysical topics. In the investigation of laboratory plasma, Stark broadening theory can help, for example, in plasma diagnostics, for the determination of the density and temperature of the plasma.

The importance of cobalt is equally present in technology as in astrophysical research. Cobalt is, for example, used in the preparation of magnetic and wear-resistant alloys. Lithium cobalt oxide as a cobalt compound is widely used in lithium ion battery cathodes. Cobalt-60 is a commercially important radioisotope, used as a radioactive tracer and as a source of high energy gamma rays.

From the perspective of astrophysical science, cobalt is important in the spectral analysis of so-called chemically peculiar (CP) stars. The main characteristic of these stars is anomalous strong or weak absorption lines in their spectra in comparison with the solar spectrum [13], so the investigations of those spectra are of particular interest for the modelling of CP star atmospheres as well as in the research of stellar evolution. The special part of these investigations is line shape modelling for comparison with actual measured spectra where lines of transition metal ions, such as singly ionized cobalt, are observed.

Thus, spectral lines of singly charged cobalt ion (Co II), for example, have been observed in Hg-Mn stars [14]. The persistence of large cobalt deficiency in the atmospheres of those objects, with metalicity of the order of −2 dex is noticed [15]. It is also very interesting to investigate another subgroup of CP stars, so-called cobalt stars (Co-stars), where an anomalous excess of cobalt abundance is observed in their spectra. Cobalt stars are mostly Ap-type, sometimes Bp-type, often having strong magnetic fields (5 kG or more). Examples of Co-stars are the Bp star HR 1094 [16], the Ap stars HD 200311 [17], HD 203932 [18] and possibly HD 208217 [19] and HR 4059.

Stellar iron, nickel and cobalt are also products of nuclear burning in a supernova event. Their strong absorption lines can be found in supernovae of types Ia and II [20] as a result of explosive nucleosynthesis. The stable form of cobalt is produced in supernovae through the so-called r-process, which occurs in their core-collapse and is responsible for the creation of approximately half of the neutron-rich atomic nuclei heavier than iron. The process entails a succession of rapid neutron captures (hence the name r-process) by heavy seed nuclei, typically $^{56}$Fe or other more neutron-rich heavy isotopes.

The first spectrum analysis of Co II was by Meggers [21], who measured the spectrum between 2150 and 5000 Å, and found eight multiplets and identified 14 lines of Co II in the solar spectrum. The analysis was extended by Findlay [22], Hagar [23], Velasco and Adames [24] and by Iglesias [25,26]. Iglesias commented that among the second spectra of the iron group elements, one of the most incompletely known spectra was that of Co II. The critical compilations of energy levels of Co II from more recent times which are also used in our calculations are from Sugar and Corliss [27] and Pickering et al. [28]. Pickering recorded high-resolution spectra of singly ionized cobalt by Fourier transform spectrometry in the region 1420–33,333 Å with cobalt-neon and cobalt-argon hollow cathode lamp as a source [28,29] and, therefore, it further contributed to the completion of the knowledge of these complex spectra.

The observed levels in Co II belong to two configuration systems. The "normal" system consists of $3d^7(^ML)nl$ subconfigurations, which are built on the parent terms $(^ML)$ in Co III, and transitions involving these levels dominate the emission spectrum of Co II. The subconfigurations $3d^6(^ML)4snl$ in the "doubly excited" system are built on the $(^ML)$ grandparent terms in Co IV, and they were not part of our interests. The Stark widths analyzed and used here [2,3] were calculated for multiplets created from a normal system of configurations, $3d^7(^ML)nl$, which is well known for $nl$ = 4 s and 4p, and according to observations those transitions are expected to be in pure LS coupling [28]. The predicted accuracy of the MSE method is around ±50 percent, but even in the cases of emitters with complex spectra, for example Xe II and Kr II, this method often gives better agreement with experiments, with relative error less than ±30 percent [30,31]. Of course, the used model also has some error bars, but our qualitative conclusions are confirmed with calculations using three different papers with model atmospheres for DA and DB white dwarfs and for A type stars. A high precision can not be achieved since we used the published models and included Stark broadening of spectral lines a posteriori. However, the presence of Stark broadening influence electron density and temperature and, consequently, on parameters of the model of atmosphere and for the best precision the Stark broadening data should be introduced a priori, during the calculation of model atmosphere.

For the purpose of this work, we chose four lines from the list of 46 Co II spectral lines for which Stark widths have already been calculated and published elsewhere [2,3], and we investigated if atmospheric layers with possible domination of Stark broadening over the thermal Doppler broadening for each of these four lines exist (Figures 1–6) To show this, different models of atmosphere of A-type star and DA and DB WD were used. Stark and Doppler broadening were presented as a function of optical depth or temperature of atmospheric layers. For investigation of this dependence, which is shown in Figure 3a,b, the Kurucz model of A spectral type of a star was used with the logarithm of surface gravity log g = 4.5 and effective temperature $T_{eff}$ = 10,000 K [32]. In the case of DA and DB dwarfs, the results of similar investigations are presented in Figures 1 and 2, using the model atmospheres from Wickramasinghe [33]. For the presentation of this dependence according to different $T_{eff}$ or log g for DB stars, appropriate model atmospheres from Koester were used [34].

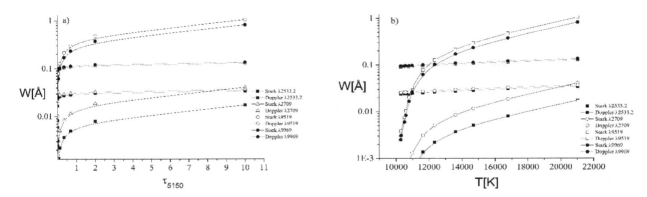

**Figure 1.** (a) Stark and Doppler broadening for spectral lines λ2533.2, λ2709, λ9519 and λ9969 as a function of optical depth in the atmosphere of a hydrogen-rich (DA) white dwarf. Model atmosphere with $T_{eff}$ = 15,000 K and log g = 8 is taken from [33]. (b) Same as Fig1a, but as a function of atmospheric layer temperature instead of optical depth.

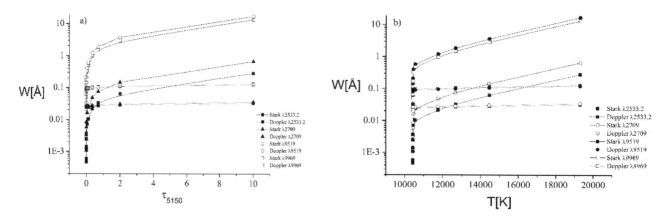

**Figure 2.** (**a**) Same as in Figure 1a, but for the model atmosphere of a helium-rich (DB) white dwarf [33], with same model parameters, $T_{eff}$ = 15,000 K and log g = 8. (**b**) Same as Figure 2a, but as a function of atmospheric layer temperature instead of optical depth.

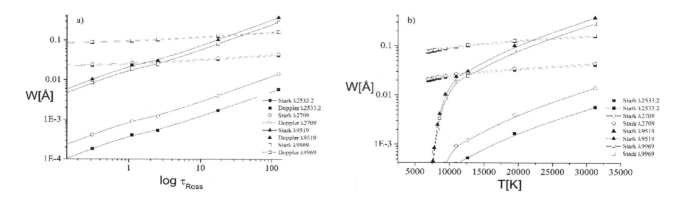

**Figure 3.** (**a**) Same as in Figures 1a and 2a, but as a function of logarithm of Rosseland optical depth, for the model atmosphere of A-type star [32] with model parameters log g = 4.5 and $T_{eff}$ = 10,000 K. (**b**) Same as Figure 3a, but as a function of atmospheric layer temperature instead of optical depth.

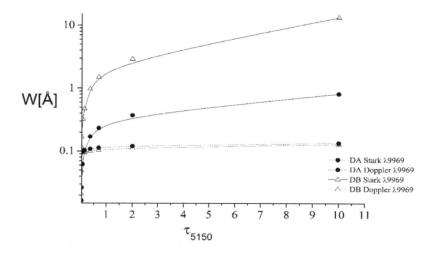

**Figure 4.** Comparison of Stark and Doppler broadening influence on Co II line λ9969 in the atmosphere of DA and DB white dwarfs, respectively, as a function of optical depth. Calculations have been performed for model atmospheres of DA and DB white dwarfs [33] with the same model parameters as in previous figures, $T_{eff}$ = 15,000 K and log g = 8.

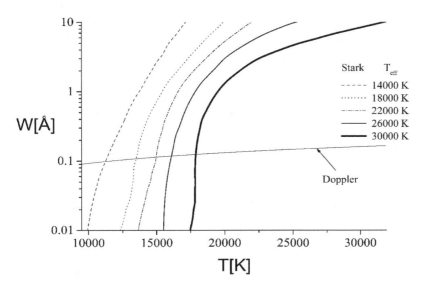

**Figure 5.** Stark and Doppler broadening of Co II spectral line λ9969 as a function of temperature of atmospheric layers in a DB white dwarf. Stark widths are shown for models [32] with five different values of effective temperature, $T_{eff}$ = 14,000–30,000 K and log g = 8.

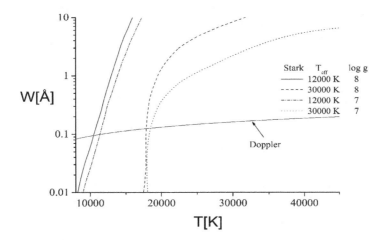

**Figure 6.** Stark and Doppler broadening of Co II spectral line λ9969 as a function of temperature of atmospheric layers in a DB white dwarf for two different values of model gravity, log g = 7 and log g = 8, each with two extremal values of effective temperatures, $T_{eff}$ = 12,000 K and $T_{eff}$ = 30,000 K.

## 3. Results and Discussion

In Figures 1–3, the comparisons of Stark widths and Doppler widths of λ2533.2, λ2709, λ9519 and λ9969 Co II spectral lines as a function of the optical depth in the white dwarf and A-star atmospheres are presented, to show in which layers of stellar atmosphere Doppler broadening caused by thermal motion of particles is dominated by Stark broadening caused by impacts of Co II ions with electrons.

In astrophysics, optical depth is a measure of the extinction coefficient or absorptivity, integrated from zero towards deeper layers up to a specific depth in stellar atmosphere. So, it is local characteristic as with electron temperature and it increases from zero towards deeper layers. Because it varies with wavelength, it is usually given for a standard wavelength of 5150 Å or as the Rosseland mean optical depth averaged over frequencies. Since we use published models of stellar atmospheres, we use optical depth as provided by authors of the models.

The first two lines, λ2533.2 and λ2709, from multiplets $(^4P)4s\ ^3P$–$(^4P)4p\ ^3D^o$ and $(^4F)4p\ ^3G^o$–$(^4F)5s$ $^3F$, respectively, are in the ultraviolet part of the spectrum, while the last two lines considered by us, λ9519 and λ9969, from multiplets $(^4F)5s\ ^5F$–$(^4F)5p\ ^5G^o$ and $(^4F)5s\ ^3F$–$(^4F)5p\ ^5F^o$, respectively, are in the infrared part of the spectrum. In Figure 1a,b, this analysis is done for DA WD model atmospheres [33] with parameters $T_{eff}$ = 15,000 K and log g = 8. Stark and Doppler broadening as a function of optical depth $\tau$ in the atmosphere at 5150 Å are shown in Figure 1a, and as a function of layer temperature in Figure 1b. The same comparisons but for DB white dwarf atmosphere model with the same parameters are shown in Figure 2a,b. In Figure 3a,b, we can see the behaviors in the function of logarithm of Rosseland optical depth and temperature in the stellar atmospheres for an A-type model atmosphere [32] with parameters log g = 4.5 and $T_{eff}$ = 10,000 K. Stark width in comparison with Doppler width increases as wavelength increases, because if a wavelength is larger than the corresponding atomic energy levels are closer and because of that, the perturbation of the emitter/absorber is larger and the emitted spectral line is broader. We notice also that Stark widths are proportional to $\lambda^2$, while Doppler widths are proportional to $\lambda$ [3]. For the last line, $\lambda9959$, the point where Stark width reaches Doppler width is deeper in the atmosphere than for the previous line, $\lambda9519$, because the Stark width values for this line are smaller since the corresponding atomic energy levels are further away than in the previous case and the perturbation is smaller.

We can see that for the hydrogen-rich (DA) type of WD, Stark broadening starts to be more significant than the Doppler broadening already in the atmospheric layers with relatively smaller optical depth, for spectral lines $\lambda9519$ and $\lambda9969$ near to $\tau \approx 0.5$. Electron-impact broadening for the line $\lambda2709$ becomes more significant for layers after $\tau \approx 10$, while for line $\lambda2533.2$ Doppler broadening is dominant for all considered values of optical depth. For the helium-rich (DB) type of WD, domination of Stark broadening for all four lines over the Doppler broadening starts before optical depth $\tau \approx 1$, where most of spectral lines are formed, so we can expect that electron-impact broadening for all three lines should be more important than thermal broadening in DB dwarf spectra. Difference between the importance of Stark broadening in comparison with Doppler broadening between DA and DB type of WD is in favor of DB type, because a helium-rich (DB) dwarf can generate more free electrons than the hydrogen-rich (DA) dwarf with the same density, causing higher perturber density [3]. This advantage in the domination of Stark width over Doppler width in the DB type in comparison with the DA type is also obvious from Figure 4, where these widths are presented as a function of optical depth.

From the same analysis for an A-type stellar atmosphere, we can see that for the spectral lines $\lambda9519$ and $\lambda9969$, Stark broadening also becomes the most significant broadening mechanism, but after reaching the deeper layers of the atmosphere, around optical depth of $\tau \approx 50$ and $\tau \approx 70$ respectively, e.g., for temperatures of atmospheric layers around 20,000 and 25,000 K, respectively. For the other two lines, Doppler broadening remains dominant even for layers with larger optical depth, e.g., higher temperatures of considered atmospheric layers.

It is obvious from Figures 1–4 that Stark broadening has larger impact on Co II spectral lines in the infrared spectral range, and this impact will be larger for DB dwarfs than on the rest of the considered objects. So, we decided to investigate how effective temperature and surface gravity of DB WD affect the relationship between Stark and Doppler widths for a particular spectral line.

In Figure 5, comparison of Stark and Doppler broadening of Co II line $\lambda9969$ in white dwarf atmospheres is presented as a function of layer temperature for five different models [34] of DB white dwarf atmospheres with effective temperatures from 14,000 to 30,000 K with a step of 4000 K, and log g = 8. The effective temperature is approximately taken as the temperature of the surface of the star. As effective temperature increases, the Stark broadening becomes more prominent in layers of the DB atmosphere, with temperatures which are more and more smaller than the effective temperature, because in these layers temperature becomes high enough to ionize helium more efficiently, so that electron density is higher. For example the difference between the effective temperature and temperature where Stark and Doppler broadening are approximately equal increases from the model with $T_{eff}$ = 14,000 K where it is several thousand kelvins to the model with $T_{eff}$ = 30,000 K, where it is larger than 10,000 K.

Finally, in Figure 6, this comparison for the same line is shown for four model atmospheres of DB white dwarfs [34] with effective temperatures $T_{eff}$ of 12,000 and 30,000 K, with two different values of log g for each temperature. We can see that electron-impact broadening becomes more important in DB white dwarf atmosphere than thermal broadening with the increase in surface gravity.

## 4. Conclusions

In this work, the usefulness and applicability of calculated set of data with Stark widths of 46 Co II lines for the investigations of spectra from atmospheres of stellar type A and hydrogen-rich (DA) and helium-rich (DB) white dwarfs are investigated. One can conclude that Stark broadening is very important for white dwarfs and for the same plasma conditions, its influence is larger for the DB than for the DA type. For A-type stars, Stark broadening may be non-negligible in comparison with thermal Doppler width, especially for higher wavelengths in the red part of the spectrum. Additionally, the influence of Stark broadening increases with the increase in the effective temperature and surface gravity analyzing as an example the DB type of WD.

We hope that the calculated set of 46 Co II Stark widths and these results will be useful for their use for hot star and WD spectroscopy, and also contribute to more accurate cobalt abundance determination. There are no other experimental or theoretical data for Stark broadening of Co II spectral lines analyzed here. As follows from our work, measurements of Stark broadening of Co II spectral lines will be of interest not only for comparison with the results obtained here but also for analysis and synthesis of stellar Co II spectral lines. This set of data, previously published as a hard copy in Ref. [3], is available online here in computer readable form. It will be implemented later and in the STARK-B database [35–38], which is also a part of the Virtual Atomic and Molecular Data Center (VAMDC) [39,40] and may be accessed through its portal [41].

**Author Contributions:** Conceptualization, M.S.D.; Formal analysis, Z.M. and V.A.S.; Validation, V.A.S.; Visualization, V.A.S.; Writing—original draft, Z.M.; Writing—review & editing, M.S.D. All authors contributed equally to this work. All authors have read and agreed to the published version of the manuscript.

## References

1. Adelman, S.J.; Golliver, A.D.; Lodén, L.O. On the cobalt abundances of early-type stars. *Astron. Astrophys.* **2000**, *353*, 335–338.
2. Majlinger, Z.; Dimitrijević, M.S.; Simić, Z. On the Stark broadening of Co II spectral lines. *Astron. Astrophys. Trans.* **2018**, *30*, 323–330.
3. Majlinger, Z.; Dimitrijević, M.S.; Srećković, V. Stark broadening of Co II spectral lines in hot stars and white dwarf spectra. *Mon. Not. R. Astron. Soc.* **2020**, *496*, 5584–5590. [CrossRef]
4. Dimitrijević, M.S.; Konjević, N. Stark widths of doubly- and triply-ionized atom lines. *J. Quant. Spectrosc. Radiat. Transf.* **1980**, *24*, 451–459. [CrossRef]
5. Chougule, A.; Przybilla, N.; Dimitrijević, M.; Schaffenroth, V. The impact of improved Stark-broadening widths on the modeling of double-ionized chromium lines in hot stars. *Contrib. Astron. Obs. Skalnaté Pleso* **2020**, *50*, 139–146. [CrossRef]
6. Popović, L.Č.; Milovanović, N.; Dimitrijević, M.S. The electron-impact broadening effect in hot star atmospheres: The case of singly- and doubly-ionized zirconium. *Astron. Astrophys.* **2001**, *365*, 656–659.
7. Baranger, M. Symplified Quantum Mechanical Theory of Pressure Broadening. *Phys. Rev.* **1958**, *111*, 481–493. [CrossRef]
8. Griem, H.R. Semiempirical Formulas for the Electron-Impact Widths and Shifts of Isolated Ion Lines in Plasmas. *Phys. Rev.* **1968**, *165*, 258–266. [CrossRef]
9. Griem, H.R. *Spectral Line Broadening BY Plasmas*; Academic Press: New York, NY, USA, 1974.

10. Sahal-Bréchot, S. Impact theory of the broadening and shift of spectral lines due to electrons and ions in a plasma. *Astron. Astrophys.* **1969**, *1*, 91–123.

11. Sahal-Bréchot, S. Impact theory of the broadening and shift of spectral lines due to electrons and ions in a plasma (continued). *Astron. Astrophys.* **1969**, *2*, 322–354.

12. Konjević, N. Plasma broadening and shifting of non-hydrogenic spectral lines: Present status and applications. *Phys. Rep.* **1999**, *316*, 339–401. [CrossRef]

13. Preston, G.W. The Chemically Peculiar Stars of the Upper Main Sequence. *Annu. Rev. Astron. Astrophys.* **1974**, *12*, 257–277. [CrossRef]

14. Bolcal, C.; Didelon, P. Comparison of Line Identifications. In *Elemental Abundance Analyses*; Adelman, S.J., Lanz, T., Eds.; Institut d'Astronomie de l'Université de Lausanne: Chavannes-des-Bois, Switzerland, 1988; pp. 152–153.

15. Smith, K.C.; Dworetsky, M.M. Elemental abundances in normal lateB and HgMn stars from co-added IUE spectra. I. Iron-peak elements. *Astron. Astrophys.* **1993**, *274*, 335–355.

16. Sadakane, K. A Chlorine-Cobalt Peculiar Star HR 1094. *Publ. Astron. Soc. Jpn.* **1992**, *44*, 125–133.

17. Adelman, S.J. The Peculiar a Star HD 200311: A Photographic Region Line-Identification Study. *Astrophys. J. Suppl. Ser.* **1974**, *28*, 51. [CrossRef]

18. Gelbmann, M.; Kupka, F.; Weiss, W.W.; Mathys, G. Abundance analysis of roAp stars. II. HD 203932. *Astron. Astrophys.* **1997**, *319*, 630–636.

19. Adelman, S.J.; Cowley, C.R.; Leckrone, D.S.; Roby, S.W.; Wahlgren, G.M. The Abundances of the Elements in Sharp-lined Early-Type Stars from IUE High-Dispersion Spectrograms. I. Cr, Mn, Fe, Ni, and Co. *Astrophys. J.* **1993**, *419*, 276–285. [CrossRef]

20. Jaschek, C.; Jaschek, M. *The Behaviour of Chemical Elements in Stars*; Cambridge University Press: Cambridge, UK, 1995.

21. Meggers, W.F. Multiplets in the Co II Spectrum. *J. Wash. Acad. Sci.* **1928**, *18*, 325–330.

22. Findlay, J.H. The Spark Spectrum of Cobalt, Co II. *Phys. Rev.* **1930**, *36*, 5–12. [CrossRef]

23. Hagar, N.E. Ph.D. thesis, Princeton University, Princeton, NJ, USA, 1951. (cited in Ref. [28]).

24. Velasco, R.; Adames, J. Nuevos niveles de energia del ion Co$^+$. *Anales* **1965**, *61A*, 269–274.

25. Iglesias, L. Niveles de energia del espectro Co II. *Opt. Pura Appl.* **1972**, *5*, 195–202.

26. Iglesias, L. Espectro Co II. *Opt. Pura Applicada* **1979**, *12*, 63–89.

27. Sugar, J.; Corliss, C. Atomic Energy Levels of the Iron-Period Elements: Potassium through Nickel. *J. Phys. Chem. Ref. Data* **1985**, *14*, 1–644.

28. Pickering, J.C.; Raassen, A.J.J.; Uylings, P.H.M.; Johansson, S. The Spectrum and Term Analysis of Co II. *Astrophys. J. Suppl. Ser.* **1998**, *117*, 261–311. [CrossRef]

29. Pickering, J.C. Analysis of 4d–4f Transitions in Co II. *Phys. Scr.* **1998**, *57*, 385–394. [CrossRef]

30. Popović, L.Č.; Dimitrijević, M.S. Stark broadening of Xe II lines. *Astron. Astrophys. Suppl. Ser.* **1996**, *116*, 359–365. [CrossRef]

31. Popović, L.Č.; Dimitrijević, M.S. Stark broadening parameters for Kr II lines from 5s–5p transitions. *Astron. Astrophys. Suppl. Ser.* **1998**, *127*, 295–297. [CrossRef]

32. Kurucz, R.L. Model atmospheres for G, F, A, B, and O stars. *Astrophys. J. Suppl. Ser.* **1979**, *40*, 1–340. [CrossRef]

33. Wickramarsinghe, D.T. Model atmospheres for DA and DB white dwarfs. *Mem. R. Astron. Soc.* **1972**, *76*, 129–179.

34. Koester, D. Model atmospheres for DB white dwarfs. *Astron. Astrophys. Suppl. Ser.* **1980**, *39*, 401–409.

35. Sahal-Bréchot, S.; Dimitrijević, M.S.; Moreau, N.; Ben Nessib, N. The STARK-B database VAMDC node: A repository for spectral line broadening and shifts due to collisions with charged particles. *Phys. Scr.* **2015**, *90*, 054008. [CrossRef]

36. Dimitrijević, M.S.; Sahal-Bréchot, S.; Moreau, N. The STARK-B Database, A Node of Virtual Atomic and Molecular Data Center (VAMDC). *Publ. Astron. Obs. Belgrade* **2018**, *98*, 285–288.

37. Sahal-Bréchot, S.; Dimitrijević, M.S.; Moreau, N. Virtual Laboratory Astrophysics and the STARK-B database VAMDC node: A resource for electron and ion impact widths and shifts of isolated lines. *J. Phys. Conf. Ser.* **2020**, *1412*, 132052. [CrossRef]

38. STARK-B. Available online: http://stark-b.obspm.fr (accessed on 12 July 2020).

39. Dubernet, M.; Boudon, V.; Culhane, J.; Dimitrijević, M.S.; Fazliev, A.; Joblin, C.; Kupka, F.; Leto, G.; Le Sidaner, P.; Loboda, P.; et al. Virtual atomic and molecular data centre. *J. Quant. Spectrosc. Radiat. Transf.* **2010**, *111*, 2151–2159. [CrossRef]

40. Dubernet, M.-L.; Antony, B.; A Ba, Y.; Babikov, Y.L.; Bartschat, K.; Boudon, V.; Braams, B.; Chung, H.-K.; Daniel, F.; Delahaye, F.; et al. The virtual atomic and molecular data centre (VAMDC) consortium. *J. Phys. B. At. Mol. Opt. Phys.* **2016**, *49*, 074003. [CrossRef]

41. VAMDC Portal. Available online: https://portal.vamdc.eu (accessed on 20 July 2020).

# 2

# Theoretical Stark Broadening Parameters for UV–Blue Spectral Lines of Neutral Vanadium in the Solar and Metal-Poor Star HD 84937 Spectra

**Cristóbal Colón** [1,*] ⬛, **María Isabel de Andrés-García** [1], **Lucía Isidoro-García** [2] and **Andrés Moya** [1]

[1] Department of Applied Physics, E.T.S.I.D. Industrial, Universidad Politécnica de Madrid, Calle Ronda de Valencia 3, 28012 Madrid, Spain; mariaisabel.deandres@upm.es (M.I.d.A.-G.); a.moya@upm.es (A.M.)

[2] Department of Industrial Chemistry and Polymers, E.T.S.I.D. Industrial, Universidad Politécnica de Madrid, Calle Ronda de Valencia 3, 28012 Madrid, Spain; lucia.isidoro@upm.es

\* Correspondence: cristobal.colon@upm.es

**Abstract:** Using Griem's semi-empirical approach, we have calculated the Stark broadening parameters (line widths and shifts) of 35 UV–Blue spectral lines of neutral vanadium (V I). These lines have been detected in the Sun, the metal-poor star HD 84937, and Arcturus, among others. In addition, these parameters are also relevant in industrial and laboratory plasma. The matrix elements required were obtained using the relativistic Hartree–Fock (HFR) method implemented in Cowan's code.

**Keywords:** atomic data; atomic processes; stark broadening

---

## 1. Introduction

Chemical elements belonging to the iron group, from Scandium to Zirconium (Z = 21–30), are critical for an understanding of nucleosynthesis in different super-nova types. In particular, Vanadium (together with Scandium and Titanium) is produced by explosive silicon burning and oxygen burning in the core-collapse supernova (SN) phase [1]. In recent works, this element has been analyzed in detail, especially in metal-poor stars [2]. In particular, Cowan et al. [3]—and references therein—find a correlation among Scandium, Titanium, and Vanadium in metal-poor stars. Therefore, the interest in Vanadium spectra has increased, and information about them is relevant for improving its abundance determination accuracy.

Among all the effects that have an impact on atomic spectral lines, the knowledge of the broadening and shift produced by charged particles is essential. Accurate measurements of Stark broadening and shift parameters are required to properly analyze astrophysical data. In the case of Vanadium, Manrique et al. [4] have provided experimental information related to the Stark broadening and shift parameters of the spectral lines of ionized Vanadium (V II).

In addition, the neutral Vanadium has been studied in detail. Infrared spectral lines of neutral vanadium (V I)—7363.1, 8027.3, 8255.8, 9037.6 Å, and so on—have been detected in the Solar spectrum, the spectrum of the metal-poor star HD 84937, and the spectrum of Arcturus ([5–7], respectively).

The levels of V I have been the subject of both experimental and theoretical studies. The earliest works were compiled by [8,9]. The most recent works [10–12] were collected in an exhaustive compilation, revision, and expansion of the V I levels that was carried out by Thorne et al. [13].

As this is an element of great interest, there are relatively recent publications in the literature with transition probabilities that are both theoretical and experimental. A critical analysis of these parameters can be found in [14].

A similar situation is found in the case of other parameters: There are excellent works devoted to the level lifetimes. The most recent are those of Hartog et al. [15], Wang et al. [16], and Holmes et al. [17].

However, despite its interest, we have not found data on the parameters of broadening and displacement of the spectral lines of neutral Vanadium by collision with electrons (Stark parameters) in the National Institute of Standards and Technology (NIST) database [18], nor in the Stark-B database [19].

In addition to its astrophysical interest, the role of vanadium in the corrosion protection of alloys of high industrial interest, as in the case of Ti-6Al-4V [20], is well known. In laser-generated plasma in Laser Sock Process (LSP) experiments with samples of the material mentioned above, intense spectral lines of neutral vanadium and single ionized vanadium can be seen [21,22].

Therefore, electron impact line widths and shifts of V I are relevant in the diagnostics of the plasma of stellar atmospheres, but they are also necessary in the analysis of the plasma in industrial processes.

In this work, electron impact line widths and shifts for 35 spectral lines of neutral vanadium have been calculated using a semi-empirical formalism [23]. The parameters are presented at an electron density of $10^{15}$ cm$^{-3}$. We have chosen that electron density because of its proximity to the densities that appear in the LSP experiments. To use them with other densities, they must be multiplied by the appropriate factor; the Stark parameters scale proportionally with the electron density in the semi-empirical approximation.

In the next section, we present our theoretical calculations, which, unlike in previous works, have used previous results of other authors. Below, we present our results for several level lifetimes (comparing them with experimental values obtained from the literature) and Stark broadening parameters. Regarding some lines of astrophysical interest, we show their Stark broadening parameters versus temperature in a graphical representation.

## 2. Materials and Methods

The semi-empirical formalism in which Griem considered the 1958 Baranger formulation [24] uses Equations (1) and (2), where $\omega_{se}$ and $d$ represent the Stark line width (half-width at half-maximum, HWHM) and the Stark line shift, respectively, in angular frequency units, $E_H$ is the hydrogen ionization energy, $E=(3/2)\,kT$ means the energy of the perturbing electron, $N_e$ is the free electron density, and $T$ is the electron temperature. The initial and final levels of the transitions are denoted by $i$ and $f$, respectively. In these equations, $g_{se}$ and $g_{sh}$ are the effective Gaunt factors proposed by Seaton [25] and Van Regemorter [26], respectively.

$$\omega_{se} \approx 8 \left(\frac{\pi}{3}\right)^{3/2} \frac{\hbar}{ma_0} N_e \left(\frac{E_H}{kT}\right)^{1/2} \left[\sum_{i'} |\langle i'|\,\vec{r}\,|i\rangle|^2\, g_{se}\left(\frac{E}{\Delta E_{i'i}}\right) + \sum_{f'} |\langle f'|\,\vec{r}\,|f\rangle|^2\, g_{se}\left(\frac{E}{\Delta E_{f'f}}\right)\right] \tag{1}$$

$$d \approx -8 \left(\frac{\pi}{3}\right)^{3/2} \frac{\hbar}{ma_0} N_e \left(\frac{E_H}{kT}\right)^{1/2} \left[\sum_{i'} \left(\frac{\Delta E_{i'i}}{|\Delta E_{i'i}|}\right) |\langle i'|\,\vec{r}\,|i\rangle|^2\, g_{sh}\left(\frac{E}{\Delta E_{i'i}}\right) - \sum_{f'} \left(\frac{\Delta E_{f'f}}{|\Delta E_{f'f}|}\right) |\langle f'|\,\vec{r}\,|f\rangle|^2\, g_{sh}\left(\frac{E}{\Delta E_{f'f}}\right)\right] \tag{2}$$

The expression $\Delta\lambda = \omega_{se}\lambda^2/\pi c$ was used to convert the units of angular frequency (obtained in Griem's expressions) into wavelength units. In this equation, $\Delta\lambda$ is the Stark broadening (full-width at half maximum, FWHM), $\lambda$ is the wavelength, and c is the light speed.

Cowan's Code [27] was used to obtain the necessary matrix elements included in the Equations (1) and (2). This code is a relativistic Hartree–Fock (HFR) approach that uses an intermediate coupling scheme (IC). It allows us to get a complete set of transition probabilities ($A_{ij}$) of the V I spectral lines by obtaining through them the required matrix elements.

Since the neutral vanadium is an element with a large number of known levels (346 odd levels and 198 even levels in the previously cited work [13]), we preferred not to repeat the very tedious calculations that had already been carried out successfully by this last author Thorne et al. [13]. We used the energies from Thorne et al. [13] and the transition probabilities from Cowan's code to determine the necessary matrix elements in Equations (1) and (2).

The calculations of the present authors were made using the following configurations: even parity, $3d^34s^2 + 3d^44s + 3d^45s + 3d^46s + 3d^44d + 3d^45d + 3d^5 + 3d^34s5s + 3d^34s6s + 3d^34s4d + 3d^34s5d + 3d^34p^2$, and odd parity, $3d^44p + 3d^45p + 3d^34s4p + 3d^34s5p + 3d^44f + 3d^34s4f + 3d^24s^24p$. The atomic parameters that were derived from these calculations and that were used (as input) in our work can be found in Table 6 of Thorne et al. [13].

As there are no previous experimental or theoretical data on the Stark broadening parameters of the spectral lines of the V I, we can only compare atomic data without the effects of the plasma environment, i.e., the intermediate results obtained for the transition probabilities or the level lifetimes. In this case, there are a lot of spectral lines with experimental transition probabilities to include in the comparison (which was made successfully) in our text. We preferred to test the lifetimes of the upper levels of the 35 spectral lines that we considered.

## 3. Results

Our results for the lifetimes of the upper levels corresponding to the 35 spectral lines with the Stark broadening parameters calculated in this work are presented in Table 1. The first three columns show the level energy, the configuration (as it appears in the NIST databases: using spin-orbit coupling notation and keeping the component with the highest weight), and the wavelengths of the transitions studied. The remaining columns show the values of the experimental level lifetimes found in the literature and the theoretical values obtained in our calculations.

As can be seen, the theoretical values obtained are close to the experimental values, with the noticeable exception of the lifetime value of $3d^44p^4I^o_{9/2}$, which is experimentally (25.8 ns) a factor of two greater than the theoretical value (12.37 ns) obtained in our calculations. This result contrasts with the results obtained for levels of the same multiplet, such as the $3d^44p^4I^o_{15/2}$ level, in which the results obtained are 12.3 and 11.9 ns, respectively.

A possible explanation is found by observing the theoretical values obtained with Cowan's code (ab initio) before adjusting the parameters with the experimental values of the energy levels. For the level $3d^44p^4I^o_{9/2}$, 17.04 ns was obtained, and for the level $3d^44p^4I^o_{15/2}$, 13.85 ns was obtained. This is a known effect: In this case, the energy-optimized Cowan method produces an undesired effect on the result of the transition probabilities for level $3d^44p^4I^o_{9/2}$. It is clearly observed that, in the 37,285 $cm^{-1}$ level, the relative weight of the vectors (in the spin–orbit coupling notation) $3d^44p^4I^o_{9/2}$, $3d^44p^2H^o_{9/2}$ and $3d^34s4p^2H^o_{9/2}$ changes slightly when adjustments have been made to the experimental levels, significantly affecting the level lifetime.

However, as noted above, our results for the transition probabilities and the oscillator strengths are close to the experimental results in almost all the experimental spectral lines. Stark broadening calculations that depend on all transition probabilities (as can be seen from Equations (1) and (2)) will certainly reflect this effect beyond the effect of any particular transition probability.

Our results for the Stark broadening parameters calculated in this work are presented in Table 2. The first column presents the wavelength and the experimental transition probability of the spectral line. Columns two and three present the transition levels' configurations. The remaining columns show the Stark line widths and shifts at 5, 10 , 20 and 30 · $10^3$K and at an electron density of $10^{15}$ $cm^{-3}$. It should be noted that Stark's change parameters could be poorly evaluated. This is due to the fact that the different addends in Equation (2) sometimes carry different signs and are, therefore, very sensitive to possible inaccuracies in the calculations of the matrix elements. This does not happen with the Stark magnification parameter when all terms are positive.

**Table 1.** Lifetimes of upper levels corresponding to spectral lines with the Stark broadening parameters calculated in this work.

| Energy Level (cm$^{-1}$) [a] | Configuration | Wavelength in Å [a] | Lifetimes (ns) | | |
|---|---|---|---|---|---|
| | | | Expt. [b] | Expt. [c] | This Work |
| 28,368 | $3d^34s4p^6D^o_{3/2}$ | 3823.21 | 35.9 | | 37.8 |
| 29,202 | $3d^34s4p^6P^o_{3/2}$ | 3690.27 3695.86 3705.03 | 6.7 | | 10.0 |
| 29,296 | $3d^34s4p^6P^o_{5/2}$ | 3692.21 3704.69 | 6.7 | 6.9(5) | 10.0 |
| 29,418 | $3d^34s4p^6P^o_{7/2}$ | 3688.06 | 6.7 | 7.0(5) | 9.98 |
| 32,846 | $3d^34s4p^4F^o_{5/2}$ | 3056.33 | 4.1 | | 4.78 |
| 32,988 | $3d^34s4p^4F^o_{7/2}$ | 3060.45 4095.47 | 4.2 | | 4.79 |
| 40,001 | $3d^34s4p^4G^o_{9/2}$ | 3207.40 | 3.3 | | 5.3 |
| 43,706 | $3d^34s4d\,^6H_{7/2}$ | 3667.73 | 5.9 | | 6.4 |
| 24,648 | $3d^44p^6P^o_{3/2}$ | 4444.20 | 28.4 | | 22.2 |
| 24,770 | $3d^44p^4P^o_{1/2}$ | 4412.14 | 24.7 | 24(1) | 23.6 |
| 24,830 | $3d^44p^6F^o_{3/2}$ | 4400.57 4421.56 | 9.2 | | 10.4 |
| 24,838 | $3d^44p^6P^o_{7/2}$ | 4419.93 4437.83 | 28.1 | | 22.5 |
| 24,898 | $3d^44p^6F^o_{5/2}$ | 4408.19 | 9.1 | | 10.38 |
| 25,253 | $3d^44p^6F^o_{11/2}$ | 4379.23 | 9.0 | | 10.26 |
| 26,122 | $3d^44p^4F^o_{7/2}$ | 4182.58 | 18.0 | | 14.2 |
| 26,249 | $3d^44p^4D^o_{3/2}$ | 3808.51 | 12.5 | | 13.5 |
| 26,397 | $3d^44p^6D^o_{1/2}$ | 4116.55 | 8.2 | | 9.1 |
| 26,505 | $3d^44p^6D^o_{5/2}$ | 4105.16 4116.47 | 8.1 | | 9.0 |
| 26,604 | $3d^44p^6D^o_{7/2}$ | 4099.78 4115.17 | 8.0 | | 9.1 |
| 26,738 | $3d^44p^6D^o_{9/2}$ | 4111.78 | 8.0 | | 9.1 |
| 38,115 | $3d^44p^4D^o_{7/2}$ | 3400.39 3533.67 | 5.5 | | 6.5 |
| 37,285 | $3d^44p^4I^o_{9/2}$ | 4468.00 | **25.8** | | 12.37 |
| 37,518 | $3d^44p^4I^o_{15/2}$ | 4452.00 | 12.3 | | 11.9 |
| 37,644 | $3d^44p^4G^o_{9/2}$ | 4560.71 | 12.3 | | 13.9 |
| 39,391 | $3d^44p^4F^o_{9/2}$ | 4232.45 | 9.5 | 5.7(4) | 8.0 |
| 41,860 | $3d^44p^4G^o_{9/2}$ | 4050.95 | 7.5 | | 7.3 |

[a]NIST , [b] Hartog et al. [15], [c] Wang et al. [16].

As an example, we present the Stark broadening parameters versus temperature for three intense spectral lines of neutral vanadium in Figure 1.

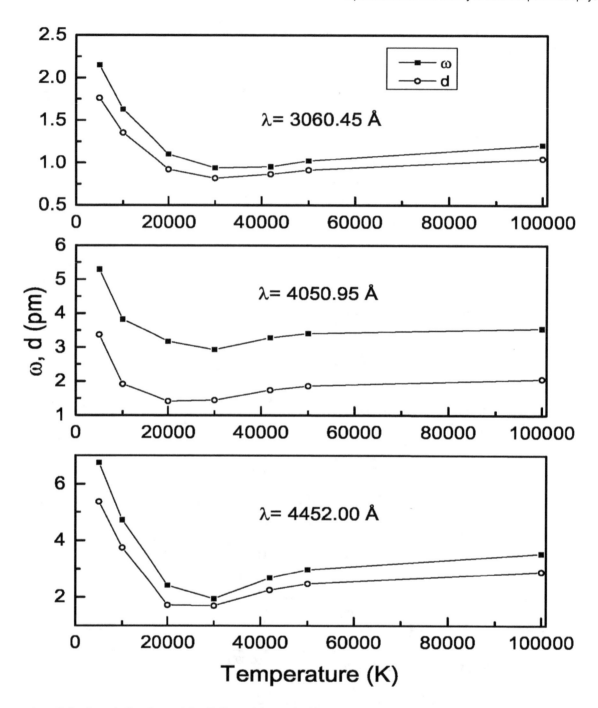

**Figure 1.** Calculated Stark width (full-width at half maximum (FWHM)) $\omega$ (pm), and shifts (d(pm)) normalized to $Ne = 10^{15}$ cm$^{-3}$ vs. temperature for spectral lines of neutral vanadium of astrophysical interest.

**Table 2.** Neutral vanadium (V I) line-widths (FWHM), $\omega$ (pm), and shifts (d(pm)) normalized to $Ne = 10^{15}$ cm$^{-3}$.

| Wavelength $\lambda$ (Å) [a] / Aij ($10^8$ s$^{-1}$) [a] | Transition Levels | | T ($10^3$K) | $\omega$ (pm) | d (pm) |
| --- | --- | --- | --- | --- | --- |
| | Upper | Lower | | | |
| 3056.33 | $3d^34s4p^4F^o_{5/2}$ | $3d^34s^{24}F_{5/2}$ | 5 | 1.61 | −1.32 |
| Aij =1.33 | | | 10 | 1.22 | −1.01 |
| | | | 20 | 0.82 | −0.69 |
| | | | 30 | 0.69 | −0.60 |
| 3060.45 | $3d^34s4p^4F^o_{7/2}$ | $3d^34s^{24}F_{7/2}$ | 5 | 2.15 | −1.76 |
| Aij =1.46 | | | 10 | 1.63 | −1.35 |
| | | | 20 | 1.10 | −0.92 |
| | | | 30 | 0.94 | −0.82 |
| 3207.40 | $3d^34s4p^4G^o_{9/2}$ | $3d^34s^{24}F_{9/2}$ | 5 | 2.94 | −2.56 |
| Aij = 0.218 | | | 10 | 2.02 | −1.74 |
| | | | 20 | 1.34 | −1.15 |
| | | | 30 | 0.92 | −0.76 |
| 3400.39 | $3d^44p^4D^o_{7/2}$ | $3d^44s^4D_{7/2}$ | 5 | 2.30 | −1.78 |
| Aij = 0.191 | | | 10 | 1.46 | −1.0 |
| | | | 20 | 0.99 | 0.60 |
| | | | 30 | 1.03 | −080 |
| 3533.67 | $3d^44p^4D^o_{7/2}$ | $3d^34s^{24}P_{5/2}$ | 5 | 2.39 | −1.83 |
| Aij = 0.63 | | | 10 | 1.51 | −1.02 |
| | | | 20 | 1.19 | −0.78 |
| | | | 30 | 1.09 | −0.84 |
| 3667.73 | $3d^34s4d\ ^6H_{7/2}$ | $3d^34s4p^6G^o_{5/2}$ | 5 | 3.34 | −2.81 |
| Aij = 1.46 | | | 10 | 1.89 | −1.52 |
| | | | 20 | 2.37 | −2.10 |
| | | | 30 | 2.24 | −2.19 |
| 3688.06 | $3d^34s4p^6P^o_{7/2}$ | $3d^44s^6D_{7/2}$ | 5 | 2.20 | −1.63 |
| Aij = 0.32 | | | 10 | 1.43 | −1.17 |
| | | | 20 | 1.01 | −0.71 |
| | | | 30 | 0.56 | −0.37 |
| 3690.27 | $3d^34s4p^6P^o_{3/2}$ | $3d^44s^6D_{1/2}$ | 5 | 0.89 | −0.60 |
| Aij = 0.44 | | | 10 | 0.57 | −0.43 |
| | | | 20 | 0.40 | −0.25 |
| | | | 30 | 0.23 | −0.14 |
| 3692.21 | $3d^34s4p^6P^o_{5/2}$ | $3d^44s^6D_{5/2}$ | 5 | 1.47 | −1.04 |
| Aij = 0.55 | | | 10 | 0.95 | −0.75 |
| | | | 20 | 0.67 | −0.44 |
| | | | 30 | 0.39 | −0.25 |
| 3695.86 | $3d^34s4p^6P^o_{5/2}$ | $3d^44s^6D_{7/2}$ | 5 | 1.11 | −0.82 |
| Aij = 0.62 | | | 10 | 0.72 | −0.59 |
| | | | 20 | 0.51 | −0.36 |
| | | | 30 | 0.27 | −0.18 |
| 3704.69 | $3d^34s4p^6P^o_{3/2}$ | $3d^44s^6D_{3/2}$ | 5 | 1.70 | −1.27 |
| Aij = 0.77 | | | 10 | 1.11 | −0.91 |
| | | | 20 | 0.78 | −0.56 |
| | | | 30 | 0.43 | −0.28 |
| 3705.03 | $3d^34s4p^6P^o_{3/2}$ | $3d^44s^6D_{5/2}$ | 5 | 1.34 | −1.05 |
| Aij = 0.418 | | | 10 | 0.89 | −0.75 |
| | | | 20 | 0.62 | −0.47 |
| | | | 30 | 0.31 | −0.21 |

**Table 2.** *Cont.*

| Wavelength $\lambda$ (Å) [a] $A_{ij}$ ($10^8$ s$^{-1}$) [a] | Transition Levels Upper | Lower | T ($10^3$K) | $\omega$ (pm) | d (pm) |
|---|---|---|---|---|---|
| 3808.51 | $3d^44p^4D^o_{3/2}$ | $3d^44s^4F_{3/2}$ | 5 | 1.54 | −1.31 |
| $A_{ij}$ = 0.143 | | | 10 | 1.01 | −0.85 |
| | | | 20 | 0.68 | −0.61 |
| | | | 30 | 0.61 | −0.55 |
| 3823.21 | $3d^34s4p^6D^o_{3/2}$ | $3d^44s^6D_{5/2}$ | 5 | 1.31 | −1.26 |
| $A_{ij}$ =0.164 | | | 10 | 0.91 | −0.88 |
| | | | 20 | 0.60 | −0.58 |
| | | | 30 | 0.21 | −0.20 |
| 4050.95 | $3d^44p^4G^o_{9/2}$ | $3d^44s^4G_{9/2}$ | 5 | 5.30 | −3.37 |
| $A_{ij}$ = 0.70 | | | 10 | 3.82 | −1.91 |
| | | | 20 | 3.18 | −1.41 |
| | | | 30 | 2.93 | −1.44 |
| 4095.47 | $3d^34s4p^4F^o_{7/2}$ | $3d^44s^4D_{5/2}$ | 5 | 3.37 | −2.67 |
| $A_{ij}$ = 0.42 | | | 10 | 2.58 | −2.08 |
| | | | 20 | 1.55 | −1.22 |
| | | | 30 | 1.50 | −1.29 |
| 4099.78 | $3d^44p^6D^o_{7/2}$ | $3d^44s^6D_{5/2}$ | 5 | 2.81 | −2.18 |
| $A_{ij}$ = 0.391 | | | 10 | 1.99 | −1.54 |
| | | | 20 | 1.08 | −0.76 |
| | | | 30 | 0.82 | −0.73 |
| 4105.157 | $3d^44p^6D^o_{5/2}$ | $3d^44s^6D_{3/2}$ | 5 | 2.03 | −1.56 |
| $A_{ij}$ = 0481 | | | 10 | 1.44 | −1.10 |
| | | | 20 | 0.77 | −0.53 |
| | | | 30 | 0.60 | −0.53 |
| 4111.778 | $3d^44p^6D^o_{9/2}$ | $3d^44s^6D_{9/2}$ | 5 | 3.92 | −3.12 |
| $A_{ij}$ = 1.0 | | | 10 | 2.77 | −2.20 |
| | | | 20 | 1.55 | −1.16 |
| | | | 30 | 1.11 | 1.0 |
| 4115.17 | $3d^44p^6D^o_{7/2}$ | $3d^44s^6D_{7/2}$ | 5 | 3.11 | −2.47 |
| $A_{ij}$ = 0.57 | | | 10 | 2.20 | −1.75 |
| | | | 20 | 1.22 | −0.91 |
| | | | 30 | 0.87 | −0.78 |
| 4116.47 | $3d^44p^6D^o_{5/2}$ | $3d^44s^6D_{5/2}$ | 5 | 2.32 | −1.84 |
| $A_{ij}$ = 0.215 | | | 10 | 1.64 | −1.30 |
| | | | 20 | 0.91 | −0.67 |
| | | | 30 | 0.64 | −057 |
| 4116.55 | $3d^44p^6D^o_{1/2}$ | $3d^44s^6D_{1/2}$ | 5 | 0.79 | −0.63 |
| $A_{ij}$ = 0.276 | | | 10 | 0.56 | −0.44 |
| | | | 20 | 0.31 | −0.23 |
| | | | 30 | 0.22 | −020 |
| 4182.58 | $3d^44p^4F^o_{7/2}$ | $3d^44s^6D_{5/2}$ | 5 | 3.06 | −2.40 |
| $A_{ij}$ =0.012 | | | 10 | 2.06 | −1.60 |
| | | | 20 | 1.33 | −1.17 |
| | | | 30 | 1.18 | −1.0 |
| 4232.45 | $3d^44p^4F^o_{9/2}$ | $3d^44s^4F_{9/2}$ | 5 | 5.37 | −3.47 |
| $A_{ij}$ = 0.69 | | | 10 | 3.37 | −2.37 |
| | | | 20 | 2.75 | −1.50 |
| | | | 30 | 2.45 | −1.34 |

**Table 2.** *Cont.*

| Wavelength $\lambda$ (Å) [a] Aij ($10^8$ s$^{-1}$) [a] | Transition Levels Upper | Transition Levels Lower | T ($10^3$ K) | $\omega$ (pm) | d (pm) |
|---|---|---|---|---|---|
| 4379.23 | $3d^44p^6F^o_{11/2}$ | $3d^44s^6D_{9/2}$ | 5 | 4.82 | −3.64 |
| Aij = 1.15 | | | 10 | 3.41 | −2.58 |
| | | | 20 | 1.75 | −1.16 |
| | | | 30 | 1.27 | −1.06 |
| 4400.57 | $3d^44p^6F^o_{3/2}$ | $3d^44s^6D_{1/2}$ | 5 | 1.41 | −1.01 |
| Aij = 0.347 | | | 10 | 1.0 | −0.72 |
| | | | 20 | 0.48 | −0.28 |
| | | | 30 | 0.39 | −0.32 |
| 4408.19 | $3d^44p^6F^o_{5/2}$ | $3d^44s^6D_{5/2}$ | 5 | 2.59 | 2.00 |
| Aij = 0.51 | | | 10 | 1.83 | −1.41 |
| | | | 20 | 0.96 | −0.66 |
| | | | 30 | 0.66 | −0.55 |
| 4412.136 | $3d^44p^4P^o_{1/2}$ | $3d^44s^6D_{1/2}$ | 5 | 0.93 | −0.71 |
| Aij = 0.0426 | | | 10 | 0.59 | −0.44 |
| | | | 20 | 0.36 | −0.33 |
| | | | 30 | 0.31 | −0.23 |
| 4419.93 | $3d^44p^6P^o_{7/2}$ | $3d^44s^6D_{5/2}$ | 5 | 3.10 | −1.73 |
| Aij = 0.0122 | | | 10 | 2.17 | −1.24 |
| | | | 20 | 1.52 | −0.38 |
| | | | 30 | 1.15 | −0.06 |
| 4421.56 | $3d^44p^6F^o_{3/2}$ | $3d^44s^6D_{5/2}$ | 5 | 2.06 | −1.66 |
| Aij = 0.133 | | | 10 | 1.46 | −1.17 |
| | | | 20 | 0.80 | −0.60 |
| | | | 30 | 0.50 | −0.42 |
| 4437.83 | $3d^44p^6P^o_{7/2}$ | $3d^44s^6D_{7/2}$ | 5 | 3.45 | −2.06 |
| Aij = 0.0836 | | | 10 | 2.42 | −1.48 |
| | | | 20 | 1.69 | −0.54 |
| | | | 30 | 1.21 | −0.11 |
| 4444.20 | $3d^44p^6P^o_{3/2}$ | $3d^44s^6D_{3/2}$ | 5 | 1.50 | −1.29 |
| Aij = 0.148 | | | 10 | 1.23 | −0.74 |
| | | | 20 | 0.85 | −0.27 |
| | | | 30 | 0.62 | −006 |
| 4452.00 | $3d^44p^4I^o_{15/2}$ | $3d^44s^4H_{13/2}$ | 5 | 6.75 | −5.37 |
| Aij = 0.81 | | | 10 | 4.72 | −3.75 |
| | | | 20 | 2.42 | −1.72 |
| | | | 30 | 1.94 | −1.70 |
| 4468.00 | $3d^44p^4I^o_{9/2}$ | $3d^44s^4H_{7/2}$ | 5 | 3.97 | −3.11 |
| Aij =0.174 | | | 10 | 2.78 | −2.18 |
| | | | 20 | 1.39 | −0.97 |
| | | | 30 | 1.16 | −1.0 |
| 4560.71 | $3d^44p^4G^o_{9/2}$ | $3d^44s^4F_{7/2}$ | 5 | 4.0 | −3.09 |
| Aij = 0.58 | | | 10 | 2.80 | −2.15 |
| | | | 20 | 1.39 | −0.96 |
| | | | 30 | 1.07 | −0.88 |

Note. A negative shift is towards the red, [a]*NIST.*

## 4. Conclusions

In this work, we presented theoretical Stark broadening parameters of 35 spectral lines of V I of astrophysical and industrial interest in the ultraviolet–blue range. It is the first time that these values, for which there are no experimental measurements, have been calculated. To get the required matrix elements, we used the transition probabilities obtained from Cowan's code (around 190,000). The theoretical lifetimes of the upper levels of these transitions are also shown.

**Author Contributions:** Conceptualization: C.C., M.I.d.A.-G., L.I.-G., and A.M.; Methodology: C.C., M.I.d.A.-G., L.I.-G., and A.M.; Investigation: C.C., M.I.d.A.-G., L.I.-G., and A.M.; Writing—review and editing: C.C. and L.I.-G. All authors have read and agreed to the published version of the manuscript.

**Acknowledgments:** This work was financially supported by the Spanish DGI project MAT2015- 63974-C4-2-R.

## References

1.    Woosley, S.E.; Weaver, A.T. The Evolution and Explosion of Massive Stars. II. Explosive Hydrodynamics and Nucleosynthesis. *Rev. Mod. Phys.* **1995**, *101*, 181.

2.    Ou, X.; Roederer, I.U.; Sneden, C.; Cowan, J.J.; Lawler, J.E.; Shectman, S.A.; Thompson, I.B. Vanadium Abundance Derivations in 255 Metal-poor Stars. *arXiv* **2020**, arXiv:2008.05500.

3.    Cowan, J.J.; Sneden, C.; Roederer, I.U.; Lawler, J.E.; Hartog, E.A.D.; Sobeck, J.S.; Boesgaard, A.M. Detailed Iron-peak Element Abundances in Three Very Metal-poor Stars. *Astrophys. J.* **2020**, *890*, 119. [CrossRef]

4.    Manrique, J.; Pace, D.M.D.; Aragón, C.; Aguilera, J.A. Experimental Stark widths and shifts of V II spectral lines. *Mon. Not. R. Astron. Soc.* **2020**. [CrossRef]

5.    Lawler, J.E.; Wood, M.P.; Hartog, E.A.D.; Feigenson, T.; Sneden, C.; Cowan, J.J. Improved V I log(gf) values and abundance determinations in the photospheres of the sun and metal-poor star hd 84937. *Astrophys. J. Suppl. Ser.* **2014**, *215*, 20. [CrossRef]

6.    Scott, P.; Asplund, M.; Grevesse, N.; Bergemann, M.; Jacques Sauval, A. The elemental composition of the Sun-II. The iron group elements Sc to Ni. *Astron. Astrophys.* **2015**, *573*, A26. [CrossRef]

7.    Wood, M.P.; Sneden, C.; Lawler, J.E.; Hartog, E.A.D.; Cowan, J.J.; Nave, G. Vanadium Transitions in the Spectrum of Arcturus. *Astrophys. J. Suppl. Ser.* **2018**, *234*, 25. [CrossRef]

8.    Moore, C.E. *Atomic Energy Levels*; NBS 467-11C; U.S.GPO: Washington, DC, USA, 1958.

9.    Sugar, P.; Corliss, C. Atomic Energy Levels of the Iron-Period Elements: Potassium through Nickel. *J. Phys. Chem. Ref. Data* **1985**, *14* (Suppl. 2), 203.

10.   Palmeri, P.; Biemont, E.; Aboussaid, A.; Godefroid, M. Hyperfine structure of infrared vanadium lines. *J. Phys. B At. Mol. Opt. Phys.* **1995**, *28*, 3741–3752. [CrossRef]

11.   Palmeri, P.; Biémont, E.; Quinet, P.; Dembczyński, J.; Szawiola, G.; Kurucz, R.L. Term analysis and hyperfine structure in neutral vanadium. *Phys. Scr.* **1997**, *55*, 586–598. [CrossRef]

12.   Lefèbvre, P.H.; Garnir, H.P.; Biémont, E. Hyperfine Structure of Neutral Vanadium Lines and Levels. *Phys. Scr.* **2002**, *66*, 363–366. [CrossRef]

13.   Thorne, A.P.; Pickering, J.C.; Semeniuk, J. The spectrum and term analysis of V I. *Astrophys. J. Suppl. Ser.* **2010**, *192*, 11. [CrossRef]

14.   Saloman, E.B.; Kramida, A. Critically Evaluated Energy Levels, Spectral Lines, Transition Probabilities, and Intensities of Neutral Vanadium (V i). *Astrophys. J. Suppl. Ser.* **2017**, *231*, 18. [CrossRef]

15.   Hartog, E.A.D.; Lawler, J.E.; Wood, M.P. Radiative lifetimes of V I and V II. *Astrophys. J. Suppl. Ser.* **2014**, *215*, 7. [CrossRef]

16.   Wang, Q.; Jiang, L.Y.; Quinet, P.; Palmeri, P.; Zhang, W.; Shang, X.; Tian, Y.S.; Dai, Z.W. TR-LIF lifetime measurements and HFR+CPOL calculations of radiative parameters in vanadium atom (V I). *Astrophys. J. Suppl. Ser.* **2014**, *211*, 31. [CrossRef]

17.   Holmes, C.E.; Pickering, J.C.; Ruffoni, M.P.; Blackwell-Whitehead, R.; Nilsson, H.; Engström, L.; Hartman, H.; Lundberg, H.; Belmonte, M.T. Experimentally measured radiative lifetimes and oscillator strengths in neutral vanadium. *Astrophys. J. Suppl. Ser.* **2016**, *224*, 35. [CrossRef]

18.   Kramida, A.; Ralchenco, Y.; Reader, J. NIST ASD Team (2013) NIST Atomic Spectra Database (v.5.3). 2013. Available online: http://physics.nist.gov/asd (accessed on 20 August 2020).

19. Sahal-Brechot, S.; Dimitrijevic, M.; Nessib, N.B.; Moreau, N. STARK-B. 2020. Available online: http://stark-b.obspm.fr/index.php/contact (accessed on 20 August 2020).

20. Iqbal, A.; Suhaimi, H.; Zhao, W.; Jamil, M.; Nauman, M.M.; He, N.; Zaini, J. Sustainable Milling of Ti-6Al-4V: Investigating the Effects of Milling Orientation, Cutterś Helix Angle, and Type of Cryogenic Coolant. *Metals* **2020**, *10*, 258. [CrossRef]

21. Carreón, H.; Barriuso, S.; González, J.A.P.; González-Carrasco, J.L.; Moreno, J.L.O. Thermoelectric assessment of laser peening induced effects on a metallic biomedical Ti6Al4V. In Proceedings of the Frontiers in Ultrafast Optics: Biomedical, Scientific, and Industrial Applications XIV, San Francisco, CA, USA, 2–5 February 2014; SPIE: Bellingham WA, USA, 2014; Volume 8972, pp. 89721Q-1–89721Q-7.

22. De Andrés García, M.I. Utilización de la línea H$\alpha$ del hidrógeno en la caracterización de los plasmas generados por láser para aplicaciones industriales (técnicas LSP) y espectroscópicas. Ph.D. Thesis, ETSIDI, Madrid, Spain, 2017. doi:10.20868/UPM.thesis.47301. [CrossRef]

23. Griem, H.T. Semiempirical formulas for the Electron-Impact widths and shifts of isolated ion lines in plasmas. *Phys. Rev.* **1968**, *165*, 258–266. [CrossRef]

24. Baranger, M. General Impact Theory of Pressure Broadening*. *Phys. Rev.* **1958**, *112*, 855–865. [CrossRef]

25. Seaton, M.J. The Theory of Excitation and Ionization by Electron Impact. In *Proceedings of the Atomic and Molecular Processes*; Bates, D.R., Ed.; Elsevier: Amsterdam, The Netherlands, 1962.

26. Van Regemorter, H. Rate of Collisional Excitation in Stellar Atmospheres. *Astrophys. J.* **1962**, *136*, 906. [CrossRef]

27. Cowan, R. *The Theory of Atomic Struture and Spectra*; University of California Press: Berkeley, CA, USA, 1981; p. 57. [CrossRef]

# On the Stark Broadening of Be II Spectral Lines

**Milan S. Dimitrijević** [1,2]🆔, **Magdalena Christova** [3,]* **and Sylvie Sahal-Bréchot** [2]🆔

[1]   Astronomical Observatory, Volgina 7, 11060 Belgrade, Serbia; mdimitrijevic@aob.rs
[2]   Sorbonne Université, Observatoire de Paris, Université PSL, CNRS, LERMA, F-92190 Meudon, France; sylvie.sahal-brechot@obspm.fr
[3]   Department of Applied Physics, Technical University–Sofia, 1000 Sofia, Bulgaria
*   Correspondence: mchristo@tu-sofia.bg

**Abstract:** Calculated Stark broadening parameters of singly ionized beryllium spectral lines have been reported. Three spectral series have been studied within semiclassical perturbation theory. The plasma conditions cover temperatures from 2500 to 50,000 K and perturber densities $10^{11}$ cm$^{-3}$ and $10^{13}$ cm$^{-3}$. The influence of the temperature and the role of the perturbers (electrons, protons and He$^+$ ions) on the Stark width and shift have been discussed. Results could be useful for plasma diagnostics in astrophysics, laboratory, and industrial plasmas.

**Keywords:** atomic data; atomic processes; line formation

## 1. Introduction

Atomic and spectroscopic data of light elements are of great importance in astrophysics and cosmology. LiBeB abundance is closely related to major questions concerning primordial nucleogenesis, stellar structure, mixing between atmosphere and interior, evolution, etc. [1–5]. It is well known that the abundance of the chemical elements versus the mass number is a notably decreasing curve [6,7]. The LiBeB trio is exceptional from this general trend in nature, because these elements are simple and rare. The mystery of lower light element abundance remains unresolved up to today [3]. What is the origin of these elements?; Are they generated in the normal course of stellar nucleosynthesis?; How they are destroyed?; these are all questions still unanswered. The three fragile light nuclei burn in the same $(p, \alpha)$ process and undergo nuclear reactions at relatively low temperatures, estimated at near 2.5, 3.5, and $5 \times 10^6$ K for densities similar to those in the Sun. In solar-type stars, these temperatures are reached not far below the convection zone and well outside the core, and circulation and destruction of the light elements can result in observable abundance changes. Observations of these changes can provide an invaluable probe of stellar structure and mixing. Be and B have been observed in extreme Population II with low Z, in a number of low metallicity halo dwarf stars [5]. Some papers suggest that observed Be and B were generated by cosmic-ray spallation in the early Galaxy, and the standard model of primordial nucleosynthesis is unable to produce significant yields of both light elements [5]. According to [3], there are no theoretical explanations for the reduction in the abundances, a trend of decreasing with effective temperature and a dip at Teff ~6600 K in F, G, and K-dwarfs that have been found in the Hyades and other old clusters. Interactions between emitting atoms and surrounding electrons and ions result in Stark broadening of spectral lines. This broadening mechanism of line profiles is usually a principle one in the case of white dwarfs, and is of interest for main sequence stars from A type and late B type, and sometimes dominant ones [8,9]. Stark broadening parameters of

Be II lines could serve effectively for the adequate modelling of stellar objects, opacity calculations, and diagnostics of astrophysical objects, laboratory and technological plasmas.

Previous Stark broadening calculations based on the semiclassical perturbation theory [10,11] of Be II lines have been performed in [12,13]. Here, an additional data set of Stark broadening parameters (widths and shifts) for Be II transitions, not included in [12,13], has been calculated. This dataset is used to compare the importance of the broadening due to electron-, proton-, and ionized helium-impacts, in function of temperature and to investigate the behavior of Stark broadening parameters within three spectral series.

## 2. Data Description and Method of Research

A dataset with new results for Stark broadening parameters of Be II spectral lines has been provided (Table S1). The calculations were performed using the semiclassical perturbation theory [10,11,14].

Within this theory, in the case of a charged emitter/absorber, full width at half intensity maximum (FWHM) $W$, and shift $d$ of an isolated spectral line originating from the transition between the initial level $i$ and the final level $f$ are given as:

$$
\begin{aligned}
W &= 2N \int_0^\infty v f(v)dv \left( \sum_{i' \neq i} \sigma_{ii'}(v) + \sum_{f' \neq f} \sigma_{ff'}(v) + \sigma_{el} \right) \\
d &= N \int_0^\infty v f(v)dv \int_{R_3}^{R_d} 2\pi \rho d\rho \sin(2\varphi_p)
\end{aligned}
\tag{1}
$$

where with $i'$ and $f'$ are denoted perturbing levels, $N$, $v$, and $f(v)$ are the perturber density, velocity, and the Maxwellian distribution of perturber velocities, respectively, and $\rho$ is the impact parameter of the perturber colliding with the emitter/absorber. The inelastic cross sections $\sigma_{kk'}(v)$, $k = i,f$, can be expressed by an integration of the transition probability $P_{kk'}(\rho,v)$:

$$
\sum_{i' \neq i} \sigma_{ii'}(v) = \frac{1}{2}\pi R_1^2 + \int_{R_1}^{R_d} 2\pi \rho d\rho \sum_{i' \neq i} P_{ii'}(\rho,v)
\tag{2}
$$

and the elastic contribution to the width is:

$$
\sigma_{el} = 2\pi R_2^2 + \int_{R_2}^{R_d} 2\pi \rho d\rho (\sin^2 \delta) + \sigma_r
\tag{3}
$$

$$
\delta = \left( \varphi_p^2 + \varphi_q^2 \right)^{1/2}
\tag{4}
$$

In the above Equations, $\sigma_{el}$ is the elastic cross section, while $\varphi_p$ and $\varphi_q$ are phase shifts due to the polarization and quadrupolar potential (see Section 3 of Chapter 2 in [10]). For the cut-offs $R_1$, $R_2$, and $R_D$, see Section 1 of Chapter 3 in [11]. The quantity $\sigma_r$ denotes the contribution of Feshbach resonances (see [14] and references therein), which concerns only electron-impact widths.

For isolated lines, the profile $F(\omega)$ has Lorentzian form:

$$
F(\omega) = \frac{W/(2\pi)}{\left(\omega - \omega_{if} - d\right)^2 + (W/2)^2}
\tag{5}
$$

where $\omega_{if} = E_i - E_f/\hbar$ and $E_i$, $E_f$ are the energies of the initial and final state. Therefore, if we know the Stark broadening parameters, width $W$ and shift $d$, we can determine the spectral line profile.

Here, we calculated Stark broadening parameters for three spectral series: $1s^24d-1s^2np$; $1s^24d-1s^2nf$ and $1s^24f-1s^2nd$, where $n = 6-8$. The temperature varied from 2500 K to 50,000 K and the perturber density was $10^{11}$ cm$^{-3}$ and $10^{13}$ cm$^{-3}$. The values of energy levels were taken from [15], and the oscillator strengths were calculated using [16]. For atoms such as beryllium an error around 20% was expected [17]. The dataset (Table S1) contains full Stark widths at half intensity maximum and shifts of Be II spectral lines due to collisions with electrons, protons and ionized helium, the main constituents of stellar atmospheres.

## 3. Results and Discussion

### 3.1. Series $1s^2 4d$–$1s^2 np$

#### 3.1.1. Temperature Dependence

Previous calculations [12,13] of Stark broadening parameters for the same series include results for one transition: $1s^2 4d$–$1s^2 5p$. In order to complete the data, using available energy values from [15], results for transitions from higher levels have been obtained. With the increase in the principal quantum number of the upper atomic energy level within a spectral series, the maximal perturber density for which the impact approximation is valid, decreases. For a density of $10^{13}$ cm$^{-3}$, it is valid for all electron-impact broadening parameters within the considered data set, but not for all other widths and shifts in the case of collisions with heavier particles, protons and helium ions. Therefore, we performed calculations for $10^{11}$ cm$^{-3}$ too, where impact approximation was valid for all perturbers within the considered data set. Moreover, these electron densities are typical for stellar atmospheres and for lower ones the linear extrapolation could be applied. In Figure 1a,b the temperature ($T$) dependence of Stark width and shift due to collisions with electrons, protons and ionized helium for $1s^2 4d$–$1s^2 6p$ transition have been illustrated.

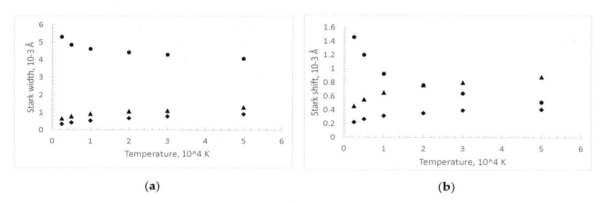

|      (a)      |      (b)      |

**Figure 1.** Stark broadening width (**a**) and shift (**b**) for multiplet $1s^2 4d^2 D$–$1s^2 6p^2 P°$ (6638.3 Å) versus temperature from different types of perturbers: electrons—circle; protons—rhombus; ionized helium ions—triangle. Perturber density was $1 \times 10^{13}$ cm$^{-3}$.

The dominant width was from the impacts with electrons. It decreased very slowly with the temperature. The broadening due to interactions with protons and He$^+$ ions were almost the same and they had the same trend, very slow increases with $T$. These values were notably lower (within an order) than the electron impact width. We observed the same trends with temperature for impact shifts, but the values for the three perturbers were much closer except for temperatures below 10,000 K, where the electron-impact shift started to dominate. For 20,000 K, the shifts due to electron- and He$^+$ ion-impacts were the same. For higher temperatures, the He$^+$ shifts were highest.

The Stark broadening parameters for $1s^2 4d$–$1s^2 8p$ transition are depicted in Figure 2a,b.

The trends of widths due to different perturbers with temperature are the same as in the previous case. The values converged with increasing temperature, particularly electron width and He$^+$ width. Comparing the shifts for the two transitions, we observed analogy in the behavior and differences in the values for different components. The impact component from He$^+$ ions was dominant for practically all temperatures. For temperatures above 15,000 K, proton shift also overcame the electron shift.

If we compare Stark broadening parameters for $n = 6$ and $n = 8$, we can see that the influence of collisions with protons and helium ions, compared to the collisions with electrons, increases for higher $n$, so that the shift becomes dominant with the increase in temperature.

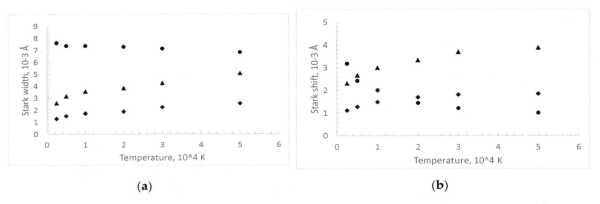

**Figure 2.** Stark broadening width (**a**) and shift (**b**) for multiplet 1s24d2D–1s28p2Po (4874.1 Å) versus temperature from different types of perturbers: electrons—circle; protons—rhombus; ionized helium ions—triangle. Perturber density was $1 \times 1013$ cm$^{-3}$.

### 3.1.2. Dependence on Principal Quantum Number

In Figure 3a,b dependence of Stark broadening parameters on the principal quantum number of the upper atomic energy level of the spectral series $1s^2 4d^2D–1s^2 np^2 P^o$, for $n = 5$–8 have been presented for electron-impact broadening. Stark broadening parameters, widths, and shifts have been expressed in angular frequency units and as decimal logarithms. The values for $n = 5$, have been extrapolated linearly to the electron density of $10^{11}$ cm$^{-3}$. Namely, from references [12,13], where Stark broadening parameters for perturber densities from $10^{13}$ cm$^{-3}$ to $10^{19}$ cm$^{-3}$ are provided, it follows that the dependence on the electron density is linear towards the lower densities. Their increasing with principal quantum number is very regular.

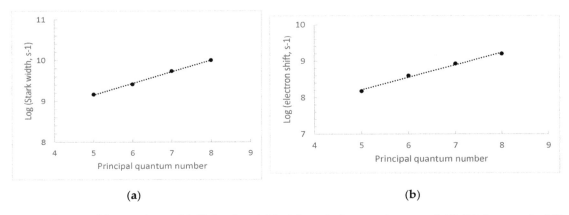

**Figure 3.** Decimal logarithm of full Stark width (**a**) and electron-impact shift (**b**) for spectral lines within the $2s^2 4d–2s^2 np$ (n = 5–8) spectral series versus principal quantum number. Electron density was $1 \times 10^{11}$ cm$^{-3}$ and temperature was $1 \times 10^4$ K. The Stark full width at half intensity maximum (FWHM) values for $n = 5$ have been taken from [13].

In the case of proton-, and helium ion-impacts, the behavior with the increase in $n$ was the same, while for the shift, we had the same behavior for $n = 6$–8. For $n = 5$ the proton-impact shift was negative (−0.101 Å [13]), as was the helium ion-impact shift (−0.0865 Å [13]).

We confirmed that for all three spectral series the behavior of widths and shifts with the increase in the principal quantum number was regular, and for all three considered series the result was the same and the behavior was very similar. Therefore, we can conclude that for all three Be II spectral series, this regular behavior can be used for interpolation and extrapolation to estimate the missing values within a considered series, and for confirmation of experimental and theoretical results: for all considered perturbations concerning Stark widths, and for electron-impact shifts.

*3.2. Series $1s^2 4d$–$1s^2 nf$*

Comparison of electron impact widths and shifts versus temperature for three transitions is illustrated in Figure 4a,b. The behavior of electron widths was the same: they decreased notably for lower temperatures up to 20,000 K and slowly for higher *T*. In accordance with the theory, the greater values corresponded to the higher transition. It is visible that the width's gradient slightly increased with principal quantum number for lower temperatures. All shifts were positive, red shifts. Figure 4b shows that shift trends were the same. Between temperatures 2500 K and 5000 K there was a big jump followed by decrease towards the higher temperatures. For temperatures above 10,000 K, electron shifts had practically the same values. As a difference from the line width, where contributions of virtual transitions to all energy levels were positive, in the case of shift we had a sum of positive and negative contributions, so that the behavior with temperature could be more complicated. In the considered series, the closest perturbing level to *nf* levels was *nd* level, with negative contribution to the shift. With the increase in temperature the role of a particular level decreases, therefore we first had an increase due to a relative decrease in negative contribution with an increasing role of levels contributing positively, and then the characteristic decreased with the increase in temperature.

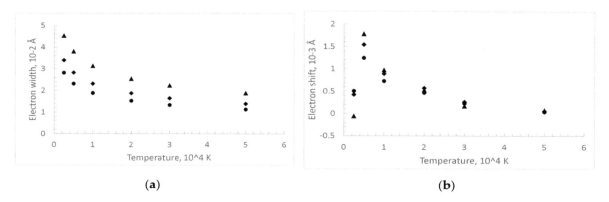

(a)                                                                                        (b)

**Figure 4.** Stark broadening parameters for electron-impacts: width (**a**) and shift (**b**) versus temperature for the $2s^2 4d$–$2s^2 nf$ series: $n = 6$—circle; $n = 7$—rhombus; $n = 8$—triangle. Perturber density was $1 \times 10^{13}$ cm$^{-3}$.

*3.3. Series $1s^2 4f$–$1s^2 nd$*

The temperature dependence of Stark broadening parameters for corresponding transitions is demonstrated in Figure 5a,b, respectively.

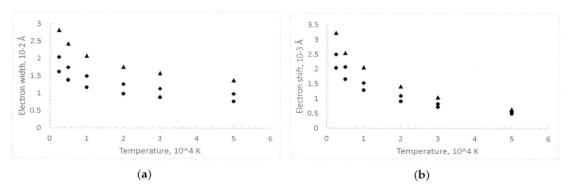

(a)                                                                                        (b)

**Figure 5.** Electron-impact width (**a**) and shift (**b**) versus temperature for the $2s^2 4f$–$2s^2 nd$ series: $n = 6$—circle; $n = 7$—rhombus; $n = 8$—triangle. Electron density is $1 \times 10^{13}$ cm$^{-3}$.

Both width and shift decreased with *T*. The width's decreasing was uniform for all three transitions, while shifts decreased and converged to practically the same value for 50,000 K.

If we compare the behavior of Stark broadening parameters within the spectral series $1s^24d$–$1s^2nf$ and $1s^24f$–$1s^2nd$ we can see that the behavior of widths due to collisions with electrons is very similar and that for all considered temperatures they increased with the increasing of the principal quantum number in a similar manner, and decreased uniformly with the increase in temperature. On the other hand, the shifts within these two series had different behavior, especially at a low temperature limit. However, in the case of the $1s^24f$–$1s^2nd$ series, shifts uniformly decreased with the increase in temperature; in the $1s^24d$–$1s^2nf$ series they considerably increased from $T = 2500$ K to $T = 5000$ K, and then uniformly decreased. A common characteristic for both series was a shift convergence with temperature. The width values did not converge.

## 4. Conclusions

Stark broadening parameters, FWHM and shifts due to impacts with electrons, protons and helium ions have been obtained for three spectral series in the Be II spectrum, by using semiclassical perturbation theory [10,11,14]. The calculations were performed for a temperature interval from 2500 to 50,000 K, and electron densities of $10^{11}$ and $10^{13}$ cm$^{-3}$. The dependence of Stark broadening parameters with temperature and the role of different perturbers (electrons, protons and He$^+$ ions) on the Stark width and shift have been discussed. Additionally, the regularity of behavior of Stark broadening parameters within the three considered spectral series was confirmed, and it was found that such regularities can be used for the interpolation and extrapolation of missing values and for a confirmation of experimental and theoretical results.

The obtained results may be of interest in astrophysics, especially for investigations of stellar atmospheres, and in particular for the problem of LiBeB abundances, closely related to major questions concerning primordial nucleogenesis, stellar structure, mixing between atmosphere and interior, as well as the stellar evolution, but also for stellar opacities, radiative transfer, stellar atmospheres modelling and analysis, and synthesis of stellar spectra. These data may be also useful for laboratory plasma diagnostics, modelling and investigation, and for the examination of regularities and systematic trends of Stark broadening parameters.

**Author Contributions:** Conceptualization, M.S.D. and M.C.; software, S.S.-B.; validation, M.S.D. and S.S.-B.; formal analysis, M.C. and M.S.D.; writing—original draft preparation, M.C. and M.S.D.; writing—review and editing, M.S.D. and M.C.; supervision, M.S.D. and S.S.-B. All authors have read and agreed to the published version of the manuscript.

**Acknowledgments:** Thanks to the Scientific and Research Sector of TU-Sofia for partial support.

## References

1.    Duncan, D.K.; Peterson, R.C.; Thorburn, J.A.; Pinsonneault, M.H. Boron abundances and internal mixing in stars. I the Hyades giants. *Astrophys. J.* **1998**, *499*, 871–882.
2.    Vangioni-Flam, E.; Cassé, M.; Audouze, J. Lithium-beryllium-boron: Origin and evolution. *Phys. Rep.* **2000**, *333*, 365–387.
3.    Lyubimkov, L.S. Light Chemical Elements in Stars: Mysteries and Unsolved Problems. *Astrophysics* **2018**, *61*, 262–285.
4.    Delbourgo-Salvador, P.; Vangioni-Flam, E. Primordial abunances of Be and B from standard Big-Bang nucleosynthesis. In *Origin and Evolution of Elements*; Prantzos, N., Vangioni-Flam, E., Casse, M., Eds.; Cambridge University Press: London, UK, 1993; pp. 132–138.
5.    Thomas, D.; Schramm, D.N.; Olive, K.A.; Fields, B.D. Primordial nucleosynthesis and the abundance of beryllium and boron. *Astrophys. J.* **1993**, *406*, 569–579.
6.    Burbige, E.M.; Burbidge, G.R.; Fowler, W.A.; Hoyle, F. Synthesis of the Elements *in* Stars. *Rev. Mod. Phys.* **1957**, *29*, 547–650.
7.    Vangioni-Flam, E.; Cassé, M. Cosmic Lithium-Beryllium-Boron story. *Astrophys. Space Sci.* **1999**, *265*, 77–86.
8.    Popović, L.Č.; Dimitrijević, M.S.; Ryabchikova, T. The electron-impact broadening effect in CP stars: The case of La II, La III, Eu II, and Eu III lines. *Astron. Astrophys.* **1999**, *350*, 719–724.

9.  Tankosić, D.; Popović, L.C.; Dimitrijević, M.S. The electron-impact broadening parameters for Co III spectral lines. *Astron. Astrophys.* **2003**, *399*, 795–797.
10. Sahal-Bréchot, S. Impact Theory of the Broadening and Shift of Spectral Lines due to Electrons and Ions in a Plasma. *Astron. Astrophys.* **1969**, *1*, 91–123.
11. Sahal-Bréchot, S. Broadening of ionic isolated lines by interactions with positively charged perturbers in the quasistatic limit. *Astron. ˇAstrophys.* **1969**, *2*, 322–354.
12. Dimitrijević, M.S.; Sahal-Bréchot, S. Stark broadening parameters tables for Be II lines for astrophysical purposes. *Bull. Astron. Belgrade* **1992**, *145*, 65–81.
13. Dimitrijević, M.S.; Sahal-Bréchot, S. Stark broadening of Be II spectral lines. *JQSRT* **1992**, *48*, 397–403.
14. Sahal-Bréchot, S.; Dimitrijević, M.S.; Ben Nessib, N. Widths and shifts of isolated lines of neutral and ionized atoms perturbed by collisions with electrons and ions: An outline of the semiclassical perturbation (SCP) method and of the approximations used for the calculations. *Atoms* **2014**, *2*, 225–252.
15. Kramida, A.E. Critical Compilation of Wavelengths and Energy Levels of Singly Ionized Beryllium (BeII). *Phys. Scr.* **2005**, *72*, 309–319.
16. Bates, D.R.; Damgaard, A. The calculation of absolute strength of spectral lines. *Philos. Trans. R. Soc. Lond. Ser. A* **1949**, *242*, 101–122.
17. Dimitrijević, M.S.; Sahal-Bréchot, S. Comparison of measured and calculated Stark broadening parameters for neutral-helium lines. *Phys. Rev. A* **1985**, *31*, 316–320.

# 4

# Monopole Contribution to the Stark Width of Hydrogenlike Spectral Lines in Plasmas

**Eugene Oks**

Physics Department, 380 Duncan Drive, Auburn University, Auburn, AL 36849, USA; goks@physics.auburn.edu

**Abstract:** One of the most reliable and frequently used methods for diagnosing various laboratory and astrophysical plasmas is based on the Stark broadening of spectral lines. It allows for determining from the experimental line profiles important parameters, such as the electron density and temperature, the ion density, the magnetic field, and the field strength of various types of the electrostatic plasma turbulence. Since, in this method, radiating atoms or ions are used as the sensitive probes of the above parameters, these probes have to be properly calibrated. In other words, an accurate theory of the Stark broadening of spectral lines in plasmas is required. In the present paper, we study, analytically, the monopole contribution to the Stark width of hydrogen-like spectral lines in plasmas. For this purpose, we use the formalism from paper by Mejri, Nguyen, and Ben Lakhdar. We show that the monopole contribution to the width has a non-monotonic dependence on the velocity of perturbing electrons. Namely, at relatively small electron velocities, the width decreases as the velocity increases. Then it reaches a minimum and (at relatively large electron velocities), as the velocity further increases, the width increases. The non-monotonic dependence of the monopole contribution to the width on the electron velocity is a *counter-intuitive result*. The outcome that at relatively large electron velocities, the monopole contribution to the width increases with the increase in the electron velocity is in a *striking distinction* to the dipole contribution to the width, which decreases as the electron velocity increases. We show that, in the situation encountered in various areas of plasma research (such as in magnetically-controlled fusion), where there is a relativistic electron beam (REB) in a plasma, the monopole contribution to the width due to the REB exceeds the corresponding dipole contribution by four orders of magnitude and practically determines the entire Stark width of hydrogenic spectral lines due to the REB.

**Keywords:** plasma spectroscopy; plasma diagnostics; Stark broadening; Stark width; monopole contribution

## 1. Introduction

One of the most reliable and frequently used methods for diagnosing various laboratory and astrophysical plasmas is based on the Stark broadening of spectral lines. It allows determining from the experimental line profiles such important parameters as, for example, the electron density and temperature, the ion density, the magnetic field, and the field strength of various types of the electrostatic plasma turbulence. Since, in this method, radiating atoms or ions are used as the sensitive probes of the above parameters, these probes have to be properly calibrated. In other words, an accurate theory of the Stark broadening of spectral lines in plasmas is required.

In particular, the theory of the Stark broadening of hydrogenlike spectral lines by plasma electrons was initially developed by Griem and Shen [1] (later being presented also in books [2,3]). In the literature it is frequently called the Conventional Theory (hereafter CT), sometimes also referred to as the standard theory. The assumption made in the CT was that the motion of the perturbing electron can be described in frames of a two-body problem—the perturbing electron moves along a hyperbolic trajectory around a "particle" of the charge $Z - 1$ (in atomic units).

In paper [4], the authors took into account that, actually, it is a three-body problem: the perturbing electron, the nucleus, and the bound electron, so that trajectories of the perturbing electrons are more complicated. They showed analytically by examples of the electron broadening of the Lyman lines of He II that this effect increases with the growth of the electron density $N_e$, becomes significant already at $N_e \sim 10^{17}$ cm$^{-3}$ and very significant at higher densities.

There were analytical advances beyond the CT, including the development of the so-called generalized theory of the Stark broadening of hydrogen-like spectral lines by plasma electrons [5]. Details can be found also in books [6,7] and references therein.

In all of the above works, the authors focused on the *dipole* interaction of the radiating ion with perturbing electrons. In distinction, in paper [8] the authors analytically studied the *shift* in hydrogen-like spectral lines due to the *monopole* interaction with plasma electrons.

In the present paper we use the formalism from paper [8] to analytically study the monopole contribution to the width of hydrogen-like spectral lines. We demonstrate that the monopole contribution to the width has a non-monotonic dependence on the velocity of perturbing electrons. Namely, at relatively small electron velocities, the width decreases as the velocity increases. Then it reaches a minimum and (at relatively large electron velocities), as the velocity further increases, the width increases.

## 2. Analytical Results

The monopole interaction potential can be represented as follows (Equation (3b) from paper [8])

$$V^{(0)}(t) = -e^2[1/|R(t) - 1/r] \, E[R(t) < r], \tag{1}$$

where R and r are the absolute values of the radii-vectors of the perturbing electron and of the bound electron, respectively; E[ ... ] is the Heaviside function manifesting the fact that the monopole interaction vanishes for $R(t) > r$.*/ According to Equation (17) from paper [8], for the Lyman lines, the monopole contribution to the shift, caused by $N_e$ electrons/cm$^3$ of velocity v, is given by

$$d_{nl \to 1s} = 2\pi N_e v \int_0^{\rho_{max}} \rho \, \sin[< nl|\Phi_0|nl> - <1s|\Phi_0|1s>] \, d\rho, \tag{2}$$

where the matrix elements of the electron broadening operator have the form

$$<nlm|\Phi_0|nl'm'> = -[e^2/(\hbar v)](1 + u_0)\{\ln[(1 + x)/(1 - x)] - 2x\} \, E[R(t) < r_{nl}] \, \delta_{ll'}\delta_{mm'}. \tag{3}$$

Here

$$x = [1 - (u^2 + u_0^2)/(1 + u_0)^2]^{1/2}, \, u = \rho/r_{nl}, \, u_0 = \rho_0/r_{nl}. \tag{4}$$

*/ The Heaviside function (also known as the step-function) in paper [8] and in the present paper is the mathematical embodiment of the vanishing monopole contribution for $R(t) > r$. It has been also previously used in other papers for the same purpose. For example, it was also employed in papers [9,10] devoted to the effect of penetrating collisions (corresponding to $R(t) < r$) on the shift of hydrogenic lines. As a result, the authors of papers [9,10] eliminated a huge discrepancy (up to an order of magnitude) between the theoretical shift and the shift observed from the laboratory and astrophysical sources. This means that the usage of the Heaviside function for this purpose provides a sufficient accuracy of the results.

In Equation (4), $\rho$ is the impact parameter and

$$\rho_0 = (Z-1)e^2/(m_e v^2), \quad r_{nl} = (<nl|r^2|nl>)^{1/2} = (a_0 n/Z)\{[5n^2 + 1 - 3l(l+1)]/2\}^{1/2}, \tag{5}$$

where $r_{nl}$ is the root-mean-square size of the radiating ion in the state of the quantum numbers $n$ and $l$, $a_0$ is the Bohr radius. Here are some useful practical formulas from paper [8]:

$$e^2/(\hbar v) = [13.605/kT_e(eV)]^{1/2}, \quad u_0 = [Z(Z-1)/n^2]\,[13.605/kT_e(eV)]. \tag{6}$$

As noted in paper [8], from the condition $R < r_{nl}$ it follows that

$$u_{max} = \rho_{max}/r_{nl} = (1 + 2u_0)^{1/2}. \tag{7}$$

Equation (7) is equivalent to

$$\rho_{max} = (r_{nl}^2 + 2\,r_{nl}\rho_0)^{1/2}. \tag{8}$$

In paper [8], it was noted that for relatively high temperatures, such that $e^2/(\hbar v) << 1$, one has $|<nlm|\Phi_0|nlm>| < 1$. In the opposite limit of relatively low temperatures, such that $u_0 >> 1$, the authors of paper [8] estimated that $|<nlm|\Phi_0|nlm>|$ does not exceed $(4/3)2^{1/2}n/[Z(Z-1)]^{1/2}$. Then, by limiting themselves to the range of parameters where $Z$ is no less than 5 and $n$ is no more than 4, the authors of paper [8] replaced, in Equation (2), $\sin[\ldots]$ by its argument.

In the present paper we are interested in the monopole contribution to the width $w^{(0)}$. For the Lyman lines, it can be represented by Equation (2) with $\sin[\ldots]$ replaced by $\cos[\ldots]$:

$$w^{(0)}{}_{nl \to 1s} = 2\pi N_e v \int_0^{\rho_{max}} \rho \, \cos[< nl|\Phi_0|nl > - < 1s|\Phi_0|1s >]\, d\rho, \tag{9}$$

In distinction to paper [8] we focus on the situation where $n >> 1$ (or practically $n > 4$), so that the contribution of the ground level can be disregarded, and Equation (9) simplifies to

$$w^{(0)}{}_{nl \to 1s} = 2\pi N_e v \int_0^{\rho_{max}} \rho \, \cos(< nl|\Phi_0|nl >)\, d\rho. \tag{10}$$

In a further distinction to paper [8], we do not limit ourselves by the case where $|<nlm|\Phi_0|nlm>| < 1$. Therefore, we keep the corresponding trigonometric function ($\cos[..]$) in the integrand in Equation (10).

By using the relation between $x$ and $\rho$ from Equation (4), we now, in Equation (10), proceed from the integration over $\rho$ to the integration over $x$:

$$w^{(0)}{}_{nl \to 1s} = 2\pi N_e v r_{nl}^2 (1 + u_0)^2 \int_0^y x \, \cos\{[e^2(1 + u_0)/(\hbar v)][\ln((1 + x)/(1 - x)) - 2x]\}, \tag{11}$$

where

$$y = (1 + 2u_0)^{1/2}/(1 + u_0). \tag{12}$$

We denote

$$A = (Z-1)a_B/r_{nl}, \quad B = \hbar v/e^2, \tag{13}$$

$B$ being the scaled dimensionless velocity of the perturbing electrons. Then, the width $w^{(0)}{}_{nl \to 1s}$ can be represented in the following final form

$$w^{(0)}{}_{nl \to 1s} = (2\pi N_e r_{nl}^2 e^2/\hbar)\, F[A, B], \tag{14}$$

where

$$F[A, B] = B\left(1 + A/B^2\right)^2 \int\limits_0^{y(A, B)} x \cos\{(1/B + A/B^3)[\ln((1 + x)/(1 - x)) - 2x]\}\, dx. \qquad (15)$$

The upper limit of the integration in Equation (15) is

$$y(A, B) = (1 + 2 A/B^2)^{1/2}/(1 + A/B^2) \qquad (16)$$

Thus, the dependence of the width $w^{(0)}_{nl \rightarrow 1s}$ on the scaled dimensionless electron velocity B is given by the function F(A, B). Figure 1 shows a three-dimensional plot of this function.

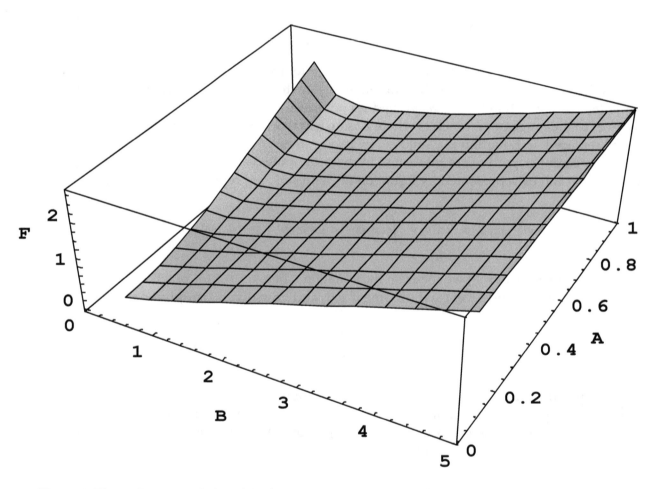

**Figure 1.** Three-dimensional plot of the function F(A, B) representing the dependence of the monopole contribution to the width $w^{(0)}_{nl \rightarrow 1s}$ (from Equation (14)) of the scaled dimensionless electron velocity B (defined in Equation (13)). The function F(A, B) is defined by Equations (15) and (16).

Figure 2 presents the dependence of the function F(A, B) on the scaled dimensionless electron velocity B for three values of the parameter A: A = 1 (solid line), A = 0.6 (dashed line), and A = 0.3 (dash-dotted line). Both from Figures 1 and 2, it is seen that the width $w^{(0)}_{nl \rightarrow 1s}$ has a non-monotonic dependence on B. Namely, at relatively small electron velocities, as B increases, the width decreases. Then, it reaches a minimum and (at relatively large electron velocities), as B further increases, the width increases. The non-monotonic dependence of the monopole contribution to the width on the electron velocity is a *counter-intuitive result*.

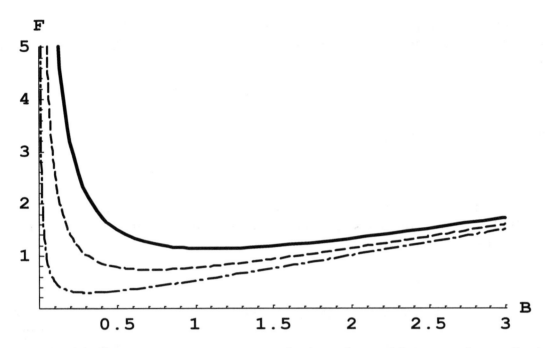

**Figure 2.** Plot of the function F(A, B), representing the dependence of the monopole contribution to the width $w^{(0)}_{nl \to 1s}$ (from Equation (14) on the scaled dimensionless electron velocity B (defined in Equation (13)) for three values of the parameter A: A = 1 (solid line), A = 0.6 (dashed line), and A = 0.3 (dash-dotted line). The function F(A, B) is defined by Equations (15) and (16).

For relatively large electron velocities—i.e., when B >> max(A, 1), the integration in Equation (15) becomes trivial and we get F(A, B) = B/2, so that

$$w^{(0)}_{nl \to 1s} = \pi r_{nl}^2 N_e v. \tag{17}$$

Physically this means that the corresponding optical cross-section—i.e., the cross-section for the line broadening collisions, becomes equal to the "geometrical" cross section $\pi r_{nl}^2$.

It is remarkable that, at relatively large electron velocities, the monopole contribution to the width increases with increasing velocity. This is in a *striking distinction* to the dipole contribution to the width, which decreases as the electron velocity increases.

For thermal velocities of plasma electrons, the parameter B (defined in Equation (13)) does not reach the range of B >> 1. However, there are situations where there is a relativistic electron beam (REB) in a plasma. Some examples are inertial fusion, heating of plasmas by a REB, acceleration of charged particles in plasmas, and generation of high-intensity coherent microwave radiation—see papers [11–13] and references therein.

Last but not least: in magnetic fusion research, one has sometimes to deal with a REB developing in the plasma. Namely, in some discharges in tokamaks, due to the phenomenon of runaway electrons, there occurs a decay in the plasma current and is partial replacement by runaway electrons that reach relativistic energies. This situation endangers the performance of the next generation tokamak ITER—see papers [14–16] and references therein.

In paper [17], the authors calculated, analytically, the *dipole* contribution $w_d$ to the Stark width of hydrogenic spectral lines due to a REB. Based on the results of paper [17] it can be estimated as follows:

$$w_d \sim N_{beam} c (n^2/Z)^2 \lambda_{Comp}^2 / (1 - 1/\gamma^2)^{1/2}, \qquad \lambda_{Comp} = \hbar/(m_e c), \qquad \gamma = 1/(1 - v^2/c^2)^{1/2}. \tag{18}$$

In Equation (18), $N_{beam}$ is the REB density, $\lambda_{Comp}$ is the Compton wavelength, and $\gamma$ is the relativistic factor. Using Equations (5), (17) and (18), we can estimate the ratio of the corresponding *monopole* contribution $w_m$ due to the REB to $w_d$ as follows:

$$w_m/w_d \sim (a_0/\lambda_{Comp})^2/(1 - 1/\gamma^2)^{1/2}. \tag{19}$$

For an ultra-relativistic REB—i.e., for $\gamma \gg 1$—Equation (19) simplifies to

$$w_m/w_d \sim (a_0/\lambda_{Comp})^2 = (\hbar c/e^2)^2 \sim 10^4 \gg 1. \tag{20}$$

It shows that the monopole contribution to the width due to the REB exceeds the corresponding dipole contribution by four orders of magnitude and practically determines the entire Stark width of hydrogenic spectral lines due to the REB.

## 3. Conclusions

By using the formalism from paper [8], we studied analytically the monopole contribution to the width of hydrogen-like spectral lines. We demonstrated that the monopole contribution to the width has a non-monotonic dependence on the velocity of perturbing electrons. Namely, at relatively small electron velocities, as the velocity increases, the width decreases. Then, it reaches a minimum and (at relatively large electron velocities), as the velocity further increases, the width increases. The non-monotonic dependence of the monopole contribution to the width on the electron velocity is a *counter-intuitive result*.

We showed, analytically, that at relatively large electron velocities, the so-called optical cross-section—i.e., the cross-section for the line broadening collisions—becomes equal to the "geometrical" cross section. We underscored that at relatively large electron velocities, the monopole contribution to the width increases with increasing velocity. This is in a *striking distinction* to the dipole contribution to the width, which decreases as the electron velocity increases.

Finally, we studied the situation, encountered in various areas of plasma research, where there is a relativistic electron beam (REB) in a plasma. We showed that the monopole contribution to the Stark width due to the REB exceeds the corresponding dipole contribution by four orders of magnitude and practically determines the entire Stark width of hydrogenic spectral lines due to the REB.

## References

1.   Griem, H.R.; Shen, K.Y. Stark broadening of hydrogenic ion lines in a plasma. *Phys. Rev.* **1961**, *122*, 1490. [CrossRef]
2.   Griem, H.R. *Plasma Spectroscopy*; McGraw-Hill: New York, NY, USA, 1964.
3.   Griem, H.R. *Spectral Line Broadening by Plasmas*; Academic Press: Cambridge, MA, USA, 1974.
4.   Sanders, P.; Oks, E.J. Allowance for more realistic trajectories of plasma electrons in the Stark broadening of hydrogenlike spectral lines. *Phys. Commun.* **2018**, *2*, 035033. [CrossRef]
5.   Oks, E.; Derevianko, A.; Ispolatov, Y.; Quant, J. A generalized theory of stark broadening of hydrogen-like spectral lines in dense plasmas. *Spectrosc. Rad. Transfer.* **1995**, *54*, 307. [CrossRef]
6.   . Oks, E. *Stark Broadening of Hydrogen and Hydrogenlike Spectral Lines in Plasmas: The Physical Insight*; Alpha Science International: Oxford, UK, 2006.
7.   Oks, E. *Diagnostics of Laboratory and Astrophysical Plasmas Using Spectral Lines of One-, Two-, and Three-Electron Systems*; World Scientific: Hackensack, NJ, USA, 2017.
8.   Mejri, M.A.; Nguyen, H.; Lakhdar, Z.B. Shifts of one-electron ions lines from n= 0 interactions with electrons in hot and dense plasma. *Europ. Phys. J. D* **1998**, *4*, 125. [CrossRef]

9.    Sanders, P.; Oks, E.E.J. Estimate of the Stark shift by penetrating ions within the nearest perturber approximation for hydrogenlike spectral lines in plasmas. *Phys. B Atom. Mol. Opt. Phys.* **2017**, *50*, 245002. [CrossRef]

10.   Oks, E.J. New source of the red shift of highly-excited hydrogenic spectral lines in astrophysical and laboratory plasmas. *Astrophys. Aerosp. Technol.* **2017**, *5*, 143.

11.   Guenot, D.; Gustas, D.; Vernier, A.; Beaurepaire, B.; Böhle, F.; Bocoum, M.; Losano, M.; Jullien, A.; Lopez-Martins, A.; Lifschitz, A.; et al. Relativistic electron beams driven by kHz single-cycle light pulses. *Nat. Photonics* **2017**, *11*, 293. [CrossRef]

12.   Kurkin, S.A.; Hramov, A.E.; Koronovskii, A.A. Microwave radiation power of relativistic electron beam with virtual cathode in the external magnetic field. *Appl. Phys. Lett.* **2013**, *103*, 043507. [CrossRef]

13.   de Jagher, P.C.; Sluijter, F.W.; Hopman, H.J. Relativistic electron beams and beam-plasma interaction. *Phys. Rep.* **1988**, *167*, 177. [CrossRef]

14.   Boozer, A.H. Runaway electrons and ITER. *Nucl. Fusion* **2017**, *57*, 056018. [CrossRef]

15.   Decker, J.; Hirvijoki, E.; Embreus, O.; Peysson, Y.; Stahl, A.; Pusztai, I.; Fülöp, T. Numerical characterization of bump formation in the runaway electron tail. *Plasma Phys. Control. Fusion* **2016**, *58*, 025016. [CrossRef]

16.   Smith, H.; Helander, P.; Eriksson, L.-G.; Anderson, D.; Lisak, M.; Andersson, F. Runaway electrons and the evolution of the plasma current in tokamak disruptions. *Phys. Plasmas* **2006**, *13*, 102502. [CrossRef]

17.   Oks, E.; Sanders, P.J. Stark broadening of hydrogen/deuterium spectral lines by a relativistic electron beam: Analytical results and possible applications to magnetic fusion edge plasmas. *Phys. Commun.* **2018**, *2*, 015030. [CrossRef]

# 5

# Stark-Zeeman Line Shape Modeling for Magnetic White Dwarf and Tokamak Edge Plasmas: Common Challenges

Joël Rosato [1,*], Ny Kieu [1], Ibtissem Hannachi [1], Mohammed Koubiti [1], Yannick Marandet [1], Roland Stamm [1], Milan S. Dimitrijević [2] and Zoran Simić [2]

[1] Laboratoire PIIM, Aix-Marseille Université, CNRS, 13397 Marseille Cedex 20, France; missny0909@gmail.com (N.K.); ibtissem.hannachi@yahoo.fr (I.H.); mohammed.koubiti@univ-amu.fr (M.K.); yannick.marandet@univ-amu.fr (Y.M.); roland.stamm@univ-amu.fr (R.S.)

[2] Astronomical Observatory, Volgina 7, 11060 Belgrade 38, Serbia; mdimitrijevic@aob.rs (M.S.D.); zsimic@aob.rs (Z.S.)

[*] Correspondence: joel.rosato@univ-amu.fr

Academic Editor: Ulrich D. Jentschura

**Abstract:** The shape of atomic spectral lines in plasmas contains information on the plasma parameters, and can be used as a diagnostic tool. Under specific conditions, the plasma located at the edge of tokamaks has parameters similar to those in magnetic white dwarf stellar atmospheres, which suggests that the same line shape models can be used. A problem common to tokamak and magnetic white dwarfs concerns the modeling of Stark broadening of hydrogen lines in the presence of an external magnetic field and the related Zeeman effect. In this work, we focus on a selection of issues relevant to Stark broadening in magnetized hydrogen plasmas. Various line shape models are presented and discussed through applications to ideal cases.

**Keywords:** line shapes; Stark broadening; Zeeman effect; tokamaks; white dwarfs

## 1. Introduction

Tokamaks are devices of toroidal shape employed in controlled nuclear fusion research for the confinement of a burning hydrogen plasma, with ion and electron temperatures in the order of $10^8$ K. The ITER project (www.iter.org) aims to demonstrate the principle of producing more energy from the fusion process than is used to initiate it; this large-scale tokamak is presently under construction in France, and it should be operational for first plasma discharge by the end of 2025 [1]. In order to support the operation of the machine, an extensive set of spectroscopic measurements is planned [2]; for example, passive spectroscopy of the Balmer series will provide information on the isotopic proportion of the fueling gas ($N_T/N_D$, $N_H/N_D$) in the divertor region [3], where a large amount of neutrals is expected. Under specific conditions (plasma in "detached" regime), the electron density in the divertor can attain values larger than $10^{14}$ cm$^{-3}$, so that the Stark effect due to the plasma microfield becomes the dominant line-broadening mechanism on lines with a high principal quantum number. Recent observations of D$\varepsilon$ (transition $7 \rightarrow 2$ of deuterium) in the divertor of ASDEX-Upgrade have been used for the determination of the electron density, using the Stark broadening of the line [4]. Balmer lines with a low principal quantum number like D$\alpha$ are also affected by the microfield and Stark broadening enters into competition with both Doppler broadening and Zeeman splitting [5]. The need for accuracy in spectroscopic diagnostics of tokamak edge plasmas has prompted an interest in the development of line broadening models accounting for the simultaneous action of electric and magnetic fields on atomic energy levels, e.g., [6]. Recently, a line shape database for the first Balmer lines accounting for the Stark and Zeeman effects has been devised for tokamak spectroscopy applications using computer simulations [7]. Similar databases for Stark broadening exist in astrophysics for stellar atmosphere diagnostic applications, e.g., STARK-B (stark-b.obspm.fr) [8]. Magnetic white dwarfs are of particular interest because they present Balmer spectra with Zeeman triplets [9,10] just as in tokamaks, and atmosphere models indicate zones where the electron density value is comparable to that in detached divertor plasmas [11]. Within this context, modeling work by tokamak spectroscopists in conjunction with astrophysicists aimed at improving Stark line shape databases is presently ongoing. In this paper, we report on models and techniques used for the description of Stark-Zeeman line shapes. A special emphasis is put on ion dynamics and collision operator models.

## 2. Stark Broadening Formalism

The formalism used in Stark line shape modeling involves atomic physics and statistical physics. An atom immersed in a plasma and emitting a photon is considered; during photon emission, the energy levels are perturbed due to the presence of the charged particles surrounding the atom, and this perturbation results in a splitting and broadening of spectral lines. The Doppler effect, which stems from the atom's motion, is not considered, hereafter, for the sake of clarity (it can be accounted for through convolution). According to standard textbooks (e.g., [12]), a line shape $I(\omega)$ is given by the Fourier transform of the atomic dipole autocorrelation function, a quantity described in the framework of quantum mechanics

$$I(\omega) = \frac{1}{\pi} \mathrm{Re} \int_0^\infty dt C(t) e^{i\omega t} \tag{1}$$

$$C(t) = \sum_{\alpha\alpha'\beta\beta'\varepsilon} \rho_{\alpha\alpha} (\mathbf{d}_{\alpha\beta} \cdot \varepsilon)(\mathbf{d}^*_{\alpha'\beta'} \cdot \varepsilon) \left\{ U_{\alpha'\alpha}(t) U^*_{\beta'\beta}(t) \right\} \tag{2}$$

Here, $\rho$ is the restriction of the density operator to the atomic Hilbert space evaluated at initial time, $\mathbf{d} \cdot \varepsilon$ is the dipole projected onto the polarization vector $\varepsilon$, the indices $\alpha$, $\alpha'$ (resp. $\beta$, $\beta'$) are used for matrix elements and denote states in the upper (resp. lower) level, the brackets { ... } denote an average over the perturber trajectories, and $U(t)$ is the evolution operator. It obeys the time-dependent Schrödinger equation

$$i\hbar \frac{dU}{dt}(t) = [H_0 + V(t)] U(t) \tag{3}$$

Here $H_0$ is the Hamiltonian accounting for the atomic energy level structure and $V(t) = -\mathbf{d}.\mathbf{F}(t)$ is the time-dependent Stark effect term resulting from the action of the microscopic electric field $\mathbf{F}(t)$ and written in the Schrödinger picture, here. When this term is neglected, the Schrödinger equation has the trivial solution $U(t) = \exp(-iH_0 t/\hbar)$, which shows, using Equations (2) and (1), that $I(\omega)$ reduces to a set of delta functions (or Lorentzian functions if the natural broadening is retained). By contrast, the case where $\mathbf{F}(t)$ is significant is much trickier, because there is no general exact analytical solution. The time-dependent perturbation theory yields a formal expansion (Dyson series), which is not applicable in explicit calculations because of the non-commutation of the interaction term at different times (time-ordering problem). This concerns, in particular, the microfield due to ions. Several models, based on suitable approximations, have been developed in such a way to provide an analytical expression for the line shape (e.g., the impact and static approximations, the model microfield method, etc.). On the other hand, computer simulations involving the numerical integration of the Schrödinger Equation (3) [13,14] provide reference line profiles and can serve as a benchmark for testing models. In the following, we focus on the ion dynamics issue, which is important for hydrogen lines with a low principal quantum number.

## 3. Ion Broadening at the Impact Limit

Modeling work has been done over the last decade, in order to set up an analytical formula for the Lyman $\alpha$ and Lyman $\beta$ line shapes for opacity investigations in tokamak divertor plasmas ([15] and Refs. therein). Kinetic simulations of radiation transport using the EIRENE code (www.eirene.de) have indicated the possibility of an alteration of the ionization-recombination balance due to the Lyman opacity [16]. The code employs a Monte Carlo method with random walks and related estimators; line shapes in the atomic frame of reference, i.e., without Doppler broadening, are involved in the sampling of photon frequencies. Because of the low principal quantum numbers ($n = 2$ and 3) involved in Lyman $\alpha$ and Lyman $\beta$, the ion dynamics is important, and a collision operator can be used for ions as a first approach for an estimate of the Stark broadening. The model of Griem, Kolb, and Shen (GKS, [17]), developed for electron broadening in hydrogen plasmas, can be adapted in a simple way for ions. Consider a single ion perturbing the emitter (binary interaction); the collision operator $K$ reads

$$K = N \int_0^\infty dv f(v) v \int_0^{\lambda_D} db 2\pi b \{1 - S\}_{\text{angle}} \tag{4}$$

where the integrals are done over the velocity (module) $v$ and the impact parameter (module) $b$, $N$ and $f$ stand for the perturber's density and velocity distribution function, the brackets denote an angular average and $S$ is the scattering matrix corresponding to a binary collision. The emitter motion is accounted for through a reduced mass model. The upper bound $\lambda_D$ is the Debye length, which accounts phenomenologically for the screening of the Coulomb field due to the electrons. In the GKS model, the integration domain is restricted in such a way that $S$ can be expanded at the second order in the Stark perturbation $V(t) = -\mathbf{d}.\mathbf{F}(t)$ ("weak collisions"); the integrals can be performed explicitly in the case of a Coulomb field and the resulting collision operator reads (notations are different from those in [17])

$$K_{\text{weak}} = N\pi b_W^2 v_0 \times \frac{2}{\sqrt{\pi}} \frac{r^2}{n^4} E_1 \left[ \left( \frac{b_W}{\lambda_D} \right)^2 \times \frac{r^2}{n^4} \right] \tag{5}$$

Here, $v_0 = \sqrt{(2k_B T / m_{\text{red}})}$ is the thermal velocity, accounting for reduced mass; $b_W = (2/3)^{1/2} \hbar n^2 / m_e v_0$ is the Weisskopf radius, which discriminates between weak ($b > b_W$) and strong ($b < b_W$) collisions (sometimes this is defined without the $(2/3)^{1/2}$ factor); $r^2$ is the square of the atomic electron position operator, in atomic units, restricted to the subspace relative to quantum number $n$, and can be formally replaced by its diagonal matrix element $(9/4)n^2(n^2 - l^2 - l - 1)$; and $E_1$ is the exponential integral function. Equation (5) represents the contribution of weak collisions only; these are usually dominant with respect to strong collisions, that the impact approximation is valid. An estimate of the strong collision contribution to $K$ can be done by setting formally $S \equiv 0$ in Equation (4) (this function strongly oscillates with respect to $b$ and $v$) and adjusting the integration domain in such a way that only impact parameters smaller than the Weisskopf radius are retained. An explicit calculation yields

$$K_{\text{strong}} = N\pi b_W^2 v_0 \times \frac{2}{\sqrt{\pi}} \frac{r^2}{n^4} \tag{6}$$

and the final expression for the collision operator is $K = K_{\text{weak}} + K_{\text{strong}}$. A Stark line shape described within the impact approximation can be written as a sum of Lorentzian functions with a half-width at half-maximum given by the matrix elements of the collision operator. In the framework of radiative transfer Monte Carlo simulations, this renders the sampling of photon frequencies straightforward. Figure 1 shows a plot of Lyman $\alpha$ at conditions of tokamak edge plasmas, calculated without Doppler broadening and using the GKS collision operator. The result is in contrasted to the result of a computer simulation, where the Schrödinger equation is solved numerically for a given set of microfield histories. These histories are simulated using particles moving in a cube with periodic boundary conditions (see [7] for details on the code). As can be seen, the impact approximation agrees well with the simulation result. Two essential features of the impact model are the assumptions of binary collision and short collision time; these features are characterized by the ratios $b_W/r_0$ and $\tau_c/t_i$, which must be much smaller than unity. Here, $r_0 \sim N^{-1/3}$ is the mean interparticle distance, $\tau_c \sim r_0/v_0$ is the collision time, and $t_i$ is the time of interest (which is estimated as the inverse line width). Here, the values of

$3 \times 10^{12}$ cm$^{-3}$ and 1 eV, assumed for density and temperature, respectively, yield values of 4% and 0.6% for these ratios.

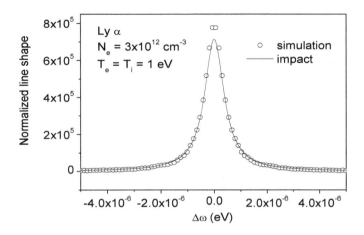

**Figure 1.** Deuterium Lyman $\alpha$ line profile in the atomic frame of reference, i.e., without Doppler broadening, in conditions such that the ratios $b_W/r_0$ and $\tau_c/t_i$ are much smaller than unity (see text). As can be seen, the impact theory agrees well with the simulation.

## 4. Influence of Zeeman Effect on Line Broadening

In tokamaks, the magnetic field is strong enough that hydrogen lines exhibit a Zeeman triplet structure. This structure is a feature of energy degeneracy removal relative to the magnetic quantum number $m$; it can alter the Stark broadening if the Zeeman perturbation $\hbar\omega_Z = |\mu_B B|$ ($\mu_B$ being the Bohr magneton) exceeds the characteristic Stark perturbation, estimated either by $n^2 ea_0 F_0$ (static picture, $F_0$ being the Holtsmark field) or $\hbar N \pi b_W^2 v_0 \ln(\lambda_D/b_W)$ (impact limit, the logarithm stems from the behavior of the exponential integral for small argument). An extension of the GKS impact collision operator that accounts for this degeneracy removal can be devised through the use of an alternative $S$ operator in Equation (4) involving the Zeeman Hamiltonian. The method employs an adaptation of the GBKO model (Griem, Baranger, Kolb, Oertel [18]), initially developed for helium lines, for hydrogen lines with Zeeman effect; details can be found in [6,19]. The result is a collision operator with a dependence on the $m$ quantum number. The following relation holds

$$\langle \alpha|K|\alpha \rangle = N \int_0^\infty dv f(v) v \sigma_\alpha(v) \tag{7}$$

$$\sigma_\alpha = \pi(b_\alpha^{\text{st}})^2 + 2\pi b_{W\alpha}^2 \left\{ K_\alpha^{//} \ln\left(\frac{\lambda_D}{b_\alpha^{\text{st}}}\right) + \left[K_\alpha^\perp a(s) - iK_\alpha'^\perp b(s)\right]_{s=\lambda_D/b_m}^{s=b_\alpha^{\text{st}}/b_m} \right\} \tag{8}$$

where $\alpha$ stands for the quantum numbers $n$, $l$, $m$; the constants $K_\alpha^{//} = \Sigma_{\alpha'} |z_{\alpha'\alpha}|^2/n_\alpha^4$, $K_\alpha^\perp = \Sigma_{\alpha'}[|x_{\alpha'\alpha}^+|^2 + |x_{\alpha'\alpha}^-|^2]/n_\alpha^4$, and $K'_\alpha^\perp = \Sigma_{\alpha'}[|x_{\alpha'\alpha}^+|^2 - |x_{\alpha'\alpha}^-|^2]/n_\alpha^4$ involve the atomic electron position operator (here, by convention, $x^\pm = (x \pm iy)/\sqrt{2}$); $a(s)$ and $b(s)$ are the GBKO functions writable in terms of the modified Bessel functions; $b_\alpha^{\text{st}}$ is the strong collision radius; and $b_m = v/\omega_Z$ is a distance relative to Zeeman degeneracy removal. The latter mitigates the characteristic strength of the Stark perturbation. If the ordering $b_{W\alpha} << b_m$ is satisfied, the $a$ function evaluated at the upper bound can be replaced by a logarithm, and the length $b_m$ plays a role similar to the Debye length. Note that the Weisskopf radius here is a function of $v$; the definition given in the previous section corresponds to evaluation at the thermal velocity $v = v_0$. In Equation (8), the dependence on velocity has not been written explicitly for the sake of simplicity. The so-called "strong collision radius"—here, $b_\alpha^{\text{st}}$—coincides with the Weisskopf radius at the limit B $\to$ 0, and is different at finite magnetic field. It can be determined by the solving of an integral equation; see [6,19] for details. The mitigation of Stark broadening due to the Zeeman effect is all the more important so that the Zeeman energy sublevels are separated. This effect is particularly important on the lateral components of Lyman $\alpha$ because they are not affected by $\Delta m = 0$ couplings (i.e., $K_\alpha{}^{//}$ vanishes identically in Equation (8)). Figure 2 shows an illustration of the Stark broadening mitigation on the lateral blue component. The same plasma conditions as in the previous section are considered, and a magnetic field of 2 T is assumed. The adaptation of the GBKO model yields a reduction of the line width by a factor of three. A slight shift towards the blue side is also visible; this corresponds to the imaginary part of the collision operator matrix element, and its amplitude is determined by the GBKO $b$ function.

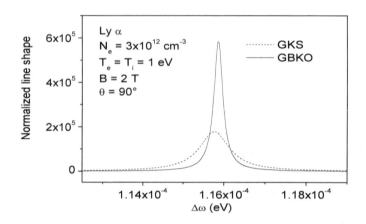

**Figure 2.** Lateral blue component of the Lyman $\alpha$ Zeeman triplet in the atomic frame of reference. A magnetic field of 2 T is assumed and the observation is assumed perpendicular to $B$ ($\theta = 90°$). "GBKO" refers to the Griem-Baranger-Kolb-Oertel 1962 model, adapted to hydrogen in the presence of Zeeman effect; "GKS" refers to the original Griem-Kolb-Shen 1959 model for hydrogen without degeneracy removal. As can be seen, the Zeeman degeneracy removal, retained in the GBKO model, strongly reduces the line width. A slight shift is also visible.

## 5. An Extension of the Impact theory to Non-Binary Interactions

The impact approximation for ions can be inaccurate under high density or low temperature conditions, namely, when the dimensionless parameters $<\alpha \mid K \mid \alpha> \times \tau_c$ and $b_W/r_0$ are significant. Comparisons with ab initio simulations have indicated that the use of the impact approximation can result in a systematic overestimate of the Stark width, e.g., [6]. Recently, a model has been developed in such a way to account for the deviations analytically [20]. The model uses kinetic theory techniques and is inspired by the "unified theory", a formalism proposed in the late sixties as an extension of the impact approximation for electrons [21,22]. The main result of the unified theory is a frequency-dependent collision operator, which accounts for incomplete collisions (i.e., collisions not achieved during the time of interest) and provides a correct asymptotic line shape formula. If we apply the unified theory to ions, the collision operator reads

$$K(\omega) = N \int_0^\infty dt e^{i\omega t} \left\{ V(t) e^{-iH_0 t/\hbar} Q(t) V(0) \right\} \tag{9}$$

where $V$ stands for the Stark interaction corresponding to a binary collision with an ion, N is the number of ions, the brackets { ... } denote statistical average, and $Q(t)$ is the evolution operator of the atom under the influence of one collision in the interaction picture. It obeys the Schrödinger equation and admits the Dyson series as a formal expression. Several techniques exist for the evaluation of the integral, either based on the exact solution or on the perturbation theory, as done for the impact collision operator. The behavior of the collision operator at small and large frequency detuning is obtained by asymptotic analysis. This shows that $K(\omega)$ behaves as $\mid \Delta\omega \mid^{-1/2}$ in the wings, and reduces to the impact collision operator at the center. For applications in tokamak edge and divertor conditions, it has been necessary to modify the original model Equation (9) in such a way as to account for N body interaction effects, referred to as "correlated collisions". Such correlations occur if, during one collision, the atom "feels" the presence of the other perturbers. Correlated collisions are important when the characteristic collision frequency $<\alpha | K | \alpha>$ becomes of the same order as, or larger than, the inverse correlation time of the emitter-perturber interaction potential [23,24]. The latter is of the order of the inverse plasma frequency $\omega_{pi}^{-1}$ (due to Debye screening). In the model reported in [20], correlated collisions are retained through a resummation procedure applied to kinetic equations of BBGKY-type. The resulting collision operator obeys a nonlinear equation with a structure similar to Equation (9)

$$K(\omega) = N\int_0^{\infty} dt e^{i\omega t}\left\{ V(t)e^{-i(H_0 - iK_0)t/\hbar}Q(t)V(0)\right\}\qquad(10)$$

In the exponential, $K_0 = K(\omega_0)$ is the collision operator evaluated at the central frequency of the line under consideration. Its presence denotes a non-Hermitian part in the atomic Liouvillian, which can be interpreted as a renormalization or "dressing" of the atomic energy levels, induced by the presence of other perturbers during a single collision and their correlation with the collision under consideration. This model presents similarities to the result of the resonance broadening theory used for plasma turbulence [25,26]. In practice, a calculation of the collision operator from Equation (10) should be done by iterations. Tests have indicated that such a procedure yields a fast convergence, with typically no more than ten iterations. Figure 3 presents a plot of Lyman $\alpha$ broadened due to ions at $N_e = 10^{15}$ cm$^{-3}$, $T_e = T_i = 1$ eV, which was obtained using impact approximation, the unified theory Equation (9), and its extension Equation (10) applied to ions. No magnetic field is retained here. A numerical result from a simulation is also shown in the figure. As can be seen, the impact approximation strongly overestimates the width. This stems from the inadequacy of this model as a regime where incomplete and correlated collisions are present. The extended unified theory gives a much better result, with an overestimate of the width no larger than 20%. We can also see that the standard unified theory (i.e., not accounting for correlated collisions) is not sufficient here, and yields a different (and incorrect) shape structure, with a dip at the center. Correlated collisions are important in this region, because they govern the average atomic evolution operator at long times; hence, by virtue of the Fourier transform, at small frequency detuning. The typical range for the dip corresponds to frequencies smaller than the matrix elements of $K_0$.

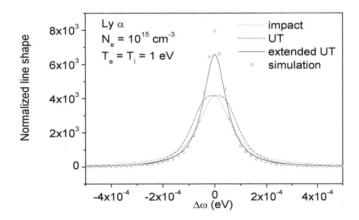

**Figure 3.** Profile of the Lyman $\alpha$ line at conditions such that neither the impact approximation nor the unified theory (UT) for ions are valid. The extension of the unified theory (UT), which accounts for correlated collisions, provides a good estimate of the line width.

## 6. Stark-Zeeman Line Shapes in Magnetic White Dwarfs

About 10% of white dwarfs are known to have a magnetic field strength of $10^5$ to $5 \times 10^8$ G (10 to $5 \times 10^4$ T), as indicated by spectroscopic observations and models [9,10,27]. An interpretation of the shape of absorption lines requires the Zeeman effect be accounted for in line broadening models, as done in tokamak edge plasma spectroscopy. Work is currently ongoing in order to improve already-existing databases by implementing the Zeeman effect in Stark line shape models. Figure 4 shows an illustration of the importance of the Zeeman effect in white dwarf spectra. A profile of H$\alpha$ has been calculated with a numerical simulation assuming $N_e = 10^{17}$ cm$^{-3}$, $T_e = T_i = 0.5$ eV, $B = 500$ T and observation perpendicular to $B$. These conditions are characteristic of the atmosphere of a white dwarf of DAH (magnetic DA) type, presenting hydrogen absorption lines. The spectrum exhibits a triplet structure characteristic to the Zeeman effect. The description of Stark broadening can be more challenging than in tokamak devices because particle correlations are important (here $r_0/\lambda_D$ is of the order of unity) and the quasiparticle model used in the simulation may not be accurate enough. An extension of this work would consist of using full molecular dynamic simulations. Features specific to dense plasmas, such as quadrupole interactions, could also be important, and should be addressed in detail; see [28] for recent works. The presence of a strong magnetic field changes the plasma dynamics through the gyromotion effect and this can result in an alteration of electron collision operator formulas [29,30]. Specific modifications to the atomic energy level structure (quadratic Zeeman effect, Lorentz electric field) could also be important [31].

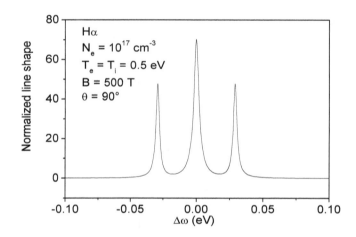

**Figure 4.** Balmer $\alpha$ line shape in conditions relevant to DAH white dwarf atmospheres. The spectrum exhibits a triplet structure due to the strong magnetic field.

## 7. Non-Hydrogen Species

Tokamak plasmas contain a sufficient amount of impurities, due to plasma-wall interactions or vacuum chamber conditioning, to be observable on spectra (e.g., oxygen, nitrogen, carbon [32]). Specific atomic species of high Z can be found in some, but not all, tokamaks (e.g., tungsten [33]). Helium is also expected to be present in high amounts in ITER during the nuclear phase, where discharges with tritium will be done. As a rule, an analysis of the position and the intensity of impurity lines provides information on the plasma composition, the recycling and sputtering processes, which helps in the setting up of tokamak discharge scenarios. The design of fast routines employs line shape codes (e.g., PPP [34]), and can also use databases available in astrophysics such as STARK-B (stark-b.obspm.fr). An issue that is specific to non-hydrogen lines concerns the modeling of collisional electron broadening, accounting for degeneracy removal with quantum number $l$. This degeneracy removal yields a reduction of the Stark effect and can induce a shift on the line under consideration. An illustration of this effect is shown in Figure 5. The He 492 nm line (1s4d $^1$D–1s2p $^1$P) has been addressed, assuming $N_e = 10^{15}$ cm$^{-3}$ and $T_e = 10,000$ K, without ion broadening and without magnetic field for the sake of clarity. Two Lorentzian curves are displayed in the graph. One corresponds to the result of a hydrogen approximation (through the GKS impact model discussed above), and the other one corresponds to the model reported in the STARK-B database, which accounts for the degeneracy removal. As can be seen, the line width is sensitive to the model used. The shift visible in the STARK-B result is a feature of the degeneracy removal. It depends on the plasma parameters, and can also be used for diagnostic purposes. An extension of already-existing databases to magnetized plasma conditions, where the Zeeman effect is visible in spectra, is under consideration. Work is presently ongoing for applications both in magnetic fusion and astrophysical plasmas.

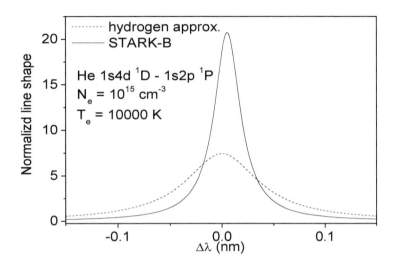

**Figure 5.** Plot of the electron impact broadening of He 1s4d $^1$D–1s2p $^1$P calculated using the STARK-B database (solid line) and a hydrogen approximation for the energy level structure (dashed line). Conditions relevant to high-density tokamak divertor plasmas are considered here, and no magnetic field is retained for the sake of clarity. As can be seen, the result significantly depends on the model used.

## 8. Conclusions

The shape of atomic spectral lines in plasmas contains information on the plasma parameters, and can be used as a diagnostic tool. In this work, we have examined a selection of problems involved in the modeling of Stark broadening for applications in magnetic fusion and in astrophysics. A specific issue concerns the description of ion dynamics on hydrogen lines, which is important if the upper principal quantum number is low as on Balmer $\alpha$. The use of a collision operator for ions is convenient because the line shape can be described analytically. In this framework, we have shown that the early model by Griem et al. (GBKO) for atoms with nondegenerate levels can be adapted to hydrogen in the presence of the Zeeman effect. It coincides with numerical simulations at the so-called impact limit, i.e., when the microfield fluctuates at a time scale much smaller than the time of interest, and when strong collisions are rare events. We have also shown that deviations in the impact limit due to non-binary interactions can be retained through an adaptation of the so-called unified theory, a formalism based on kinetic equations and related statistical concepts (BBGKY hierarchy). In astrophysics, magnetic white dwarfs present similarities with tokamak edge plasmas because the hydrogen absorption lines in atmospheres can exhibit a Zeeman triplet structure. We have illustrated this point through calculation of an absorption line in conditions relevant to a DAH stellar atmosphere. In this framework, an extension of already-existing Stark databases that is devoted to accounting for the Zeeman effect is presently under preparation. We have also shown that the description of the broadening of non-hydrogen lines is a challenging issue, both in astrophysics and in magnetic fusion. Modeling work is also ongoing in this framework in order to improve the databases.

**Acknowledgments:** This work is supported by the funding agency Campus France (Pavle Savić PHC project 36237PE). This work has also been carried out within the framework of the EUROfusion Consortium and has received funding from the Euratom research and training programme 2014–2018 under Grant agreement no 633053. The views and opinions expressed herein do not necessarily reflect those of the European Commission.

**Author Contributions:** All authors contributed equally to this work.

# References

1. ITER Press Release. 16 June 2016. Available online: https://www.iter.org/news/pressreleases (accessed on 3 July 2017).

2. Donne, A.J.H.; Costley, A.E.; Barnsley, R.; Bindslev, H.; Boivin, R.; Conway, G.; Fisher, R.; Giannella, R.; Hartfuss, H.; von Hellermann, M.G.; et al. Progress in the ITER Physics Basis, Chapter 7: Diagnostics. *Nucl. Fusion* **2007**, *47*, S337–S384. [CrossRef]

3. Sugie, T.; Costley, A.; Malaquias, A.; Walker, C. Spectroscopic Diagnostics for ITER. *J. Plasma Fusion Res.* **2003**, *79*, 1051–1061. [CrossRef]

4. Potzel, S.; Dux, R.; Müller, H.W.; Scarabosio, A.; Wischmeier, M.; ASDEX Upgrade Team. Electron density determination in the divertor volume of ASDEX Upgrade via Stark broadening of the Balmer lines. *Plasma Phys. Control. Fusion* **2014**, *56*, 025010. [CrossRef]

5. Rosato, J.; Kotov, V.; Reiter, D. Modelling of passive spectroscopy in the ITER divertor: The first hydrogen Balmer lines. *J. Phys. B At. Mol. Opt. Phys.* **2010**, *43*, 144024. [CrossRef]

6. Rosato, J.; Marandet, Y.; Capes, H.; Ferri, S.; Mossé, C.; Godbert-Mouret, L.; Koubiti, M.; Stamm, R. Stark broadening of hydrogen lines in low-density magnetized plasmas. *Phys. Rev. E* **2009**, *79*, 046408. [CrossRef] [PubMed]

7. Rosato, J.; Marandet, Y.; Stamm, R. A new table of Balmer line shapes for the diagnostic of magnetic fusion plasmas. *J. Quant. Spectrosc. Radiat. Transf.* **2017**, *187*, 333–337. [CrossRef]

8. Sahal-Bréchot, S.; Dimitrijević, M.S.; Moreau, N.; Ben Nessib, N. The STARK-B database VAMDC node: A repository for spectral line broadening and shifts due to collisions with charged particles. *Phys. Scr.* **2015**, *90*, 054008. [CrossRef]

9. Kepler, S.O.; Pelisoli, I.; Jordan, S.; Kleinman, S.J.; Kulebi, B.; Koester, D.; Peçanha, V.; Castanheira, B.G.; Nitta, A.; da Silveira Costa, J.E.; et al. Magnetic white dwarf stars in the Sloan Digital Sky Survey. *Mon. Not. R. Astron. Soc.* **2013**, *429*, 2934–2944. [CrossRef]

10. Külebi, B.; Jordan, S.; Euchner, F.; Gänsicke, B.T.; Hirsch, H. Analysis of hydrogen-rich magnetic white dwarfs detected in the Sloan Digital Sky Survey. *Astron. Astrophys.* **2009**, *506*, 1341–1350. [CrossRef]

11. Wickramasinghe, D.T. Model atmospheres for DA and DB white dwarfs. *Mon. Not. R. Astron. Soc.* **1972**, *76*, 129–179.

12. Griem, H.R. *Spectral Line Broadening by Plasmas*; Academic Press: London, UK, 1974.

13. Stamm, R.; Smith, E.W.; Talin, B. Study of hydrogen Stark profiles by means of computer simulation. *Phys. Rev. A* **1984**, *30*, 2039–2046. [CrossRef]

14. Stambulchik, E.; Maron, Y. Plasma line broadening and computer simulations: A mini-review. *High Energy Density Phys.* **2010**, *6*, 9–14. [CrossRef]

15. Rosato, J.; Reiter, D.; Kotov, V.; Marandet, Y.; Capes, H.; Godbert-Mouret, L.; Koubiti, M.; Stamm, R. Progress on Radiative Transfer Modelling in Optically Thick Divertor Plasmas. *Contrib. Plasma Phys.* **2010**, *50*, 398–403. [CrossRef]

16. Kotov, V.; Reiter, D.; Kukushkin, A.S.; Pacher, H.D.; Börner, P.; Wiesen, S. Radiation Absorption Effects in B2-EIRENE Divertor Modelling. *Contrib. Plasma Phys.* **2006**, *46*, 635–642. [CrossRef]

17. Griem, H.R.; Kolb, A.C.; Shen, K.Y. Stark Broadening of Hydrogen Lines in a Plasma. *Phys. Rev.* **1959**, *116*, 4–16. [CrossRef]

18. Griem, H.R.; Baranger, M.; Kolb, A.C.; Oertel, G. Stark Broadening of Neutral Helium Lines in a Plasma. *Phys. Rev.* **1962**, *125*, 177–195. [CrossRef]

19. Rosato, J.; Capes, H.; Godbert-Mouret, L.; Koubiti, M.; Marandet, Y.; Stamm, R. Accuracy of impact broadening models in low-density magnetized hydrogen plasmas. *J. Phys. B At. Mol. Opt. Phys.* **2012**, *45*, 1–7. [CrossRef]

20. Rosato, J.; Capes, H.; Stamm, R. Influence of correlated collisions on Stark-broadened lines in plasmas. *Phys. Rev. E* **2012**, *86*, 046407. [CrossRef] [PubMed]

21. Voslamber, D. Unified Model for Stark Broadening. *Z. Naturforsch.* **1969**, *24*, 1458–1472.

22. Smith, E.W.; Cooper, J.; Vidal, C.R. Unified Classical-Path Treatment of Stark Broadening in Plasmas. *Phys. Rev.* **1969**, *185*, 140–151. [CrossRef]

23. Capes, H.; Voslamber, D. Electron Correlations in the Unified Model for Stark Broadening. *Phys. Rev. A* **1972**, *5*, 2528–2536. [CrossRef]

24. Capes, H.; Voslamber, D. Spectral-line profiles in weakly turbulent plasmas. *Phys. Rev. A* **1977**, *15*, 1751–1766. [CrossRef]
25. Dupree, T.H. A Perturbation Theory for Strong Plasma Turbulence. *Phys. Fluids* **1966**, *9*, 1773–1782. [CrossRef]
26. Weinstock, J. Formulation of a Statistical Theory of Strong Plasma Turbulence. *Phys. Fluids* **1969**, *12*, 1045–1058. [CrossRef]
27. Landstreet, J.D.; Bagnulo, S.; Valyavin, G.G.; Fossati, L.; Jordan, S.; Monin, D.; Wade, G.A. On the incidence of weak magnetic fields in DA white dwarfs. *Astron. Astrophys.* **2012**, *545*, A30. [CrossRef]
28. Gomez, T.A.; Nagayama, T.; Kilcrease, D.P.; Montgomery, M.H.; Winget, D.E. Effect of higher-order multipole moments on the Stark line shape. *Phys. Rev. A* **2016**, *94*, 022501. [CrossRef]
29. Maschke, E.K.; Voslamber, D. Stark broadening of hydrogen lines in strong magnetic fields. In Proceedings of the Seventh International Conference, Beograd, Yugoslavia, 22–27 August 1965; Gradevinska Knjiga Publishing House : Beograd, Yugoslavia, 1966; Volume II, p. 568.
30. Rosato, J. Cases 10–11: B-induced trajectory effects. In Proceedings of the 4th Spectral Line Shapes in Plasmas (SLSP) Code Comparison Workshop, Vienna, Austria, 20–24 March 2017.
31. Kieu, N.; Rosato, J.; Stamm, R.; Kovacević-Dojcinović, J.; Dimitrijević, M.S.; Popović, L.C.; Simić, Z. A New Analysis of Stark and Zeeman Effects on Hydrogen Lines in Magnetized DA White Dwarfs. *Atoms* **2017**, submitted.
32. Koubiti, M.; Nakano, T.; Capes, H.; Marandet, M.; Mekkaoui, A.; Mouret, L.; Rosato, J.; Stamm, R. Characterization of the JT-60U Divertor Plasma Region During the Formation of a Strong Radiation. *Contrib. Plasma Phys.* **2012**, *52*, 455–459. [CrossRef]
33. Pütterich, T.; Neu, R.; Dux, R.; Whiteford, A.D.; O'Mullane, M.G. The ASDEX Upgrade Team. Modelling of measured tungsten spectra from ASDEX Upgrade and predictions for ITER. *Plasma Phys. Control. Fusion* **2008**, *50*, 085016. [CrossRef]
34. Calisti, A.; Mossé, C.; Ferri, S.; Talin, B.; Rosmej, F.; Bureyeva, L.A.; Lisitsa, V.S. Dynamic Stark broadening as the Dicke narrowing effect. *Phys. Rev. E* **2010**, *81*, 016406. [CrossRef] [PubMed]

# 6

# Using the Pairs of Lines Broadened by Collisions with Neutral and Charged Particles for Gas Temperature Determination of Argon Non-Thermal Plasmas at Atmospheric Pressure

Cristina Yubero [1], Antonio Rodero [1], Milan S. Dimitrijevic [2], Antonio Gamero [1] and Maria del Carmen García [1,*]

[1] Grupo de Física de Plasmas: Diagnosis, Modelos y Aplicaciones (FQM-136) Edificio A. Einstein (C-2), Campus de Rabanales, Universidad de Córdoba, 14071 Córdoba, Spain; f62yusec@uco.es (C.Y.); fa1rosea@uco.es (A.R.); fa1garoa@uco.es (A.G.)

[2] Astronomical Observatory, Volgina 7, 11060 Belgrade, Serbia; mdimitrijevic@aob.rs

* Correspondence: fa1gamam@uco.es

Academic Editor: Michael Brunger

**Abstract:** The spectroscopic method for gas temperature determination in argon non-thermal plasmas sustained at atmospheric pressure proposed recently by *Spectrochimica Acta Part B* 129 14 (2017)—based on collisional broadening measurements of selected pairs of argon atomic lines, has been applied to other pairs of argon atomic lines, and the discrepancies found in some of these results have been analyzed. For validation purposes, the values of the gas temperature obtained using the different pairs of lines have been compared with the rotational temperatures derived from the OH ro-vibrational bands, using the Boltzmann-plot technique.

**Keywords:** plasma spectroscopy; microwave discharges; gas temperature; stark broadening parameters; atomic emission spectroscopy

## 1. Introduction

Optical emission spectroscopy (OES) techniques based on the analysis of molecular emission spectra are commonly used for gas temperature ($T_g$) determination of plasmas at atmospheric pressure. The rotational temperature derived from them is considered as a good estimation of the kinetic temperature of the plasma heavy particles [1,2] thanks to the strong coupling between translational and rotational energy states under high-pressure conditions. The emissions of diatomic species, such as OH, $N_2$, $N_2^+$, CN, ..., have been traditionally employed with this purpose [3–10], but the use of molecular emission spectroscopy is not always easy for gas temperature measurement in plasmas: overlapping of bands, rotational population distribution of levels having a non-Boltzmann nature, wake emission of rotational bands, among others, can make it difficult to obtain reliable values of gas temperature.

Alternative OES methods for gas temperature determination are needed. In this way, the van der Waals broadening of some atomic lines has been used for this purpose due to its dependence on the plasma gas temperature. For argon plasmas, the 425.9, 522.1, 549.6, and 603.2 nm argon lines have been the most frequently employed [11–14]. This technique is based on the detection of argon lines not affected by resonance broadening (also related to $T_g$). It requires the use of additional techniques for simultaneous determination of the electron density [15], as these lines have a non-negligible Stark broadening for electron densities above $10^{14}$ cm$^{-3}$, which needs to be determined. Yubero et al. in [16] proposed a method to circumvent this dependence on electron density by considering pairs of these lines. This method allows for the determination of the gas temperature from the measurements of Lorentzian profiles of some pairs of argon atomic lines, and when applying it, no assumptions on the degree of thermodynamic equilibrium among excited states are needed (unlike methods based on rotational temperature determination).

However, the authors of [16] found a small disagreement in the results from pairs Ar I 603.2 nm/Ar I 522.1nm and Ar I 549.6 nm/Ar I 522.1 nm. In the present work, this disagreement is explained, and other pairs of lines have been employed for the gas temperature determination.

## 2. Method

This method is valid for plasmas sustained at atmospheric pressure whose line profiles can be fitted to a Voigt function-characterized by a full-width at half-maximum (FWHM), $W_V$-resulting from the convolution of a Gaussian function ($W_G$) with a Lorentzian function ($W_L$) (see, e.g., [17,18]). Indeed, the profiles of atomic lines emitted by plasmas with no presence of magnetic fields result from different broadening mechanisms leading to Gaussian or Lorentzian profile shapes, briefly described below.

The motion of emitting atoms with respect to the detector, with a continuous velocity distribution depending on their temperature, leads to the so called *Doppler broadening* and a Gaussian-shaped line profile with a FMHW $W_D$ (in nm) given by

$$W_D = 7.16 \cdot 10^{-7} \lambda \sqrt{T_g / M} \tag{1}$$

where $\lambda$, $T_g$, and $M$ are the wavelength (nm), the gas temperature (in K), and the mass of the radiating atom (in a.m.u.), respectively.

The *van der Waals broadening* is due to the dipole moment induced by neutral perturber atoms interacting with the electric field of the excited emitter atom and generates line profiles with a Lorentzian shape (with an FWHM $W_W$), according to the Lindholm–Foley theory [19].

The *resonance broadening* of spectral lines is due to dipole–dipole interactions of the emitter with ground-state atoms of the same element [20] and contributes to the Lorentzian part of the profile with an FWHM $W_R$.

The *Stark broadening* (FWHM $W_S$) of a line is due to interactions of the emitter atom with the surrounding charged particles, perturbing the electric field it experiences. In the case of a non-hydrogenic atom, the profiles of isolated spectral lines broadened by collisions with electrons have a Lorentzian shape. For thermal plasmas with a gas temperature similar to the electron one, the mobility of ions is high and the impact approximation [21] is also valid for ions, being their contribution to the broadening also being Lorentzian. In the ion impact limit, line profiles are symmetric Lorentzian. On the contrary, for plasmas where the ion mobility is small (e.g., plasmas with gas temperature relatively low), a quasistatic approximation is often needed to model the ion broadening in order to explain the slightly asymmetric shape of the profiles. The less dynamical the ions are, the more asymmetric the lines are. The finite lifetime of the excited levels gives rise to *natural broadening*, which is typically very small (~0.00001 nm) and can be neglected in the case of atmospheric pressure plasma spectroscopy.

Finally, the line profile is also affected by the instrumental function of the spectrometer used for its detection. Usually, this instrumental function can be well approximated by a Gaussian profile with an FWHM $W_I$, as shown in the next section.

Thus, the broadening contributions with a Gaussian shape will lead to a profile with an FWHM ($W_G$) given by

$$W_G = \sqrt{(W_D)^2 + (W_I)^2} \tag{2}$$

and those having a Lorentzian shape give rise to a profile that is also Lorentzian with an FWHM ($W_L$) given by

$$W_L = W_R + W_S + W_W. \tag{3}$$

46    Spectral Lines: The Theory of Line Shape in Astrophysics

The method we propose here only considers atomic lines with a negligible resonance broadening. Thus, the Lorentzian part of a line profile is only due to Stark and van der Waals broadenings, and full width at half maximum of the Lorentzian profile, $W_L$ can be written as follows:

$$W_L = W_S + W_W \tag{4}$$

Several studies of Stark broadening for atomic and singly charged ion lines [18,22–26] show that Equation (4) can be written as

$$W_S \cong \left[ w_S^e + w_S^i \right] \frac{n_e}{10^{16}} \tag{5}$$

where parameters $w_S^e$ and $w_S^i$ are electronic and ionic contributions to the full-width at half-intensity maximum given for an electron density equal to $10^{16}$ cm$^{-3}$, and $n_e$ is the electron density.

In non-thermal-plasmas where $T_g$ is much lower than $T_e$ (electron temperature), the ionic contribution $w_S^i$ can be neglected. Additionally, $W_S$ can be considered to have a weak dependence on $T_e$ in the small range of electron temperature from 5000 to 10,000 K [22]. In this way, $W_S$ depends only on $n_e$ and Equation (5) can be approximated as follows:

$$W_L(T_g, n_e) = w_S \frac{n_e}{10^{16}} + W_W(T_g) \tag{6}$$

where $w_S = w_S^e$.

On the other hand, the van der Waals broadening has a full width at half maximum $W_W$ in nm given by Griem [18], which, considering the ideal gas equation $N = P/K_B T_g$ for the density of perturbers, and where $K_B$ is the Boltzmann constant and $P$ is the pressure, can be written as

$$W_W(T_g) = \frac{C}{T_g^{\frac{7}{10}}} \text{(nm)} \tag{7}$$

with $C$ being determined by the type of gas in the discharge and the nature of the atom emitters:

$$C = \frac{8.18 \cdot 10^{-19} \lambda^2 \left( \alpha \left\langle \overline{R^2} \right\rangle \right)^{2/5} P}{k_\beta \mu^{3/10}} \text{(nm·K}^{7/10}) \tag{8}$$

where

$$\left\langle \overline{R^2} \right\rangle = \left\langle \overline{R_U^2} \right\rangle - \left\langle \overline{R_L^2} \right\rangle \tag{9}$$

is the difference of the squares of coordinate vectors (in $a_0$ units) of the upper and lower level, $\lambda$ is the wavelength of the observed line in nm, $\alpha$ is the polarizability of perturbers interacting with the excited radiator in cm$^3$, $T_g$ is the temperature of the emitters (coincident with the gas temperature) in K, and $\mu$ is the reduced mass of the emitter–perturber pair in a.m.u.

For argon plasma at atmospheric pressure, when considering the van der Waals broadening of argon atomic lines ($\mu = 19.97$ and $\alpha = 16.54 \cdot 10^{-25}$ cm$^3$), $C$ can be written as [16]

$$C = 7.5 \cdot 10^{-7} \lambda^2 \left( \left\langle \overline{R^2} \right\rangle \right)^{2/5} \text{(nm·K}^{7/10}). \tag{10}$$

Thus, the formula for $W_L$ given by Equation (3) can be approximately expressed as

$$W_L(T_g, n_e) = w_S \frac{n_e}{10^{16}} + \frac{C}{T_g^{\frac{7}{10}}}. \tag{11}$$

The method for gas temperature determination proposed by Yubero et al. in [16] is based on the measurement of the Lorentzian contribution to the entire FWHM for two atomic lines, $L1$ and $L2$:

$$W_L^{L1}(T_g, n_e) = w_S^{L1}\frac{n_e}{10^{16}} + \frac{C^{L1}}{T_g^{\frac{7}{10}}}$$
$$W_L^{L2}(T_g, n_e) = w_S^{L2}\frac{n_e}{10^{16}} + \frac{C^{L2}}{T_g^{\frac{7}{10}}} \qquad (12)$$

The dependence on the electron density can be eliminated from these expressions, and a linear dependence between the FWHM of the Lorentzian part of the total profile of these two lines, $W_L^{L1}$ and $W_L^{L2}$, is obtained:

$$W_L^{L1} = \frac{w_S^{L1}}{w_S^{L2}}W_L^{L2} + \left(C^{L1} - C^{L2}\frac{w_S^{L1}}{w_S^{L2}}\right)T_g^{-0.7}. \qquad (13)$$

Finally, the gas temperature is given by

$$T_g = \left(\frac{C^{L1} - C^{L2}\frac{w_S^{L1}}{w_S^{L2}}}{W_L^{L1} - \frac{w_S^{L1}}{w_S^{L2}}W_L^{L2}}\right)^{1/0.7}. \qquad (14)$$

Thus, as long as the Stark parameters for two lines and their $C$ parameters are known, the gas temperature can be determined from full Lorentzian FWHMs of these lines. Nevertheless, the applicability of this method relies on the knowledge of these parameters, with a certain degree of accuracy. The theoretical estimation of these parameters in the literature can lead to significant errors in gas temperature values.

Concerning this matter, Yubero et al. [16] gather the values of FWHM of the Lorentzian profile of the lines Ar I 603.2 nm, Ar I 549.6 nm, and Ar I 522.1 nm These lines correspond to transitions from high energy levels and exhibit slightly asymmetric profiles [13]. The effect of this asymmetry, not considered in the previous work [16], has been removed from the profiles according to the procedure in [13] (so, only considering the symmetric electron contribution) in order to improve the results of the method. According to this reference, the effect of ions is more important in the right shape of profile. Therefore, only the left part of the profile has been considered to generate the entire profile. Figure 1 shows an example of the symmetrization procedure for the Ar I 522.1 nm line.

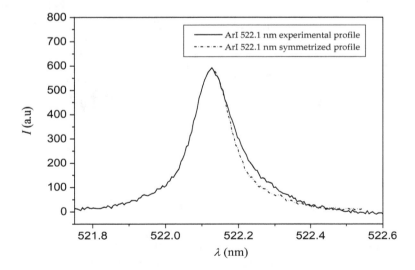

**Figure 1.** Symmetrization of Ar I 522.1 nm.

In addition, new Ar I lines at 560.7 nm and 518.8 nm have been included in this work.

Measured values of $W_L$ for these lines allow us to obtain $T_g$ using Equation (14), provided the $C$ coefficients and Stark broadening parameters are known. Table 1 includes values of the $C$ coefficients calculated for these lines from Equation (8), and the Stark broadening parameters theoretically determined by Dimitrijević et al. [22] for an electron temperature of 10,000 K.

**Table 1.** $C$ coefficients calculated from Equation (8), and the Stark broadening parameters due to electron impacts theoretically determined by Dimitrijević et al. [22] for an electron temperature of 10,000 K and an electron density of $10^{16}$ cm$^{-3}$.

| Ar I Line (nm) | $C$ | $w_S = w_S^e$ (nm) |
|---|---|---|
| 603.2 | 4.2 | 0.149 |
| 549.6 | 4.9 | 0.305 |
| 522.1 | 5.9 | 0.588 |
| 560.7 | 3.6 | 0.145 |
| 518.8 | 4.1 | 0.104 |

### 3. Experimental Set-Up

In this work, we measured the gas temperature of an argon microwave (2.45 GHz) induced plasma column sustained at atmospheric pressure and generated inside a quartz tube (with the inner and outer diameters 1.5 mm and 4 mm, respectively), described elsewhere [27]. A similar plasma had been previously characterized in [28], its electron density being of the order of $10^{14}$ (cm$^{-3}$). The electron temperature was estimated to be close to 10,000 K from observed relative populations of the argon excited levels assuming a partial local thermodynamic equilibrium [28].

Figure 2 includes a scheme of the optical detection assembly and data acquisition system to process spectroscopic measurements. A *surfaguide* was employed as a coupling device, injecting a microwave power of 100 W to the plasma. The argon flow rate was set at 0.5 slm (standard liters per minute) and adjusted with a calibrated mass flow controller.

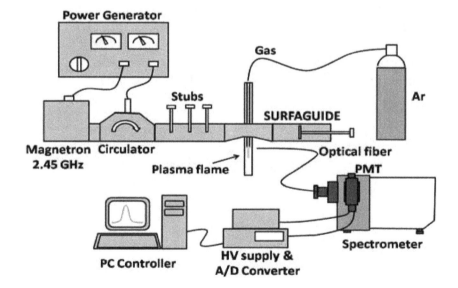

**Figure 2.** Scheme of experimental set-up.

Light emission from the plasma was analyzed by using a Czerny–Turner type spectrometer with a 1 m focal length, equipped with a 2400 grooves/mm holographic grating and a photomultiplier (spectral output interval of 200–800 mm) as a detector. The light emitted by the plasma was collected side-on using an optical fiber at different axial positions along the plasma column ($z = 4, 8,$ and 12 cm measured from the end of the column).

The instrumental function of the spectrometer was measured from the FWHM of the line Ne I 632.8 nm emitted by a helium–neon laser (this line is a good choice to make this estimate, as it has a wavelength close to those of the Ar I lines considered in this work). When using equal entrance and exit spectrometer slit widths, this function had an approximately triangular shape, which could be well-fitted with a Gaussian function. In this way, an instrumental broadening width $W_I = (0.032 \pm 0.001)$ nm was measured when using slit widths of 100 µm.

On the other hand, measurements of the light absorption have shown that the plasma studied can be considered as optically thin in the direction of observation chosen (transversally) for the Ar I lines detected [28,29].

Each Ar I line was measured experimentally several times and was fitted to a Voigt profile. Therefore, the uncertainty of each Voigt FWHM corresponds to the dispersion. The Lorentzian contribution to the entire broadening in each case was obtained from the Voigt FWHM measured for each line using the formula [30,31]

$$W_V \approx \frac{W_L}{2} + \sqrt{\left(\frac{W_L}{2}\right)^2 + W_G^2} \qquad (15)$$

assuming that $W_G \approx W_I$, since, according to Equation (2), the Doppler contribution can be considered as negligible when compared to the instrumental one under the experimental conditions in the plasma studied ($T_g \leq 2500$ K, $W_D{}^{ArI} \leq 0.003$ nm).

## 4. Results

Table 2 shows the values of Lorentzian FWHM measured for the different Ar I lines at different positions. For each pair of lines, these values allow for $T_g$ determination using Equation (14), provided $C$ coefficients and Stark broadening parameters given in Table 1.

Table 2. Lorentzian FWHM of lines Ar I 603.2 nm, Ar I 549.6 nm, Ar I 522.1 nm, Ar I 560.7 nm, and Ar I 518.8 nm measured at different axial plasma positions.

| z (cm) | z = 4 cm | z = 8 cm | z = 12 cm |
|---|---|---|---|
| $W_L{}^{603}$ (nm) | $0.0411 \pm 0.0014$ | $0.0437 \pm 0.0016$ | $0.0459 \pm 0.0014$ |
| $W_L{}^{549}$ (nm) | $0.0594 \pm 0.0019$ | $0.0626 \pm 0.0016$ | $0.0731 \pm 0.0012$ |
| $W_L{}^{522}$ (nm) | $0.0958 \pm 0.0024$ | $0.1020 \pm 0.0019$ | $0.122 \pm 0.002$ |
| $W_L{}^{560}$ (nm) | $0.0342 \pm 0.0024$ | $0.0377 \pm 0.0024$ | $0.0429 \pm 0.0012$ |
| $W_L{}^{518}$ (nm) | $0.0372 \pm 0.0021$ | $0.0418 \pm 0.0016$ | $0.0472 \pm 0.0018$ |

Table 3 shows the $T_g$ values obtained using these parameters for an electron temperature of 10,000 K. These values are also compared with the values obtained using OH ro-vibrational band [13]. Uncertainties in $T_g$ have been obtained from Equation (14) by considering $C$ and $W_S$ as theoretical constants and only taking into account uncertainties in the broadenings experimentally measured and errors of approximations used in this method.

Overall, results from the method of pairs of argon lines proposed in this work are in good agreement with those derived from the analysis of OH ro-vibrational band. However, there are some results affected by a very high uncertainty (higher than 50%) corresponding, in some cases, to the Ar I 560.7 nm line. Additionally, there are values lower than the ones obtained from rotational temperature technique corresponding mainly to the 518.8 line.

**Table 3.** Gas temperature obtained using Equation (14) with theoretical Stark broadening parameters at electron temperatures of 10,000 K, and comparison with the one obtained from OH ro-vibrational bands, using the well known Boltzmann plot technique.

| $z = 4$ cm $T^{BP}_g$ (K) = 1390 $\pm$ 70 | Ar I 549 nm | Ar I 522 nm | Ar I 560 nm | Ar I 518 nm |
|---|---|---|---|---|
| Ar I 603 nm | 1300 $\pm$ 300 | 1420 $\pm$ 180 | 570 $\pm$ 240 | 1100 $\pm$ 500 |
| Ar I 549 nm | - | 1800 $\pm$ 900 | 2200 $\pm$ 1300 | 1200 $\pm$ 230 |
| Ar I 522 nm | | - | 2000 $\pm$ 700 | 1300 $\pm$ 200 |
| Ar I 560 nm | | | - | 900 $\pm$ 300 |
| $z = 8$ cm $T^{BP}_g$ (K) = 1330 $\pm$ 70 | Ar I 549 nm | Ar I 522 nm | Ar I 560 nm | Ar I 518 nm |
| Ar I 603 nm | 1100 $\pm$ 220 | 1300 $\pm$ 200 | 700 $\pm$ 600 | 760 $\pm$ 220 |
| Ar I 549 nm | - | 1700 $\pm$ 500 | 1400 $\pm$ 500 | 920 $\pm$ 120 |
| Ar I 522 nm | | - | 1600 $\pm$ 500 | 1030 $\pm$ 110 |
| Ar I 560 nm | | | - | 800 $\pm$ 300 |
| $z = 12$ cm $T^{BP}_g$ (K) = 1520 $\pm$ 70 | Ar I 549 nm | Ar I 522 nm | Ar I 560 nm | Ar I 518 nm |
| Ar I 603 nm | 1600 $\pm$ 400 | 1680 $\pm$ 220 | 3000 $\pm$ 3000 | 500 $\pm$ 130 |
| Ar I 549 nm | - | 1800 $\pm$ 500 | 1400 $\pm$ 400 | 810 $\pm$ 110 |
| Ar I 522 nm | | - | 1500 $\pm$ 300 | 930 $\pm$ 110 |
| Ar I 560 nm | | | - | 640 $\pm$ 130 |

## 5. Discussion and Conclusions

In this paper, the OES tool proposed by Yubero et al. in [16] for the determination of the gas temperature in non-thermal plasmas, based on the measurement of the Lorentzian part of the profile of a pair of atomic emission lines, has been improved, as asymmetries of the profiles of these lines have been removed. Compared with the results of [16], better agreement with gas temperatures obtained from OH ro-vibrotional band has been found. We can conclude that symmetrization of the line profile is advisable to obtain good results from this method. Moreover, results from new lines Ar I 560.7 nm and Ar I 518.8 nm have been also included in this paper. As can be seen in Table 3, there are some pairs that do not give accurate results or $T_g$ values with higher uncertainty. They correspond to those cases in which the denominator in Equation (14) is very small, so gas temperature determination becomes very sensitive, giving rise to large errors. Examples of pairs of line giving large errors are Ar I 560.7 nm/Ar I 603.2 nm and Ar I 560.7 nm/Ar I 549.6 nm. This fact does not explain other values that are lower than those obtained from OH ro-vibrotional band. Examples of these pairs of lines are Ar I 518.8 nm/Ar I 603.2 nm, Ar I 518.8 nm/Ar I 549.6 nm, Ar I 518.8 nm/Ar I 522.1 nm, and Ar I 518.8 nm/Ar I 560.7 nm. This could be explained by errors in the theoretical broadening constants given in Table 1.

Although the method has been developed for some pairs of argon lines, it also applies for any pair of atomic lines, as long as they fulfill the following conditions:

(i) Their Stark parameters must be as accurate as possible; (ii) They have a negligible resonance broadening (or they do not have any at all). Choosing lines with an upper or lower level that do not have an electric dipole transition (resonance line) to the ground state is a way to ensure that this condition is satisfied. Additionally, analytical expressions for resonant FWHM in the literature [32,33] can be used to evaluate the importance of this broadening; (iii) Their Stark and van der Waals broadenings should not be very different from each other, so as to avoid large errors in $T_g$ determination (see Equation (14)); this condition applies to plasmas with a relatively low gas temperature (van der Waals broadening not negligible) and moderate electron densities (significant Stark broadening).

The method we propose in this work can be considered a good alternative to the traditional ones based on the measurement of rotational temperatures. In the application of this method, no assumptions on the degree of thermodynamic equilibrium for excited states existing in the plasma are needed, which is its main advantage compared to other methods.

Additionally, as other optical emission spectroscopy techniques, this is a non-plasma diagnosis method, easy to implement.

**Acknowledgments:** The authors thank the European Regional Development Funds program (EU-FEDER) and the Research Spanish Agency (Agencia Española de Investigación-AEI) of MINECO (project MAT2016-79866-R) for financial support. The authors are also grateful to the *Física de Plasmas: Diagnosis, Modelos y Aplicaciones* (FQM 136) research group of the Regional Government of Andalusia for technical and financial support.

**Author Contributions:** All authors contributed equally.

# References

1.  Nassar, H.; Pellerin, S.; Mussiol, K.; Martinie, O.; Pellerin, M.N.; Cormier, J.M. N2(+)/N2 ratio and temperature 2 measurements based on the first negative N-2(+) and second positive N-2 overlapped molecular emission spectra. *J. Phys. D Appl. Phys.* **2004**, *37*, 1904. [CrossRef]

2.  Britun, N.; Gaillard, M.; Ricard, A.; Kim, Y.M.; Kim, K.S.; Han, H.G. Determination of the vibrational, rotational and electron temperatures in N2 and Ar-N2 rf discharge. *J. Phys. D Appl. Phys.* **2007**, *40*, 1022. [CrossRef]

3.  Zhu, X.M.; Chen, W.C.; Pu, Y.K. Gas temperature, electron density and electron temperature measurement in a microwave excited microplasma. *J. Phys. D Appl. Phys.* **2008**, *41*, 105212. [CrossRef]

4.  Wang, Q.; Koleva, I.; Donnelly, V.M.; Economou, D.J. Spatially resolved diagnostics of an atmospheric pressure direct current helium microplasma. *J. Phys. D Appl. Phys.* **2005**, *38*, 1690. [CrossRef]

5.  Abdallah, M.H.; Mermet, J.M. The Behavior of Nitrogen Excited in an Inductively Coupled Argon Plasma. *J. Quant. Spectrosc. Radiat. Transf.* **1978**, *19*, 83–91. [CrossRef]

6.  Laux, C.O.; Spence, T.G.; Kruger, C.H.; Zare, R.N. Optical diagnostics of atmospheric pressure air plasmas. *Plasma Sources Sci. Technol.* **2003**, *12*, 125. [CrossRef]

7.  Fantz, U. Emission spectroscopy of molecular low pressure plasmas. *Contrib. Plasma Phys.* **2004**, *44*, 508–515. [CrossRef]

8.  Mora, M.; García, M.C.; Jiménez-Sanchidrián, C.; Romero-Salguero, F.J. Transformation of light paraffins in a microwave-induced plasma-based reactor at reduced pressure. *Int. J. Hydrog. Energy* **2010**, *35*, 4111–4122. [CrossRef]

9.  Iza, F.; Hopwood, J.A. Rotational, vibrational, and excitation temperatures of a microwave-frequency microplasma. *IEEE Trans. Plasma Sci.* **2004**, *32*, 498–504. [CrossRef]

10. Lombardi, G.; Benedic, F.; Mohasseb, F.; Hassouni, K.; Gicquel, A. Determination of gas temperature and C-2 absolute density in Ar/H-2/CH4 microwave discharges used for nanocrystalline diamond deposition from the C-2 Mulliken system. *Plasma Sources Sci. Technol.* **2004**, *13*, 375. [CrossRef]

11. Christova, M.; Castaños-Martínez, E.; Calzada, M.D.; Kabouzi, Y.; Luque, J.M.; Moisan, M. Electron density and gas temperature from line broadening in an argon surface-wave-sustained discharge at atmospheric pressure. *Appl. Spectrosc.* **2004**, *58*, 1032–1037. [CrossRef] [PubMed]

12. Christova, M.; Gagov, V.; Koleva, I. Analysis of the profiles of the argon 696.5 nm spectral line excited in non-stationary wave-guided discharges. *Spectrochim. Acta B* **2000**, *55*, 815–822. [CrossRef]

13. Yubero, C.; Dimitrijevic, M.S.; García, M.C.; Calzada, M.D. Using the van der Waals broadening of the spectral atomic lines to measure the gas temperature of an argon microwave plasma at atmospheric pressure. *Spectrochim. Acta B* **2007**, *62*, 169–176. [CrossRef]

14. Muñoz, J.; Dimitrijevic, M.S.; Yubero, C.; Calzada, M.D. Using the van der Waals broadening of spectral atomic lines to measure the gas temperature of an argon-helium microwave plasma at atmospheric pressure. *Spectrochim. Acta B* **2009**, *64*, 167–172. [CrossRef]

15. Gigosos, M.A.; Cardeñoso, V. New plasma diagnosis tables of hydrogen Stark broadening including ion dynamics. *J. Phys. B At. Mol. Opt. Phys.* **1996**, *29*, 4795. [CrossRef]

16. Yubero, C.; Rodero, A.; Dimitrijevic, M.; Gamero, A.; García, M.C. Gas temperature determination in an argon non-thermal plasma at atmospheric pressure from broadenings of atomic emission lines. *Spectrochim. Acta Part B* **2017**, *129*, 14–20. [CrossRef]

17. Zaghloul, M.R. On the calculation of the Voigt line profile: A single proper integral with a damped sine integrand MNRAS. *Mon. Not. R. Astron. Soc.* **2007**, *375*, 1043–1048. [CrossRef]

18. Griem, H.R. *Spectral Line Broadening by Plasmas*; Academic Press: New York, NY, USA, 1974.

19.  Allard, N.; Kielkopf, J. The effect of neutral non resonant collisions on atomic spectral lines. *Rev. Mod. Phys.* **1982**, *54*, 1103. [CrossRef]

20.  Griem, H.R. Stark broadening of isolated spectral lines from heavy elements in a plasma. *Phys. Rev.* **1962**, *128*, 515–531. [CrossRef]

21.  Konjevic, N. Plasma broadening and shifting of non-hydrogenic spectral lines: Present status and applications. *Phys. Rep.* **1999**, *316*, 339–401. [CrossRef]

22.  Dimitrijević, M.S.; Konjević, N. Stark broadenings of isolated spectral-lines of heavy-elements in plasmas. *J. Quant. Spectrosc. Radiat. Transf.* **1983**, *30*, 45–54. [CrossRef]

23.  Dimitrijević, M.S.; Christova, M.; Sahal-Bréchot, S. Stark broadening of visible Ar I spectral lines. *Phys. Scr.* **2007**, *75*, 809–819. [CrossRef]

24.  Christova, M.; Dimitrijević, M.S.; Sahal-Bréchot, S. Stark broadening of Ar I spectral lines emitted in surface wave sustained discharges. *Mem. Della Soc. Astron. Ital. Suppl.* **2005**, *7*, 238.

25.  Dimitrijević, M.S. A programme to provide Stark broadening data for stellar and laboratory plasma investigations. *Zh. Prikl. Spektrosk.* **1996**, *63*, 810.

26.  Sahal-Bréchot, S. Théorie de l'élargissement et du déplacement des raies spectrales sous l'effect des chocs avec les électrons et les ions dans l'approximation des impacts. *Astron. Astrophys.* **1959**, *1*, 91–123.

27.  Moisan, M.; Etermandi, E.; Rostaing, J.C. Excitation System for a Gas Plasma Surface Wave, and Associated Gas Processing System—Has Electromagnetic Material Sleeve Surrounding Gas Circulating Tube (European Patent EP 0874 537 A1). French Patent N. 2,762,748, 1998.

28.  García, M.C.; Rodero, A.; Sola, A.; Gamero, A. Spectroscopic study of a stationary surface-wave sustained argon plasma column at atmospheric pressure. *Spectrochim. Acta Part B* **2000**, *55*, 1733–1745. [CrossRef]

29.  Santiago, I.; Christova, M.; García, M.C.; Calzada, M.D. Self-absorbing method to determine the population of the metastable levels in an argon microwave plasma at atmospheric pressure. *Eur. Phys. J. Appl. Phys.* **2004**, *28*, 325–330. [CrossRef]

30.  Temme, N.M. Voigt function. In *NIST Handbook of Mathematical Functions*; Olver Frank, W.J., Lozier, D.M., Boisvert, R.F., Eds.; Cambridge University Press: New York, NY, USA, 2010; ISBN 978-0521192255.

31.  Olivero, J.J.; Longbothum, R.L. Empirical fits to the Voigt line width: A brief review. *J. Quant. Spectrosc. Radiat. Transf.* **1977**, *17*, 233–236. [CrossRef]

32.  Ali, A.W.; Giem, H.R. Theory of Resonance Broadening of Spectral Lines by Atom-Atom Impacts. *Phys. Rev.* **1965**, *140*, 1044. [CrossRef]

33.  Ali, A.W.; Giem, H.R. Theory of Resonance Broadening of Spectral Lines by Atom-Atom Impacts (ERRATA). *Phys. Rev.* **1966**, *144*, 366. [CrossRef]

# 7

# Quasar Black Hole Mass Estimates from High-Ionization Lines: Breaking a Taboo?

Paola Marziani [1,*], Ascensión del Olmo [2,*], Mary Loli Martínez-Aldama [2], Deborah Dultzin [3], Alenka Negrete [3], Edi Bon [4], Natasa Bon [4] and Mauro D'Onofrio [5]

[1]    Osservatorio Astronomico di Padova, Istituto Nazionale di Astrofisica (INAF), IT 35122 Padova, Italy

[2]    Instituto de Astrofisíca de Andalucía (IAA-CSIC), E-18008 Granada, Spain; maryloli@iaa.es

[3]    Instituto de Astronomía, Universidad Nacional Autónoma de México (UNAM), México D.F. 04510, Mexico; deborah@astro.unam.mx (D.D.); alenka@astro.unam.mx (A.N.)

[4]    Astronomical Observatory, Volgina 7, 11060 Belgrade 38, Serbia; ebon@oab.rs (E.B.); nbon@oab.rs (N.B.)

[5]    Dipartimento di Fisica & Astronomia "Galileo Galilei", Università di Padova, IT35122 Padova , Italy; mauro.donofrio@unipd.it

[*]    Correspondence: paola.marziani@oapd.inaf.it (P.M.); chony@iaa.es (A.d.O.)

Academic Editor: Robert C. Forrey

**Abstract:** Can high ionization lines such as CIV$\lambda$1549 provide useful virial broadening estimators for computing the mass of the supermassive black holes that power the quasar phenomenon? The question has been dismissed by several workers as a rhetorical one because blue-shifted, non-virial emission associated with gas outflows is often prominent in CIV$\lambda$1549 line profiles. In this contribution, we first summarize the evidence suggesting that the FWHM of low-ionization lines like H$\beta$ and MgII$\lambda$2800 provide reliable virial broadening estimators over a broad range of luminosity. We confirm that the line widths of CIV$\lambda$1549 is not immediately offering a virial broadening estimator equivalent to the width of low-ionization lines. However, capitalizing on the results of Coatman et al. (2016) and Sulentic et al. (2017), we suggest a correction to FWHM CIV$\lambda$1549 for Eddington ratio and luminosity effects that, however, remains cumbersome to apply in practice. Intermediate ionization lines (IP $\sim$ 20–30 eV; AlIII$\lambda$1860 and SiIII]$\lambda$1892) may provide a better virial broadening estimator for high redshift quasars, but larger samples are needed to assess their reliability. Ultimately, they may be associated with the broad-line region radius estimated from the photoionization method introduced by Negrete et al. (2013) to obtain black hole mass estimates independent from scaling laws.

**Keywords:** ionization processes; emission line formation; atomic spectroscopy; supermassive black holes; emission line profiles; quasars

## 1. Introduction: Statement of the Problem

A defining property of type-1 quasars is the presence of broad and narrow optical and UV lines emitted by ionic species over a wide range of ionization potentials (IPs, [1]): high-ionization lines (HILs) involving IP > 50 eV, and low-ionization lines (LILs) from ionic species with IP < 20 eV. Over the years, it has turned expedient to consider the UV resonance line CIV$\lambda$1549 as a representative of broad HILs. Lines do not all show the same profiles, and redshifts measured on different lines often show significant differences. Internal line shifts involving both broad lines in the optical and UV spectra of quasars have offered a powerful diagnostic tool of the quasar innermost structure since a few years after the discovery of quasars [2].

The CIV$\lambda$1549 is a resonant doublet ($^2P^o_{\frac{3}{2},\frac{1}{2}} \rightarrow^2 S_{\frac{1}{2}}$) emitted by gas which is, at least in part, flowing out from a region within $\sim$1000 gravitational radii from the central black hole (e.g., [3], for a review).

The occurrence of CIV$\lambda$1549 large shifts constrains the suitability of the CIV$\lambda$1549 profile broadening as a virial black hole mass ($M_{\mathrm{BH}}$) estimator (see, e.g., [4,5], for reviews). Results at low-redshift obtained in the mid-2010s suggest that the CIV$\lambda$1549 line is unsuitable for, at least, part of the quasar Population A sources (Section 4 [6]). A similar conclusion was reached at $z \approx 2$ on a sample of 15 high-luminosity quasars [7]. More recent work tends to confirm that the CIV$\lambda$1549 line width is not straightforwardly related to virial broadening (e.g., [8]). However, the CIV$\lambda$1549 line is strong and observable up to $z \approx 6$ with optical spectrometers. It is so highly desirable to have a consistent virial broadening estimate up to the highest redshifts that various attempts (e.g., [9]) have been done at "rehabilitating" CIV$\lambda$1549 line width estimators to bring them in agreement with the width of LILs such as H$\beta$ and MgII$\lambda$2800 [10].

In this contribution, we briefly stress the importance of black hole mass estimates (Section 2) and recapitulate the basic method of $M_{\mathrm{BH}}$ estimates applied to large samples of quasars under the virial assumption (Section 3). To improve estimates that have been known to be plagued by systematic and statistical uncertainties as large as a factor $\sim$100, we adopt the "rehabilitating" power of the quasar eigenvector 1 (E1; Section 4). Our analysis is then focused on virial broadening estimators (VBEs; Section 5) from: (a) LILs (IP < 15 eV: H$\beta$, MgII 2800); (b) HILs (IP > 40 eV: CIV$\lambda$1549); (c) intermediate-ionization lines (IILs: SiIII]$\lambda$1892, Al III$\lambda$1860), for which we provide a brief summary of preliminary results. Reported results from our group were published during the last decade [11–14]. We finally suggest that "photoioionization" computations of $M_{\mathrm{BH}}$ (Section 6) may offer a solution to some of the problems associated with the use of an average scaling law.

## 2. Importance of Black Hole Mass Determination

The relevance of supermassive black hole $M_{\mathrm{BH}}$ estimates is not limited to the sake of a better understanding of quasars interpreted as accreting system. Black hole masses are a key parameter in the evolution of the galaxies and in cosmology as well. Massive, fast outflows are affected by the ratio of radiation to gravitational forces. They provide feedback effects to the host galaxy, and are even invoked to account for the $M_{\mathrm{BH}}$-bulge velocity dispersion correlation [15–17]. The ratio between radiation and gravitation forces also influences broad-line region dynamics; lower column density material may flow out of the emitting region [14,18,19]. Black hole masses of high redshift quasars provide constraints on primordial black hole collapse (and references therein [20,21]). Overestimates by a factor as large as $\sim$100 for supermassive black holes at high $z$ may even pose a spurious challenge to concordance cosmology. Present-day estimates constrain the relation between the formation of the seed black holes and the collapse of the protogalaxy. Collapse of dark matter clumps yielding massive seeds that then rapidly grows by super-Eddington accretion appears necessary to explain the occurrence of supermassive black holes at very high redshifts.

## 3. Virial Black Hole Mass Estimates

The virial expression for the mass can be written as

$$M_{\mathrm{BH}} = f r_{\mathrm{BLR}} \delta v^2 / G, \tag{1}$$

where $f$ is a factor of order unity reflecting geometry and dynamics of the broad line emitting regions, and $r_{\mathrm{BLR}}$ is a representative distance of the broad-line emitting region (BLR).[1] The virial broadening $\delta v$ is provided by a measurement of the line profile width, which can be velocity dispersion $\sigma$, FWHM, or FWZI. The FWHM is by far the most handy measure employed as a VBE although some authors claim that a better estimator may be offered by the velocity dispersion [22]. The latter measure is, however, not defined in the case of Lorentzian profiles and not of easy interpretation in the case of

---

[1]    The following analysis pertains only type 1 quasars showing broad (FWHM > 1000 km s$^{-1}$) emission in permitted lines.

shifted profiles. The virial assumption had a spectacular confirmation (albeit limited to few cases) by type-1 sources for which several lines were monitored: the response time of the BLR line and the line width were found to be correlated exactly as expected for Keplerian motion around a massive central object, with $\delta v \propto r^{-\frac{1}{2}}$ [23].

We can subdivide the estimates of the radius of the BLR ($r_{BLR}$) as primary and secondary. Primary determinations are obtained as the peak or centroid of the cross-correlation function between continuum and line light curves. Secondary determinations stem from the correlation between $r_{BLR}$ and luminosity that has been derived from reverberation-mapped sources (e.g., [24–27]): $r_{BLR} \propto L^a$, $a \approx 0.5$–$0.65$ [24,28]. The relation takes different form for different lines, and has been defined for H$\beta$ and CIV. It then becomes possible to derive scaling laws for the black hole mass: $M_{BH} = M_{BH}(L, FWHM) = kL^a FWHM^b$. If $a = 0.5$, $b = 2$, the relation is consistent with the virial assumption (e.g., [29,30]), and the $r_{BLR}$ scaling laws. If $a \neq 0.5$, $b \neq 2$ (e.g., [31–33]), the mass scaling law still provides $M_{BH}$ estimates, but the accuracy of these estimates is likely to be sample-dependent (see Section 5.3).

The $M_{BH}$ scaling laws provide a simple recipe usable with single-epoch spectra of large sample of quasars. Estimates of the Eddington ratio are derived by applying a bolometric correction to the observed luminosity. The correction is typically assumed a factor 10–13 from the flux $\lambda f_\lambda$ at 5100 Å (measured in erg s$^{-1}$) and 3.4–5 from $\lambda f_\lambda$ measured at 1450 Å in the UV, [34,35]). For the sake of the present review, we will stay with these constant corrections without forgetting that different bolometric corrections should be defined along the quasar "main sequence" (introduced in Section 4), and that the correction is most likely luminosity as well as orientation dependent [36].

*Caveats*

The estimate of the $M_{BH}$ is based on several assumptions underlying reverberation mapping studies: among them, a compact region wetting the quasar continuum, and a fairly monotonic response of the line emitting gas. This latter assumption has been apparently challenged by the unpredicted behavior of NGC 5548 in 2014, where a time delay $\tau$ much shorter than expected was found when the source was in a high-luminosity state [37–39]. The physical origin of this behavior is not yet clear: in principle, shielding, optically thin gas, changing size of the continuum source could also give rise to a shorter (or a lack of) response to continuum change in the emitting line regions. In addition, periodic signals (in photometric and spectroscopic measurements) have been detected in a number of sources [40], including NGC 5548 (e.g., [41–45]). The origin of the periodicity is as yet unclear as well. A second, supermassive binary black hole has been suggested in some cases [41,46] but, in principle, black holes of masses much smaller than the primary could produce periodic photometric properties without leaving a detectable trace of their gravitational pull on the line profiles. As a matter of fact, the $r_{BLR}$-$L$ scaling relation has a non-negligible intrinsic dispersion, which may be, at least in part, accounted for by a dependence of $r_{BLR}$ estimates on the dimensionless accretion rate found in recent work [27].

The basic assumption underlying the search of virial broadening estimators is that a line (or line component) is symmetric and unshifted with respect to the quasar rest frame. Most line profiles are asymmetric, and shifted, but this has been ignored until recent years. If CIV FWHM is employed as a virial broadening estimator, the loss of information is so severe that the $M_{BH}$ distribution as a function of redshift cannot be distinguished if masses obtained from random values of the FWHM (e.g., [47]) are used in place of the actual estimates.

A single value of the structure factor is obtained by scaling the $M_{BH}$ to agree with the dynamical masses [48–52]. We expect that the geometry and dynamics of the BLR is related to the accretion mode, although it is as yet not unclear in which way. The structure factor is therefore likely to be different for different type-1 quasar populations. For Populations A and B (defined in Section 4), Collin et al. [53] find $f(\text{FWHM}) \approx 2$ and 0.5, respectively.

## 4. The Rehabilitating Power of Eigenvector 1

The eigenvector 1 (E1) was originally defined by a principal component analysis of ≈80 Palomar–Green quasars, and is associated with an anti-correlation between the strength of the FeII$\lambda$4570 blend, measured by the flux ratio $R_{\rm FeII} = F({\rm FeII}\lambda4570)/F({\rm H}\beta)$ and the FWHM of H$\beta$. The E1 is (at the very least) an useful tool to organize quasar diversity through a sequence (the main sequence, MS) in the so-called optical plane of the E1, defined by FWHM(H$\beta$) vs. $R_{\rm FeII}$. The 4D E1 parameters space includes two more parameters: the soft X-ray photon index $\Gamma_{\rm soft}$ associated with the accretion status, and the centroid at half maximum of CIV$\lambda$1549 c(1/2) CIV measuring the amplitude of the high ionization outflow originating from the BLR [54]. More parameters, however, correlate with E1 (see Table 1 of Sulentic et al. [55] and Fraix-Burnet et al. [56]). Since the analysis of Boroson and Green [57], MS trends have been found in works dealing with a number of phenomenological and physical correlates [58–65]. Recently, large Sloan Digital Sky Survey (SDSS) data samples were involved [9,66–69]. The E1 "rehabilitating" power stems, in this context, by the ability to identify systematic trends that may be missed and confused as statistical errors if structurally different sources are dumped together in a single sample.

The trends along the MS allow for the definition of spectral types [70] in bins of $\Delta R_{\rm FeII} = 0.5$ (A1, A2, A3, A4 from $R_{\rm FeII} = 0$ to $R_{\rm FeII} = 2$, tracing a change of $L/L_{\rm Edd}$ convolved with orientation, [71]) and FWHM (A for FWHM H$\beta < 4000$ km s$^{-1}$, B1, B1$^+$, B1$^{++}$ for broader sources in steps of 4000 km s$^{-1}$, tracing mainly a change in orientation). Quasars can be classified as belonging to two quasar Populations, A and B [72]. Population A (FWHM H$\beta < 4000$ km/s) includes NLSy1s. Population A and B(roader) sources are most likely associated with a different accretion mode, since Population A are at $L/L_{\rm Edd} > 0.2$–0.3 which is the theoretical limit above and is an optically and geometrically thick advection-dominated accretion disk forms [73]. Population B of low $L/L_{\rm Edd}$ may be consistent with a flat $\alpha$ disk. The distinction between Populations A and B separates sources that are also called wind- and disk-dominated by Richards et al. [66], or Populations 1 and 2 by Collin et al. [53]. It supersedes the distinction between NLSy1s and rest of type-1 AGNs, which is based on a FWHM limit that has played an important historical role but is physically arbitrary. At low $z$ (<0.7), Population A show low $M_{\rm BH}$, high $L/L_{\rm Edd}$ Population B, high $M_{\rm BH}$, low $L/L_{\rm Edd}$, a reflection of the "downsizing" of nuclear activity: practically no black hole with very large mass ($M_{\rm BH} \sim 10^9$–$10^{10}$ M$_\odot$) is accreting super-Eddington in the local Universe.

## 5. Virial Broadening Estimators along the Quasar MS

### 5.1. LILs: H$\beta$ and MgII$\lambda$2800

The profiles of LILs like H$\beta$ and MgII$\lambda$2800 change along the quasar MS. If we want to extract a VBE, we have to consider the behavior of H$\beta$ and MgII$\lambda$2800 from extreme Population B to extreme Population A.

In Population B, the broad profiles of H$\beta$ and MgII$\lambda$2800 are most frequently redward asymmetric. Composite spectra of individual spectral types can be modeled by a broad Gaussian component (the BC, symmetric and unshifted, assumed to be the virialized component), and a redshifted very-broad Gaussian component. This latter component presumably associated with "perturbed" virialized motions or gravitational redshift in the inner BLR [74] systematically increases the line width with respect to the BC that provides the VBE.

At the other end of the MS extreme, Population A sources show narrower, Lorentzian-like profiles, slightly blueward asymmetric. The H$\beta$ and MgII profiles can be modeled by Lorentzian functions plus a blueshifted excess modeled with a skewed Gaussian [12,13]. Both the LIL resonance line MgII$\lambda$2800 and H$\beta$ are affected by low-ionization outflows detected in the extreme Population A corresponding to spectral types A3–A4 [13]. However, in most of Population A, the LILs are dominated by a symmetric, virialized broad component [8,13,30,75].

We can define a parameter $\zeta$ yielding a correction to the observed profile to obtain a VBE, as follows:

$$\zeta = \frac{FWHM_{VBE}}{FWHM_{obs}} \approx \frac{FWHM_{BC}}{FWHM_{obs}} \approx \frac{FWHM_{symm}}{FWHM_{obs}}, \tag{2}$$

where the VBE FWHM can be considered best estimated by the FWHM of the broad component $FWHM_{BC}$, whose proxy can be obtained by symmetrizing the observed FWHM by various corrections described below. Even if the non-virial broadening mechanism is different along the E1 sequence, $0.75 \leq \zeta \leq 1.0$ for both H$\beta$ and MgII. This implies that, even if LILs are not always asymmetric, a modest correction is, on average, sufficient to retrieve a VBE from the observed FWHM. Table 1 reports the current estimates along with bibliographic references.

**Table 1.** The LIL and IIL $\zeta$ (Equation 2) factor for the spectral types of the quasar MS, listed for the spectral types (SpT) of highest occupation.

| SpT | H$\beta$ | MgII | AlIII | Ref. |
|:---:|:---:|:---:|:---:|:---:|
| A3–A4 | 0.8/0.9 | 0.75/0.8 | 1.0 | [12,76] |
| A1–A2 | 1.0 | 1.0 | 1.0 | [12,76] |
| B1 | 0.8 | 0.9 | 1.35 | [13,45] |
| B1+/B1++ | 0.8 | 0.9 | 1.35 | [13,45] |

### 5.2. LIL VBE at High L

At low $z$, the quasar population reaches bolometric luminosity $\sim 10^{47}$ erg s$^{-1}$ at most. It is interesting to consider the LIL VBE behavior if we add to a local sample sources more luminous than this limit. Composite spectra are helpful to outline the LIL H$\beta$ behavior over a wide luminosity range. Marziani et al. [11] computed composites in step of 1 dex for 6 dex in luminosity, joining an HE/ISAAC high-$L$ sample (52 sources) and an SDSS sample from Zamfir et al. [77]. The H$\beta$ line becomes broader with increasing $L$ (over $43 < \log L < 48.5$ [erg/s]), but shapes are similar: the Population A/Population B differences are preserved at high $L$ [11]. At the same time, the minimum FWHM(H$\beta$) increases with luminosity. This result is consistent with the virial assumption and with the $r_{BLR}$ scaling law:

$$M_{BH} \propto r_{BLR} FWHM^2 \propto L^a FWHM^2 \propto (L/M)^{-1} L^{(1-2a)/2}. \tag{3}$$

The minimum, luminosity-dependent FWHM is obtained for a limiting Eddington ratio $\approx 1$. As a consequence, the Population A limit is also luminosity dependent.

Extracting a VBE from LIL H$\beta$ at high-$L$ is possible by employing different symmetrization methods:

- substitution of the BC in place of the full H$\beta$ profile. This method requires a multicomponent maximum likelihood fitting, and is fairly reliable for Population A (even for extreme Population A) and Population B spectral types where the VBC is creating a profile inflection;
- symmetrization of the profile: FWHM$_{symm}$ = FWHM$-2$ c(1/2);
- correction based on spectral type. In practice, this means to correct H$\beta$ for Population B sources by a factor $\zeta$ as reported in Table 1.

All symmetrization methods were found to be equivalent at low- and high-$L$ [45]. In other words, the H$\beta$ profile shapes at high $L$ are consistent with those at low-$z$, lower $L$ (with some caveats, [11]).

### 5.3. CIV

Virial broadening estimators extracted from the HIL CIV$\lambda$1549 along the E1 sequence are unfortunately not easy to define. If we scale the H$\beta$ emission and overlay it on CIV$\lambda$1549 (Figure 1), we obtain a strong excess of blueshifted emission. The profile can be interpreted as an almost symmetric, unshifted "virialized" emitting region + an outflow/wind component that dominates in A3/A4 spectral types (e.g., [78–81]) and at high luminosity. The largest shifts of CIV$\lambda$1549 centroid at 1/2 along the MS are found in Population A [4,6,14].

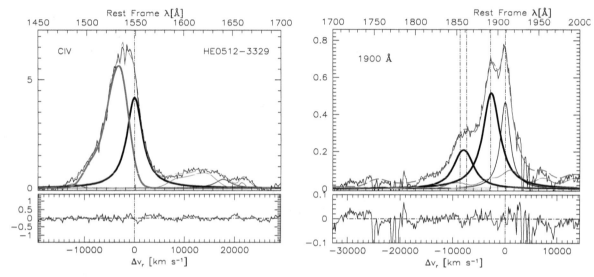

**Figure 1. Left**: the CIV profile of HE0512, a Population A quasar in the sample of Sulentic et al. [14]. The profile has been decomposed into a symmetric, unshifted component (the virialized component, thick black line) and a blue shifted component modeled by a skewed Gaussian (thick blue line). The flat-topped emission on the red side of CIV is mainly HeII$\lambda$1640; **Right**: the 1900 blend for HE0512. The AlIII and SiIII] lines at 1860 Å and 1892 Å are unshifted and symmetric, and their FWHMs are in agreement with the H$\beta$ and of the CIV virialized component profile. The thick black components can be thought as scaled H$\beta$ profiles, as in the case of CIV. Note that AlIII$\lambda$1860 lacks the blue-shifted excess of CIV. The thin line shows the CIII] $\lambda$1909 profile whose intensity remains highly uncertain because of the heavy blending with FeIII emission.

The Vestergaard and Peterson [29] scaling laws

$$\log M_{\mathrm{BH}} = c + a \log L + b \log \mathrm{FWHM},\tag{4}$$

assumed that the width of CIV and H$\beta$ are equivalent, with $a \approx 0.5$, $b = 2$, and $c$ an average constant difference between the 5100 and the 1450 luminosity. Considering the FWHM CIV as a proxy of FWHM H$\beta$ causes a bias along the E1 sequence. Especially for extreme Population A, errors can be as large as 2 dex. Figure 2 is the statement of the ensuing BH mass taboo for CIV.

Figure 2 (based on the data of Sulentic et al. [6]) emphasizes a systematic trend for which a corrective could be estimated. To this aim, we defined a sample of RQ quasars with CIV$\lambda$1549 observations including 70 sources at $z < 1$ [6], and 25 high-luminosity sources at $z > 1.4$ [14]. Matching H$\beta$ data are available from our previous observations [11,82] or from published spectra. At $L > 10^{47}$ erg s$^{-1}$, high amplitude CIV 1549 blueshifts in both Population A and B are observed with median $\approx 3000$ km s$^{-1}$ for Population A; two extreme cases involve CIV c(1/2) blueshift amplitude larger than 5000 km s$^{-1}$. Widespread powerful outflows are affecting both Population A and B sources [14,83] with worrisome implications for $M_{\mathrm{BH}}$ estimates from CIV$\lambda$1549 FWHM.

Population A sources are more frequently selected at high $z$, high $L$. Figure 2 of [84] depicts what may be called an Eddington ratio bias: for a fixed mass, higher Eddington ratio sources are preferentially better sampled. In this way, the FWHM CIV (if left uncorrected) may lead to systematic over-estimates of $M_{\mathrm{BH}}$ by even one/two orders of magnitude, with the potential of creating the (spurious) result of a population of extremely massive black holes at high redshift ($z > 2$).

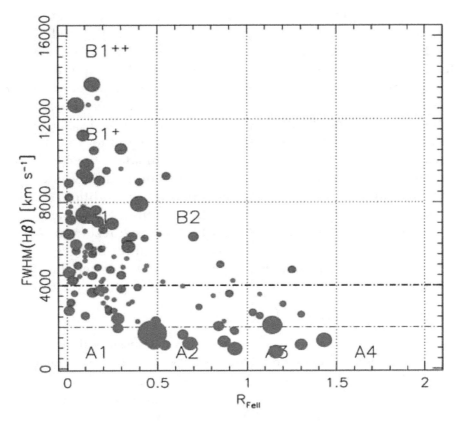

**Figure 2.** Bias of the $M_{\mathrm{BH}}$ estimates in the optical E1 plane FWHM(H$\beta$) vs. $R_{\mathrm{FeII}}$, for the full sample of Sulentic et al. [6]. Grey symbols represent differences $|\Delta \log M| = |\log M(CIV) - \log M(H\beta)| \leq 0.3$, blue dots $\Delta \log M > 0.3$, red circles $\Delta \log M < -0.3$. The size of the circles is proportional to the amplitude of the difference, with the largest blue circles showing CIV masses in excess by a factor 100. Mid-sized circles indicate a disagreement of a factor $\approx 10$. The labels identify the spectral types as described in Section 4 and the thick dot-dashed line at 4000 km s$^{-1}$ shows the boundary between Population A and Population B.

### 5.4. A CIV$\lambda$1549 VBE

The centroid shift at half-maximum c(1/2) is correlated with the FWHM in the CIV line, implying that the the FWHM excess with respect to H$\beta$ is associated with a blueshifted component, as shown in Figure 1.[2] The comparison of Figure 1 has been possible because we have H$\beta$ observations for this source. However, H$\beta$ observations are available only for a few hundred high-redshift quasars since the line is shifted in the IR.

Nonetheless, there have been several attempts in the last few years to define scaling laws that corrected for the non/virial contamination of CIV emission. A scaling law that assumes $M_{\mathrm{BH}} \propto$ FWHM$^{0.5}$ [32] accounts for the over-broadening of Population A sources, but overcorrects for Population B [45]. In general, corrections dependent on $L/L_{\mathrm{Edd}}$ (which is correlated with FWHM of CIV, [85]) or an $L/L_{\mathrm{Edd}}$ proxy such as the SiIV + OIV]1400 blend/CIV 1549 ratio are promising [9] but tend not to work well for Population B. Empirical corrections based on CIV blueshift work fairly well for high-$L$ sources only [83,86].

It is important to consider that there is a threshold in CIV shift amplitude (c(1/2)) at $L/L_{\mathrm{Edd}} \approx 0.2$ [14]. There is a strong correlation with $L/L_{\mathrm{Edd}}$ if blueshifts are significant. There is also a weak but significant correlation with luminosity in the sample of Sulentic et al. [14]: the partial

---

[2]   This happens also for MgII, but with much lower blueward displacements (few hundreds vs. few thousands km s$^{-1}$).

correlation coefficient between c(1/2) and $L$ ($L/L_{Edd}$ hidden) is significant at about a $2\sigma$ confidence level. A multivariate analysis confirms the blueshift dependence on both $L$ and $L/L_{Edd}$ in that sample. The blueshift dependences are consistent with a radiation-driven outflow, with a slope $\approx 0.15$ for $\log L$, and slope $\approx 0.5$ for $L/L_{Edd}$. A pure dependence on $L$ arises for $L/L_{Edd}$ in a small range. A strong dependence on $L/L_{Edd}$ and a weak dependence on $L$ can be obtained under a variety of physical scenarios. We expect that strength and form of $L$ and $L/L_{Edd}$ correlations are sample dependent [14].

Considering the threshold and the dependence on both $L$ and $L/L_{Edd}$, using c(1/2) as a proxy for $L/L_{Edd}$, HIL CIV corrections based on c(1/2) and $L$ reduce scatter to 0.33 dex with respect to H$\beta$ estimates, providing an unbiased $M_{BH}$ estimator. Unfortunately, the correction coefficients are different for Pops. A and B. For Population B, the correction is highly uncertain in the sample of Sulentic et al. [14], and a larger sample of sources is needed. In addition, (1) corrections derived by Coatman et al. [83] may be sample dependent because of the Eddington ratio bias; and (2) any correction based on c(1/2) requires a precise estimate of the quasar rest frame of reference, which is not easily obtained from UV data. A theoretical correction requires that c(1/2) CIV and ionization conditions in the BLR are accounted for, and has not been competed as yet. It is therefore not obvious whether $M_{BH}$ estimates based on CIV can be reliable for large samples of quasars.

*5.5. Intermediate Ionization Lines (IILs)*

An alternative to the use of CIV is to resort to the emission lines of the 1900 Å blend, whose constituents are mainly the resonant doublet of AlIII at $\lambda$1860 Å, and the intercombination lines SiIII] and CIII] at $\lambda$ 1892 and $\lambda$1909 Å, respectively (Table 1 of Negrete et al. [87] provides detailed information of the atomic transitions involved). We base our analysis on $\approx 80$ objects of the CIV sample described in Section 5.3 for which the 1900 blend has been covered. CIII]$\lambda$1909 measurements have serious problems: in Population A, CIII] is faint, and blended with strong and diffuse FeIII emission. In Population B, the CIII]$\lambda$1909 line is affected by VBC. However, considering AlIII and SiIII] it is possible to obtain a VBE consistent with H$\beta$. To this aim is necessary to assume that FWHM AlII 1860 = FWHM SiIII] 1892 i.e., to anchor the FWHM AlII to FWHM SiIII]. A source of concern is that AlIII is a resonance doublet ($^2P^o_{\frac{3}{2},\frac{1}{2}} \rightarrow ^2 S_{\frac{1}{2}}$) and part of its emission may originate in an outflow, where the AlIII, as does CIV, acts as a resonant scatterer of continuum photons. However, measured AlIII shifts are $< 0.2$ CIV shifts and the AlIII and SiIII] profiles looks fairly symmetric in the wide majority of cases (Figure 1 shows a typical case).

If we compare the FWHM of the IILs to H$\beta$, we find that it is in very good agreement with Population A. Population B IILs are narrower than H$\beta$, but a modest correction is needed to enforce an unbiased agreement: FWHM H$\beta_{BC}$ = 1.35 FWHM AlIII$_{BC}$. In other words, $\xi \approx 1.35$ (Table 1) is needed to rescale the line widths to the one of H$\beta$ BC assumed as a reference VBE.

## 6. Photoionization Masses

The ionization parameter $U$ can be written as

$$U = \frac{Q(H)}{4\pi r_{BLR}^2 cn},$$ (5)

where $Q(H)$ is the number of ionizing photons, $n$ the hydrogen density, and $c$ the speed of light. The expression for $U$ can be inverted to obtain the BLR radius

$$r_{BLR} = \left( \frac{Q(H)}{4\pi Ucn} \right)^{1/2}.$$ (6)

The radius $r_{BLR}$ estimates from photo-ionization agree with $c\tau$ from reverberation mapping for a sample of 12 sources [76]. The photon flux $n \cdot U$ is estimated using diagnostic ratios involving AlIII 1860, SiIII] 1892, SiII 1816, CIV 1549, SiIV + OIV] 1400. The photoionization method provides an unbiased estimator of $r_{BLR}$ in the sample of Negrete et al. [76], but $M_{BH}$ estimates at high $L$ remain largely untested [88]. The photoionization method, in principle, could avoid the dispersion intrinsic to the scaling laws with $L$. To obtain accurate individual estimates of $M_{BH}$ is then necessary to have an accurate knowledge of $f$ and of orientation effects that influence the FWHM. Both are still poorly known at the time of writing.

## 7. Conclusions

Retrieving a VBE that is representative of the broadening due virial motions in the low-ionization part of the BLR is a major goal in order to reduce systematic and statistical effects in single-epoch $M_{BH}$ determinations applied to large samples of quasars. The following remarks emerge from the present review.

- Low-ionization lines (H$\beta$, MgII 2800) provide reliable virial broadening estimators by applying corrections to the observed line width. The corrections depend on the spectral type along the E1 MS, but they are relatively small (less than 30%), and work up to the highest $L$ of quasars.
- The HIL CIV$\lambda$1549 is not immediately providing a reliable virial broadening estimator. The profile is broadened by an excess emission on its blue side. The shift amplitude depends on both $L/L_{Edd}$ and $L$. Large shifts are observed in Population A, with Eddington ratio above a critical $L/L_{Edd} \approx 0.2$.

- It is possible to apply corrections to the observed CIV broadening, but they remain cumbersome even for Population A. Population B sources at low Eddington ratio require a different correction (still ill-defined by the analysis of Marziani et al. [45] as Population B sources are most affected by the Eddington ratio bias mentioned in Section 5.3).

- Preliminary results on the 1900 blend indicate that the IILs lines could provide a better choice than CIV; IIL FWHM measurements appear intrinsically more robust than those of CIV since they do not require corrections based on shift measurement with respect to rest frame. However, more data are needed to assess their reliability.

The ultimate solution may be to abandon scaling laws altogether and to attempt $M_{BH}$ estimates on an individual basis, considering $r_{BLR}$ from photoionization and $f = f(L/L_{Edd}, L, \ldots)$ (where $\ldots$ indicates any relevant physical parameter, for example the spin parameter of the black hole) as well as orientation effects on the VBE along the E1 sequence.

**Acknowledgments:** A.d.O., and M.L.M.A. acknowledge financial support from the Spanish Ministry for Economy and Competitiveness through grant AYA2016-76682-C3-1-P. D. D. and A. N. acknowledge support from grants PAPIIT108716, UNAM, and CONACyT221398. This research is part of the projects 176003 "Gravitation and the large scale structure of the Universe" and 176001 "Astrophysical spectroscopy of extragalactic objects" supported by the Ministry of Education and Science of the Republic of Serbia.

**Author Contributions:** This review is partly based on the results of several research papers by the authors: [45,76] (A.N.), [12,13,45] (A.D.O.), [11,45,76,76] (D.D.), [45] (M.D.O.), [45] (M.L.M.A.), [11–13,45,76] (P.M.). For the present review, E.B., N.B., A.D.O., D.D., M.D.O. sent comments and suggestions. P.M. wrote the paper.

## Abbreviations

The following abbreviations are used in this manuscript:

| | |
|---|---|
| BC | Broad component |
| BLR | Broad line region |
| E1 | Eigenvector 1 |
| HE | Hamburg-ESO |
| HIL | High ionization line |
| IIL | Intermediate ionization line |

| | |
|---|---|
| IP | Ionization potential |
| ISAAC | Infrared Spectrometer and Array Camera |
| LIL | Low ionization line |
| FWHM | Full width half maximum |
| FWHM | Full width zero intensity |
| MS | Main sequence |
| NLSy1 | Narrow-line Seyfert 1 |
| SDSS | Sloan digital sky survey |
| VBE | Virial broadening estimator |

## References

1. Berk, D.E.V.; Richards, G.T.; Bauer, A.; Strauss, M.A.; Schneider, D.P.; Heckman, T.M.; York, D.G.; Hall, P.B.; Fan, X.; Knapp, G.R.; et al. Composite Quasar Spectra from the Sloan Digital Sky Survey. *Astron. J.* **2001**, *122*, 549–564.

2. Burbidge, G.R.; Burbidge, E.M. *Quasi-Stellar Objects*; Freeman: San Francisco, CA, USA, 1967.

3. Marziani, P.; Sulentic, J.W. Quasar Outflows in the 4D Eigenvector 1 Context. *Astron. Rev.* **2012**, *7*, 33–57.

4. Marziani, P.; Sulentic, J.W. Estimating black hole masses in quasars using broad optical and UV emission lines. *New Astron. Rev.* **2012**, *56*, 49–63.

5. Shen, Y. The mass of quasars. *Bull. Astron. Soc. India* **2013**, *41*, 61–115.

6. Sulentic, J.W.; Bachev, R.; Marziani, P.; Negrete, C.A.; Dultzin, D. Civ$\lambda$1549 as an Eigenvector 1 Parameter for Active Galactic Nuclei. *Astrophys. J.* **2007**, *666*, 757–777.

7. Netzer, H.; Lira, P.; Trakhtenbrot, B.; Shemmer, O.; Cury, I. Black Hole Mass and Growth Rate at High Redshift. *Astrophys. J.* **2007**, *671*, 1256–1263.

8. Mejía-Restrepo, J.E.; Trakhtenbrot, B.; Lira, P.; Netzer, H.; Capellupo, D.M. Active galactic nuclei at z ~ 1.5: II. Black Hole Mass estimation by means of broad emission lines. *Mon. Not. R. Astron. Soc.* **2016**, *460*, 187–211.

9. Brotherton, M.S.; Runnoe, J.C.; Shang, Z.; DiPompeo, M.A. Bias in C IV-based quasar black hole mass scaling relationships from reverberation mapped samples. *Mon. Not. R. Astron. Soc.* **2015**, *451*, 1290–1298.

10. Brotherton, M.S.; Runnoe, J.C.; Shang, Z.; Varju, M. Further Rehabilitating CIV-based Black Hole Mass Estimates in Quasars. In Proceedings of the 228th American Astronomical Society Meeting, San Diego, CA, USA, 12–16 June 2016; Abstracts, p. 400.05.

11. Marziani, P.; Sulentic, J.W.; Stirpe, G.M.; Zamfir, S.; Calvani, M. VLT/ISAAC spectra of the H$\beta$ region in intermediate-redshift quasars. III. H$\beta$ broad-line profile analysis and inferences about BLR structure. *Astron. Astrophys.* **2009**, *495*, 83–112.

12. Marziani, P.; Sulentic, J.W.; Plauchu-Frayn, I.; del Olmo, A. Low-Ionization Outflows in High Eddington Ratio Quasars. *Astrophys. J.* **2013**, *764*, 150.

13. Marziani, P.; Sulentic, J.W.; Plauchu-Frayn, I.; del Olmo, A. Is Mg II 2800 a Reliable Virial Broadening Estimator for Quasars? *Astron. Astrophys.* **2013**, *555*, A89.

14. Sulentic, J.W.; del Olmo, A.; Marziani, P.; Martínez-Carballo, M.A.; D'Onofrio, M.; Dultzin, D.; Perea, J.; Martínez-Aldama, M.L.; Negrete, C.A.; Stirpe, G.M.; et al. What does Civ$\lambda$1549 tell us about the physical driver of the Eigenvector Quasar Sequence? *arXiv* **2017**, arXiv:1708.03187.

15. Fabian, A.C. Observational Evidence of Active Galactic Nuclei Feedback. *Annu. Rev. Astron. Astrophys.* **2012**, *50*, 455–489.

16. Kormendy, J.; Ho, L.C. Coevolution (Or Not) of Supermassive Black Holes and Host Galaxies. *Annu. Rev. Astron. Astroph.* **2013**, *51*, 511–653.

17. King, A.; Pounds, K. Powerful Outflows and Feedback from Active Galactic Nuclei. *Annu. Rev. Astron. Astroph.* **2015**, *53*, 115–154.

18. Ferland, G.J.; Hu, C.; Wang, J.; Baldwin, J.A.; Porter, R.L.; van Hoof, P.A.M.; Williams, R.J.R. Implications of Infalling Fe II-Emitting Clouds in Active Galactic Nuclei: Anisotropic Properties. *Astrophys. J. Lett.* **2009**, *707*, L82–L86.

19. Marziani, P.; Carballo, M.A.M.; Sulentic, J.W.; Del Olmo, A.; Stirpe, G.M.; Dultzin, D. The most powerful quasar outflows as revealed by the Civ $\lambda$1549 resonance line. *Astrophys. Space Sci.* **2016**, *361*, 29.

20. Smith, A.; Bromm, V.; Loeb, A. The first supermassive black holes. *arXiv* **2017**, arXiv:1703.03083.

21. Trakhtenbrot, B.; Urry, C.M.; Civano, F.; Rosario, D.J.; Elvis, M.; Schawinski, K.; Suh, H.; Bongiorno, A.; Simmons, B.D. An over-massive black hole in a typical star-forming galaxy, 2 billion years after the Big Bang. *Science* **2015**, *349*, 168–171.

22. Denney, K.D.; Pogge, R.W.; Assef, R.J.; Kochanek, C.S.; Peterson, B.M.; Vestergaard, M. C IV Line-width Anomalies: The Perils of Low Signal-to-noise Spectra. *Astrophys. J.* **2013**, *775*, 60.

23. Peterson, B.M.; Wandel, A. Keplerian Motion of Broad-Line Region Gas as Evidence for Supermassive Black Holes in Active Galactic Nuclei. *Astrophys. J. Lett.* **1999**, *521*, L95–L98.

24. Kaspi, S.; Smith, P.S.; Netzer, H.; Maoz, D.; Jannuzi, B.T.; Giveon, U. Reverberation Measurements for 17 Quasars and the Size-Mass-Luminosity Relations in Active Galactic Nuclei. *Astrophys. J.* **2000**, *533*, 631–649.

25. Kaspi, S.; Brandt, W.N.; Maoz, D.; Netzer, H.; Schneider, D.P.; Shemmer, O. Reverberation Mapping of High-Luminosity Quasars: First Results. *Astrophys. J.* **2007**, *659*, 997–1007.

26. Bentz, M.C.; Peterson, B.M.; Pogge, R.W.; Vestergaard, M. The Black Hole Mass-Bulge Luminosity Relationship for Active Galactic Nuclei From Reverberation Mapping and Hubble Space Telescope Imaging. *Astrophys. J. Lett.* **2009**, *694*, L166–L170.

27. Du, P.; Lu, K.X.; Hu, C.; Qiu, J.; Li, Y.R.; Huang, Y.K.; Wang, F.; Bai, J.M.; Bian, W.H.; Yuan, Y.F.; et al. Supermassive Black Holes with High Accretion Rates in Active Galactic Nuclei. VI. Velocity-resolved Reverberation Mapping of the H$\beta$ Line. *Astrophys. J.* **2016**, *820*, 27.

28. Bentz, M.C.; Peterson, B.M.; Pogge, R.W.; Vestergaard, M.; Onken, C.A. The Radius-Luminosity Relationship for Active Galactic Nuclei: The Effect of Host-Galaxy Starlight on Luminosity Measurements. *Astrophys. J.* **2006**, *644*, 133–142.

29. Vestergaard, M.; Peterson, B.M. Determining Central Black Hole Masses in Distant Active Galaxies and Quasars. II. Improved Optical and UV Scaling Relationships. *Astrophys. J.* **2006**, *641*, 689–709.

30. Trakhtenbrot, B.; Netzer, H. Black hole growth to $z = 2 - $ I. Improved virial methods for measuring $M_{BH}$ and $L/L_{Edd}$. *Mon. Not. R. Astron. Soc.* **2012**, *427*, 3081–3102.

31. Shen, Y.; Liu, X. Comparing Single-epoch Virial Black Hole Mass Estimators for Luminous Quasars. *Astrophys. J.* **2012**, *753*, 125.

32. Park, D.; Woo, J.H.; Denney, K.D.; Shin, J. Calibrating C-IV-based Black Hole Mass Estimators. *Astrophys. J.* **2013**, *770*, 87.

33. Shen, Y.; Brandt, W.N.; Richards, G.T.; Denney, K.D.; Greene, J.E.; Grier, C.J.; Ho, L.C.; Peterson, B.M.; Petitjean, P.; Schneider, D.P.; et al. The Sloan Digital Sky Survey Reverberation Mapping Project: Velocity Shifts of Quasar Emission Lines. *Astrophys. J.* **2016**, *831*, 7.

34. Elvis, M.; Wilkes, B.J.; McDowell, J.C.; Green, R.F.; Bechtold, J.; Willner, S.P.; Oey, M.S.; Polomski, E.; Cutri, R. Atlas of quasar energy distributions. *Astrophys. J. Suppl. Ser.* **1994**, *95*, 1–68.

35. Richards, G.T.; Lacy, M.; Storrie-Lombardi, L.J.; Hall, P.B.; Gallagher, S.C.; Hines, D.C.; Fan, X.; Papovich, C.; Berk, D.E.V.; Trammell, G.B.; et al. Spectral Energy Distributions and Multiwavelength Selection of Type 1 Quasars. *Astrophys. J. Suppl. Ser.* **2006**, *166*, 470–497.

36. Runnoe, J.C.; Shang, Z.; Brotherton, M.S. The orientation dependence of quasar spectral energy distributions. *Mon. Not. R. Astron. Soc.* **2013**, *435*, 3251–3261.

37. Horne, K.; AGN STORM collaboration. Echo Mapping of the Broad Emission-Line Region in NGC 5548. *Zenodo* **2017**, doi:10.5281/zenodo.569512.

38. Pei, L.; Fausnaugh, M.M.; Barth, A.J.; Peterson, B.M.; Bentz, M.C.; De Rosa, G.; Denney, K.D.; Goad, M.R.; Kochanek, C.S.; Korista, K.T.; et al. Space Telescope and Optical Reverberation Mapping Project. V. Optical Spectroscopic Campaign and Emission-line Analysis for NGC 5548. *Astrophys. J.* **2017**, *837*, 131.

39. Fausnaugh, M.M.; Denney, K.D.; Barth, A.J.; Bentz, M.C.; Bottorff, M.C.; Carini, M.T.; Croxall, K.V.; De Rosa, G.; Goad, M.R.; Horne, K.; et al. Space Telescope and Optical Reverberation Mapping Project. III. Optical Continuum Emission and Broadband Time Delays in NGC 5548. *Astrophys. J.* **2016**, *821*, 56.

40. Bon, E.; Marziani, P.; Bon, N. Periodic optical variability of AGN. In Proceedings of the IAU Symposium 324: New Frontiers in Black Hole Astrophysics, Ljubljana, Slovenia, 12–16 September 2016; pp. 176–179.

41. Bon, E.; Jovanović, P.; Marziani, P.; Shapovalova, A.I.; Bon, N.; Borka Jovanović, V.; Borka, D.; Sulentic, J.; Popović, L.Č. The First Spectroscopically Resolved Sub-parsec Orbit of a Supermassive Binary Black Hole. *Astrophys. J.* **2012**, *759*, 118.

42. Graham, M.J.; Djorgovski, S.G.; Stern, D.; Drake, A.J.; Mahabal, A.A.; Donalek, C.; Glikman, E.; Larson, S.; Christensen, E. A systematic search for close supermassive black hole binaries in the Catalina Real-time Transient Survey. *Mon. Not. R. Astron. Soc.* **2015**, *453*, 1562–1576.

43. Bon, E.; Zucker, S.; Netzer, H.; Marziani, P.; Bon, N.; Jovanović, P.; Shapovalova, A.I.; Komossa, S.; Gaskell, C.M.; Popović, L.Č.; et al. Evidence for Periodicity in 43 year-long Monitoring of NGC 5548. *Astrophys. J. Suppl. Ser.* **2016**, *225*, 29.

44. Charisi, M.; Bartos, I.; Haiman, Z.; Price-Whelan, A.M.; Graham, M.J.; Bellm, E.C.; Laher, R.R.; Márka, S. A population of short-period variable quasars from PTF as supermassive black hole binary candidates. *Mon. Not. R. Astron. Soc.* **2016**, *463*, 2145–2171.

45. Marziani, P.; Bon, E.; Bon, N.; Dultzin, D.; Del Olmo, A.; D'Onofrio, M. **2017**, in preparation.

46. Li, Y.R.; Wang, J.M.; Ho, L.C.; Lu, K.X.; Qiu, J.; Du, P.; Hu, C.; Huang, Y.K.; Zhang, Z.X.; Wang, K.; et al. Spectroscopic Indication of a Centi-parsec Supermassive Black Hole Binary in the Galactic Center of NGC 5548. *Astrophys. J.* **2016**, *822*, 4.

47. Croom, S.M. Do quasar broad-line velocity widths add any information to virial black hole mass estimates? *arXiv* **2011**, arXiv:1105.4391.

48. Woo, J.H.; Treu, T.; Barth, A.J.; Wright, S.A.; Walsh, J.L.; Bentz, M.C.; Martini, P.; Bennert, V.N.; Canalizo, G.; Filippenko, A.V.; et al. The Lick AGN Monitoring Project: The $M_{\mathrm{BH}}$-$\sigma_*$ Relation for Reverberation-mapped Active Galaxies. *Astrophys. J.* **2010**, *716*, 269–280.

49. Gültekin, K.; Richstone, D.O.; Gebhardt, K.; Lauer, T.R.; Tremaine, S.; Aller, M.C.; Bender, R.; Dressler, A.; Faber, S.M.; Filippenko, A.V.; et al. The M-$\sigma$ and M-L Relations in Galactic Bulges, and Determinations of Their Intrinsic Scatter. *Astrophys. J.* **2009**, *698*, 198–221.

50. Onken, C.A.; Ferrarese, L.; Merritt, D.; Peterson, B.M.; Pogge, R.W.; Vestergaard, M.; Wandel, A. Supermassive Black Holes in Active Galactic Nuclei. II. Calibration of the Black Hole Mass-Velocity Dispersion Relationship for Active Galactic Nuclei. *Astrophys. J.* **2004**, *615*, 645–651.

51. Ferrarese, L.; Merritt, D. A Fundamental Relation between Supermassive Black Holes and Their Host Galaxies. *Astrophys. J. Lett.* **2000**, *539*, L9–L12.

52. Graham, A.W.; Onken, C.A.; Athanassoula, E.; Combes, F. An expanded $M_{\mathrm{bh}}$-$\sigma$ diagram, and a new calibration of active galactic nuclei masses. *Mon. Not. R. Astron. Soc.* **2011**, *412*, 2211–2228.

53. Collin, S.; Kawaguchi, T.; Peterson, B.M.; Vestergaard, M. Systematic effects in measurement of black hole masses by emission-line reverberation of active galactic nuclei: Eddington ratio and inclination. *Astron. Astrophys.* **2006**, *456*, 75–90.

54. Sulentic, J.W.; Marziani, P.; Zwitter, T.; Dultzin-Hacyan, D.; Calvani, M. The Demise of the Classical Broad-Line Region in the Luminous Quasar PG 1416-129. *Astrophys. J. Lett.* **2000**, *545*, L15–L18.

55. Sulentic, J.; Marziani, P.; Zamfir, S. The Case for Two Quasar Populations. *Open Astron.* **2011**, *20*, 427–434.

56. Fraix-Burnet, D.; Marziani, P.; D'Onofrio, M.; Dultzin, D. The Phylogeny of Quasars and the Ontogeny of Their Central Black Holes. *Front. Astron. Space Sci.* **2017**, *4*, 1.

57. Boroson, T.A.; Green, R.F. The emission-line properties of low-redshift quasi-stellar objects. *Astrophys. J. Suppl. Ser.* **1992**, *80*, 109–135.

58. Dultzin-Hacyan, D.; Sulentic, J.; Marziani, P.; Calvani, M.; Moles, M. A Correlation Analysis for Emission Lines in 52 AGN. In *IAU Colloquium 159: Emission Lines in Active Galaxies: New Methods and Techniques*; Peterson, B.M., Cheng, F.Z., Wilson, A.S., Eds.; Astronomical Society of the Pacific Conference Series; The Astronomical Society of the Pacific: San Francisco, CA, USA, 1997; Volume 113, p. 262.

59. Shang, Z.; Wills, B.J.; Robinson, E.L.; Wills, D.; Laor, A.; Xie, B.; Yuan, J. The Baldwin Effect and Black Hole Accretion: A Spectral Principal Component Analysis of a Complete Quasar Sample. *Astrophys. J.* **2003**, *586*, 52–71.

60. Kruczek, N.E.; Richards, G.T.; Gallagher, S.C.; Deo, R.P.; Hall, P.B.; Hewett, P.C.; Leighly, K.M.; Krawczyk, C.M.; Proga, D. C IV Emission and the Ultraviolet through X-Ray Spectral Energy Distribution of Radio-quiet Quasars. *Astron. J.* **2011**, *142*, 130.

61. Tang, B.; Shang, Z.; Gu, Q.; Brotherton, M.S.; Runnoe, J.C. The Optical and Ultraviolet Emission-line Properties of Bright Quasars with Detailed Spectral Energy Distributions. *Astrophys. J. Suppl. Ser.* **2012**, *201*, 38.

62. Kuraszkiewicz, J.; Wilkes, B.J.; Schmidt, G.; Smith, P.S.; Cutri, R.; Czerny, B. Principal Component Analysis of the Spectral Energy Distribution and Emission Line Properties of Red 2MASS Active Galactic Nuclei. *Astrophys. J.* **2009**, *692*, 1180–1189.

63. Mao, Y.F.; Wang, J.; Wei, J.Y. Extending the Eigenvector 1 space to the optical variability of quasars. *Res. Astron. Astrophys.* **2009**, *9*, 529–537.

64. Grupe, D. A Complete Sample of Soft X-ray-selected AGNs. II. Statistical Analysis. *Astron. J.* **2004**, *127*, 1799–1810.

65. Wang, J.; Wei, J.Y.; He, X.T. A Sample of IRAS Infrared-selected Seyfert 1.5 Galaxies: Infrared Color $\alpha(60, 25)$-dominated Eigenvector 1. *Astrophys. J.* **2006**, *638*, 106–119.

66. Richards, G.T.; Kruczek, N.E.; Gallagher, S.C.; Hall, P.B.; Hewett, P.C.; Leighly, K.M.; Deo, R.P.; Kratzer, R.M.; Shen, Y. Unification of Luminous Type 1 Quasars through C IV Emission. *Astron. J.* **2011**, *141*, 167.

67. Yip, C.W.; Connolly, A.J.; Berk, D.E.V.; Ma, Z.; Frieman, J.A.; SubbaRao, M.; Szalay, A.S.; Richards, G.T.; Hall, P.B.; Schneider, D.P.; et al. Spectral Classification of Quasars in the Sloan Digital Sky Survey: Eigenspectra, Redshift, and Luminosity Effects. *Astron. J.* **2004**, *128*, 2603–2630.

68. Shen, Y.; Ho, L.C. The diversity of quasars unified by accretion and orientation. *Nature* **2014**, *513*, 210–213.

69. Sun, J.; Shen, Y. Dissecting the Quasar Main Sequence: Insight from Host Galaxy Properties. *Astrophys. J. Lett.* **2015**, *804*, L15.

70. Sulentic, J.W.; Marziani, P.; Zamanov, R.; Bachev, R.; Calvani, M.; Dultzin-Hacyan, D. Average Quasar Spectra in the Context of Eigenvector 1. *Astrophys. J. Lett.* **2002**, *566*, L71–L75.

71. Marziani, P.; Sulentic, J.W.; Zwitter, T.; Dultzin-Hacyan, D.; Calvani, M. Searching for the Physical Drivers of the Eigenvector 1 Correlation Space. *Astrophys. J.* **2001**, *558*, 553–560.

72. Sulentic, J.W.; Marziani, P.; Dultzin-Hacyan, D. Phenomenology of Broad Emission Lines in Active Galactic Nuclei. *Annu. Rev. Astron. Astrophys.* **2000**, *38*, 521–571.

73. Abramowicz, M.A.; Straub, O. Accretion discs. *Scholarpedia* **2014**, *9*, 2408.

74. Bon, N.; Bon, E.; Marziani, P.; Jovanović, P. Gravitational redshift of emission lines in the AGN spectra. *Astrophys. Space Sci.* **2015**, *360*, 41.

75. Wang, J.; Dong, X.; Wang, T.; Ho, L.C.; Yuan, W.; Wang, H.; Zhang, K.; Zhang, S.; Zhou, H. Estimating Black Hole Masses in Active Galactic Nuclei Using the Mg II $\lambda2800$ Emission Line. *Astrophys. J.* **2009**, *707*, 1334–1346.

76. Negrete, C.A.; Dultzin, D.; Marziani, P.; Sulentic, J.W. Reverberation and Photoionization Estimates of the Broad-line Region Radius in Low-z Quasars. *Astrophys. J.* **2013**, *771*, 31.

77. Zamfir, S.; Sulentic, J.W.; Marziani, P.; Dultzin, D. Detailed characterization of H$\beta$ emission line profile in low-z SDSS quasars. *Mon. Not. R. Astron. Soc.* **2010**, *403*, 1759–1786.

78. Leighly, K.M. ASCA (and HST) observations of NLS1s. *New Astron. Rev.* **2000**, *44*, 395–402.

79. Bachev, R.; Marziani, P.; Sulentic, J.W.; Zamanov, R.; Calvani, M.; Dultzin-Hacyan, D. Average Ultraviolet Quasar Spectra in the Context of Eigenvector 1: A Baldwin Effect Governed by the Eddington Ratio? *Astrophys. J.* **2004**, *617*, 171–183.

80. Marziani, P.; Sulentic, J.W.; Negrete, C.A.; Dultzin, D.; Zamfir, S.; Bachev, R. Broad-line region physical conditions along the quasar eigenvector 1 sequence. *Mon. Not. R. Astron. Soc.* **2010**, *409*, 1033–1048.

81. Denney, K.D.; Vestergaard, M.; Watson, D.; Davis, T. Using Quasars as Standard Candles for Studying Dark Energy. In Proceedings of the American Astronomical Society Meeting, Austin, TX, USA, 8–12 January 2012; Abstracts, Volume 219, p. 440.20.

82. Marziani, P.; Sulentic, J.W.; Zamanov, R.; Calvani, M.; Dultzin-Hacyan, D.; Bachev, R.; Zwitter, T. An Optical Spectroscopic Atlas of Low-Redshift Active Galactic Nuclei. *Astrophys. J. Suppl. Ser.* **2003**, *145*, 199–211.

83. Coatman, L.; Hewett, P.C.; Banerji, M.; Richards, G.T.; Hennawi, J.F.; Prochaska, J.X. Correcting C IV-based virial black hole masses. *Mon. Not. R. Astron. Soc.* **2017**, *465*, 2120–2142.

84. Sulentic, J.W.; Martínez-Carballo, M.A.; Marziani, P.; del Olmo, A.; Stirpe, G.M.; Zamfir, S.; Plauchu-Frayn, I. 3C 57 as an Atypical Radio-Loud Quasar: Implications for the Radio-Loud/Radio-Quiet Dichotomy. *Mon. Not. R. Astron. Soc.* **2015**, *450*, 1916–1925.

85. Saito, Y.; Imanishi, M.; Minowa, Y.; Morokuma, T.; Kawaguchi, T.; Sameshima, H.; Minezaki, T.; Oi, N.; Nagao, T.; Kawatatu, N.; et al. Near-infrared spectroscopy of quasars at z ∼ 3 and estimates of their supermassive black hole masses. *Publ. Astron. Soc. Jpn.* **2016**, *68*, doi:10.1093/pasj/psv102.

86. Coatman, L.; Hewett, P.C.; Banerji, M.; Richards, G.T. C IV emission-line properties and systematic trends in quasar black hole mass estimates. *Mon. Not. R. Astron. Soc.* **2016**, *461*, 647–665.

87.  Negrete, A.; Dultzin, D.; Marziani, P.; Sulentic, J. BLR Physical Conditions in Extreme Population A Quasars: A Method to Estimate Central Black Hole Mass at High Redshift. *Astrophys. J.* **2012**, *757*, 62.

88.  Negrete, C.A.; Dultzin, D.; Marziani, P.; Sulentic, J.W. A New Method to Obtain the Broad Line Region Size of High Redshift Quasars. *Astrophys. J.* **2014**, *794*, 95.

# 8

# Stark Broadening from Impact Theory to Simulations

Roland Stamm [1,*], Ibtissem Hannachi [1,2], Mutia Meireni [1], Laurence Godbert-Mouret [1],
Mohammed Koubiti [1], Yannick Marandet [1], Joël Rosato [1], Milan S. Dimitrijević [3] and
Zoran Simić [3]

[1] Département de Physique, Aix-Marseille Université, CNRS, PIIM UMR 7345, 13397 Marseille CEDEX 20,
France; ibtissem.hannachi@univ-amu.fr (I.H.); mutia_meireni@ymail.com (M.M.);
laurence.mouret@univ-amu.fr (L.G.-M.); mohammed.koubiti@univ-amu.fr (M.K.);
yannick.marandet@univ-amu.fr (Y.M.); joel.rosato@univ-amu.fr (J.R.)
[2] PRIMALAB, Faculty of Sciences, University of Batna 1, Batna 05000, Algeria
[3] Astronomical Observatory, Volgina 7, 11060 Belgrade, Serbia; mdimitrijevic@aob.rs (M.S.D.);
zsimic@aob.rs (Z.S.)
* Correspondence: roland.stamm@univ-amu.fr

Academic Editor: Ulrich Jentschura

**Abstract:** Impact approximation is widely used for calculating Stark broadening in a plasma.
We review its main features and different types of models that make use of it. We discuss recent
developments, in particular a quantum approach used for both the emitter and the perturbers.
Numerical simulations are a useful tool for gaining insight into the mechanisms at play in
impact-broadening conditions. Our simple model allows the integration of the Schrödinger equation
for an emitter submitted to a fluctuating electric field. We show how we can approach the impact
results, and how we can investigate conditions beyond the impact approximation. The simple
concepts developed in impact and simulation approaches enable the analysis of complex problems
such as the effect of plasma rogue waves on hydrogen spectra.

**Keywords:** stark broadening; impact approximation; numerical simulation

## 1. Introduction

Stark profiles are used in astrophysics and other kinds of plasmas for obtaining information
on the charged environment of the emitting particles. Using light for conveying information on
the plasma often requires a modeling of both the plasma and the radiator. We will review different
situations requiring different modeling approaches. The impact-broadening approach considers the
emitter-plasma interaction as a sequence of brief separate collisions decorrelating the radiative dipole.
Impact models are very effective for many types of plasmas, and can be applied to different kinds
of emitters, hydrogen being an exception for most plasma conditions. Many different models using
impact approximation have been developed, and we will review the most commonly-used. One can
distinguish firstly between models keeping the quantum character of the perturbers, and those using
a classical trajectory for the charged particles. Full quantum approaches require specific calculation
techniques, which, once established, have proved to be of general interest. Another way to look at the
models is the degree of accuracy required. It is often not necessary to have an accuracy better than
about 20%, since the experimental errors are often of the same order or worse. This has enabled the
development of a semi-empirical impact model, useful especially in cases where one does not have a
sufficient set of atomic data for adequate application of more sophisticated methods with which one
can readily obtain a large number of line shapes [1], making it an effective diagnostic tool.

A typical starting point for a line shape formalism in a plasma is a full quantum formalism for the
emitter and the perturbers. It can be written as a linear response for the emitter dipole operator, and

provides the response of an emitter at a time t, knowing its state at an initial time [2]. This response in time allows the physical measurement of the spectrum to which it is linked by a Fourier transform. Quantum formalism introduces specific computational difficulties, but also brings powerful tools such as the angular averages. We will briefly discuss such approaches, and how they compare to semi-classical calculations. Classical path impact approximations have been widely developed, and exist in several levels of accuracy, depending on whether one is interested in a rapid analysis of a large number of spectra, or one asks for an accurate analysis of a few lines. We will identify situations for which other models are helpful, e.g., for the case were the emitter-perturber interactions cannot be represented by a sequence of collisions. Such models use the statistical properties of the electric field created by the perturbing particles. In astrophysics, model microfield methods provide an efficient alternative for cases where neither the impact nor the static approximation are valid. For such situations, several models have been developed and interfaced with atomic data. Their accuracy can be tested by simulation techniques avoiding some approximations, but at the expenses of computer time. Such computer simulations can be used to analyze the various physical processes involved in plasmas under arbitrary conditions. We will illustrate their use in the case of plasma rogue waves.

## 2. Impact Broadening

A detailed and accurate modeling of Stark broadening started almost sixty years ago with the development of a general impact theory having the ability of retaining the quantum character of the emitters and perturbers, and allowing both elastic and inelastic collisions between such particles [2]. The line shape is obtained by a Fourier transform of the dipole autocorrelation function (DAF), a quantity expressed as a trace over all possible states of the quantum emitter plus perturbers system:

$$C(t) = Tr[\vec{d} \cdot T^+(t) \vec{d} \, T(t) \rho], \tag{1}$$

where $\vec{d}$ is the dipole moment of the emitter, $T(t) = exp(-iHt/\hbar)$ the evolution operator and $\rho$ the density matrix, these last two quantities being dependent on the Hamiltonian $H$ for the whole system. Such an expression could be calculated by density functional theory or quantum Monte Carlo methods, taking advantage of the development of computational techniques and computer hardware [3]. Such studies have proved to be efficient for describing the properties of dense plasmas found in the interior of gaseous planets, the atmospheres of white dwarfs or the laboratory plasmas created by energetic lasers. They might be useful for understanding some features in the spectrum of such plasmas, but have not been developed yet in the context of line broadening. Probably the main reason for this is that there is no clear evidence that the dynamical effects of multiple quantum perturbers can affect a line shape. Another reason is that for most of the plasma conditions and line shapes studied, we can use the impact approximation, which assumes that the various perturbers interact separately with the emitter (binary collision assumption), and that the average collision is weak. A validity condition for the impact approximation is that the collision time is small compared with the decorrelation time of $C(t)$. If this condition and the binary collision assumptions are verified, it is possible to use a constant collision operator to account for all the effects of the perturbers on the emitter. Different approaches using impact approximation have been proposed, but we can distinguish firstly between quantum impact models that retain the quantum behavior of both emitters and perturbers, and the semiclassical impact models treating only the emitter as a quantum particle. A pictorial representation of the full quantum emitter-perturber interaction is provided by the use of wave packets for the perturbers. Each wave packet is scattered in a region within the reach of the interaction potential with the emitter. Quantum collision formalism can be applied, enabling the calculation of cross sections with the aid of scattering amplitudes. Although quantum impact calculations have been performed since the seventies [4,5], such calculations are not very numerous for line broadening due to their computational difficulty. In particular, they involve a calculation of the scattering matrix or $S$ matrix [2].

Many calculations have been applied to isolated lines of various ions, a case for which the width $w$ takes the compact form of an average over the perturbers after the use of the optical theorem [2]:

$$w = \frac{1}{2}N\{v[\sigma_i + \sigma_f + \int d\Omega |f_i(\theta, \varphi) - f_f(\theta, \varphi)|^2]\}_{Av},\qquad(2)$$

where $N$ is the perturbers' density, $v$ their velocity, $\sigma_i$ and $\sigma_f$ are inelastic cross sections, $f_i$ and $f_f$ the forward-scattering amplitudes in a direction given by $\theta$, $\varphi$ for the initial $i$ and final $f$ states, and $\{\ldots\}_{Av}$ stands for a Boltzmann thermal average.

With the advent of accurate atomic structure and $S$ matrix codes, such impact quantum calculations have been given a new life [6,7], and are most often in good agreement with other calculations. A very efficient calculation has been proposed starting from Equation (2), using a Bethe-Born approximation [8] for evaluating inelastic cross sections. This semi-empirical model uses an effective Gaunt factor, a quantity which measures the probability of an incident electron changing its kinetic energy [9]. This function has been modified and improved to develop the modified semi-empirical model which is frequently used for calculations of isolated ion lines [1].

For most plasma conditions and line shapes studied, the wave packets associated with the perturbers are small and do not spread much in time. This enables the use of classical perturbers following classical paths. Different approaches use this approximation together with the impact approximation for the electron perturbers. Early calculations of hydrogen lines with comparisons to experimental profiles proved that a profile using an impact electron broadening [10], together with a static approximation for the ion perturbers, is in overall agreement for the Balmer $H_\beta$ line in an arc plasma with a density $N = 2.2 \times 10^{22}$ m$^{-3}$ and a temperature $T = 10{,}400$ K. The remaining discrepancies concerned the central part of the line and the far line wings, two regions that required an improvement of the model.

For isolated lines of neutral atoms and ions, the semiclassical perturbation (SCP) method [11] has been successfully applied to numerous lines, and is implemented in the STARK-B database [12]. The SCP method was inspired by developments in the quantum theory of collisions between atoms and electrons or ions, and, e.g., performs the angular averages with Clebsch-Gordan coefficients. It has the ability to generate several hundred lines rapidly for a set of densities and temperatures in a single run. The accuracy of the SCP method is assessed by a comparison to experimental spectra, and is about 20 to 30% for the widths of simple spectra, but could be worse for some complex spectra. The method is continuously improving, and has been interfaced with atomic structure codes [11].

An interesting point is raised by the comparison of impact quantum and semiclassical calculations, and a comparison of those with experiments. Quantum calculations have often been found to predict narrower lines than those of semiclassical models [13]. Semiclassical calculations may be brought closer to quantum widths, e.g., by a refined calculation of the minimum impact parameter allowing the use of a classical path [14]. Surprisingly, quantum widths of Li-like and boron-like ions often show a worse agreement with experiments than semiclassical calculation, thus calling for further calculations and analysis [7,15]. As an example of such, more recent quantum calculations [15] are in fairly good agreement with experiments [15].

## 3. Simulations of Impact Theory and Ion Dynamics

The need for a model that does not assume the impact approximation arose out of the study of hydrogen lines, with the surge of accurate profile measurements in near equilibrium plasmas [16]. It appeared that a standard model using a static ion approximation, and an impact electron collision operator, showed pronounced differences with the measures near the line centers, and also in the far wings. The line wings were well reproduced by the so-called unified theory, which retains the static interaction between an electron and the atom as a strong collision occurs [17,18]. The difference in the central part of the line was linked to the use of the static ion approximation, since it depended on the reduced mass of the emitter-ion perturber system. The observation of the Lyman-$\alpha$ (L$_\alpha$) line [19] showed later that the experimental profile was a factor 2.5 broader than the theoretical line using static ions in arc plasma conditions. This was a strong motivation for developing a technique able to retain ion dynamics in a context where the electric field is created by numerous ions in motion. Since perturbative approaches were unable to account for multiple strong collisions, a computer simulation has been proposed for describing the motion of the ions. The effect on the emitter of the time dependent ion electric field is obtained by a numerical integration of the Schrödinger equation. Early calculations showed the effect of ion dynamics in the central part of the line, and were able to strongly reduce the difference between experimental and simulation profiles [20–22].

Simple hydrogen plasma simulations may be used to illustrate the behavior of an electric field component during a time interval of the order of the line shape time of interest. This time is usually taken as the DAF decorrelation time, and can also be defined as the inverse of the line width. The electric field experienced by an atom surrounded by moving charged particles can be calculated at the center of a cubic box, using particles with straight line trajectories. The edge of the cube should be assumed to be equal to a few times the Debye length $\lambda_D = \sqrt{\varepsilon_0 k_B T / (Ne^2)}$, with $T$ and $N$ the hydrogen plasma temperature and density, respectively, $e$ the electron charge, $k_B$ the Boltzmann constant, and $\varepsilon_0$ the permittivity of free space. If we simulate only the ion perturbers, we assume that each particle creates a Debye shielded electric field, in an attempt to retain ion-electron correlations. Random number generators are used to obtain the uniform positions and Maxwell-Boltzmann distributed velocities of the charged particles. If an ion leaves the cubic box, it is replaced by a new one created near the cube boundaries. For the weak coupling conditions assumed, a large number of particles (several thousand commonly) is retained in a cube with a size larger than the Debye length. Such a model provides a good approximation for the time-dependent electric field in a weakly coupled ion plasma at equilibrium, although it suffers from inaccuracies, especially if the size of the box is not large enough [23]. We show in Figure 1 the time dependence of one component of the ionic electric field calculated at the center of the box for an electron density $N_e = 10^{19}$ m$^{-3}$, and a temperature $T = 40,000$ K. The electric field is expressed in units of $E_0 = 1/(4\pi\varepsilon_0 r_0^2)$, where $r_0$ is the average distance between particles defined by $r_0^3 = 3/(4\pi N_e)$. The time interval of 5 ns used in Figure 1a is the L$_\alpha$ time of interest for such plasma conditions. The validity condition of the binary collision approximation requests that the Weisskopf radius $\rho_w = \hbar n^2 / m_e v_i$, with $n$ the principal quantum number of the L$_\alpha$ upper states ($n = 2$), and $v_i = \sqrt{2k_B T / m_p}$ the thermal ion velocity ($m_e$ and $m_p$ are resp. the electron and proton mass), is much smaller than the average distance between particles. This ratio is for L$_\alpha$ and protons of the order of 0.04, enabling the use of an impact approximation. The electric field in Figure 1a clearly exhibits several large fields that are well separated in time during the 5 ns of the L$_\alpha$ time of interest. During this time interval, only a few fields (3 in Figure 1a) have a magnitude larger than 50 $E_0$, but about 20 have a magnitude of 10 $E_0$ or more.

A piece of the same field history is shown in Figure 1b during a time interval equal to the time of interest for the Balmer-$\beta$ (H$_\beta$) line. For this time interval of 0.3 ns, the electric field shows much fewer fluctuations, the atom is no longer submitted to a sequence of sharp collisions, and we can no longer use the impact approximation. This is confirmed by a value of 0.16 for the $\rho_w/r_0$ ratio, making the use of an impact approximation for this line problematic. Looking now at Figure 1a,b, we can see a background of electric field fluctuations with a small magnitude of about $E_0$, and a typical time scale longer than the collision time $r_0/v_i$. Such fluctuations correspond to the sum of electric fields of distant particles with a magnitude on the order of $E_0$. For hydrogen lines affected by the linear Stark effect, it is well known that this effect of weak collisions is dominant in near impact regimes [10], and results from the long range of the Coulomb electric field.

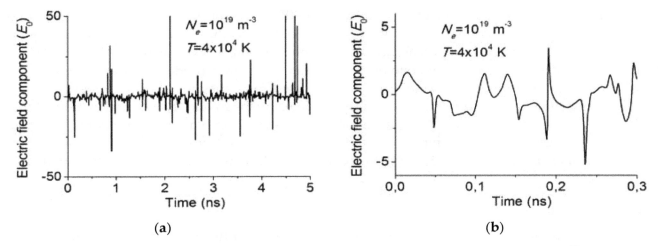

**Figure 1.** Electric field component in units of $E_0$ in a plasma with a density $N_e = 10^{19}$ m$^{-3}$, and a temperature $T = 40,000$ K, during (**a**) a time interval of the order of the time of interest for the $L_\alpha$ line, and (**b**) a time interval of the order of the time of interest for the $H_\beta$ line.

Using several thousand samples of such electric fields, it is possible to calculate the DAF for each line studied. This requires for each field history $\vec{E}(t)$ an integration of the Schrödinger equation of the emitter submitted to a dipolar interaction potential $-\vec{d}.\vec{E}(t)$. We obtain the emitter's evolution operator by finite difference computational methods, using time steps adjusted to ensure the best compromise between computer time cost and accuracy [24]. The integration time interval is provided by the time of interest for the line calculated, and a first estimate for the time step is a hundredth of the collision time. In the following, we retain only the broadening of the upper states of the line, resulting in some loss of accuracy for the first Balmer lines, but in a much faster calculation. We show in Figure 2a the DAF of $L_\alpha$ for the same plasma conditions as in Figure 1. The ab-initio DAF (solid line) is obtained by a simulation of the ions retaining also the effect of electrons with an impact approximation. We observe that this simulation is close to an impact calculation for both ions and electrons (dashed line). For the same condition and the $H_\beta$ line, the decay of the ab-initio DAF is significantly smaller than for the impact calculation, indicating again a deviation from ion impact broadening for this line.

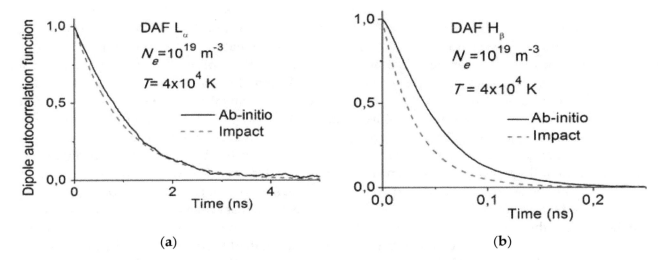

**Figure 2.** Dipole autocorrelation functions with an ab-initio simulation (solid line) and in the impact limit (dotted line) in a plasma with a density $N_e = 10^{19}$ m$^{-3}$, and a temperature $T = 40,000$ K, for (**a**) the $L_\alpha$ Lyman transition, and (**b**) the $H_\beta$ Balmer transition.

Another way of taking account of ion dynamics is with the help of stochastic processes. A stepwise constant stochastic process is used to model the electric field felt by the atom [25]. The process requires the knowledge of the microfield probability distribution function, and of a waiting time distribution function controlling the jumps from one field to the next one. Such model microfield methods are efficient for retaining ion dynamics effects, and are used for a diagnostic of hydrogen lines [26]. Stochastic processes are also used in the line shape code using the frequency fluctuation model for an inclusion of ion dynamics [27]. During the last decades, several simulations and models have been developed with the ability of retaining ion dynamics. The field is still active, with ion dynamics being one of the issues discussed in the Spectral Line Shapes in Plasmas workshop, providing many new analyses [28].

## 4. Effect of Plasma Waves

Plasmas sustain various types of waves, which behave differently in a linear and nonlinear regime. A way to distinguish between the two regimes is to calculate the ratio $W$ of the wave energy density to the plasma energy density, given by:

$$W = \varepsilon_0 E_L^2 / 4N_e k_B T \,, \tag{3}$$

where $E_L$ is the electric field magnitude of the wave. For values of $W$ much smaller than 1, we expect a linear behavior of the waves. In a linear regime, electronic Langmuir waves oscillate at a frequency close to the plasma frequency $\omega_p = \sqrt{N_e e^2 / m_e \varepsilon_0}$, and can be excited even by thermal fluctuations. We assume that the numerous emitters on the line of sight are submitted to different Langmuir waves, each with the same frequency $\omega_p$, but a different direction and phase chosen at random, and a magnitude sampled using a half-normal probability density function (PDF). In the following, we have used this half-normal PDF for the reduced electric field magnitude $F = E/E_0$:

$$P(F) = \frac{\sqrt{2}}{\sigma\sqrt{\pi}} exp\left(-\frac{F^2}{2\sigma^2}\right) \tag{4}$$

In this expression, we use the standard deviation $\sigma$ of a normal distribution, and thus obtain the mean value $E_L$ of $E$ by writing $E_L = \sigma E_0 \sqrt{2/\pi}$. Each Langmuir wave has a different electric field history, and we obtain the DAF by an average over about a thousand such field histories. For a plasma with a density $N_e = 10^{19}$ m$^{-3}$, and a temperature $T = 10^5$ K, we first calculated the L$_\alpha$ DAF for Langmuir waves with a mean electric field magnitude corresponding to $W = 0.01$ ($E_L = 15E_0$). The response of the DAF is a periodic oscillation with a period equal to $2\pi/\omega_p$, but with an amplitude much smaller than 1 for this average field magnitude of $15E_0$. After a product with an impact DAF for retaining the effect of the background electron and ion plasma, there remains no visible effect of the waves on the convolution DAF for the value $W = 0.01$. This ratio can take much larger values, however; especially if an external energy source such as a beam of charged particles is present. As $W$ increases, nonlinear phenomena start showing up, enabling, for instance, wave-wave couplings. Although only recently investigated in plasmas, the occurrence of rogue waves has been raised in various plasma conditions [29–31]. Rogue waves have been studied in many dynamical systems, and are known to the general public by the observation and study of rogue or freak waves that suddenly appear in the ocean as large isolated waves. In oceanography, rogue waves are defined as waves whose height is more than twice the mean of the largest third of the waves in a wave record. Rogue waves appear to be a unifying concept for studying localized excitations that exceed the strength of their background structures. They are studied in nonlinear optics [32], Bose-Einstein condensates [33], and many other fields outside of physics. For our line shape problem in plasmas, we postulate that nonlinear processes create rogue waves from a random background of smaller Langmuir waves. The physical mechanism at play is the coupling of the Langmuir wave with ion sound and electromagnetic waves; density fluctuations of the sound waves affect the high frequency waves through $\omega_p$. The first Zakharov equation [34] shows how density fluctuations affect Langmuir waves, and a second equation how a Langmuir wave packet can produce a density depression via the ponderomotive force [35]. We will not discuss these equations here, which are particularly useful for a study of wave collapse. Most present rogue wave studies rely on the nonlinear Schrödinger equation (NLSE), which is obtained in the adiabatic limit (slowly changing density perturbations) of the Zakharov equations [35]. A one-dimensional solution of the NLSE is commonly used to approximate the response of nonlinear media. Stable envelope solitons are possible solutions of the 1D NLSE. We will assume that there is a contribution of a stable envelope soliton for each history of the microfield, similarly to what we did for the background Langmuir wave. Using a ratio $W = 0.1$, the average peak magnitude of such solitons will be 3 times the amplitude of background Langmuir waves, fitting them in the category of rogue waves. A possible shape for the envelope is a Lorentzian, with a time dependence that bears some similarity with the celebrated Peregrine soliton [36]. We observe in Figure 3 that the DAF of $L_\alpha$ obtained with a product of the impact DAF and the Langmuir rogue wave DAF for $W = 0.1$ is affected by oscillations at the plasma frequency.

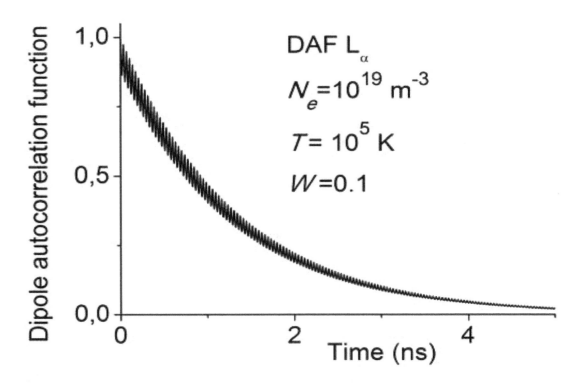

**Figure 3.** $L_\alpha$ dipole autocorrelation function in a plasma with a density $N_e = 10^{19}$ m$^{-3}$, and a temperature $T = 10^5$ K, calculated with a product of the impact DAF and the Langmuir rogue wave DAF for $W = 0.1$.

Looking at the line shape obtained with a Fourier transform, we can see in Figure 4 that the peak of the line including the wave effect is about 10% lower than the impact line peak, but this with almost no effect on the line width. Not shown in Figure 4, we noticed that the wing of the line affected by rogue waves had a slightly slower decay than the impact profile, indicating a transfer of intensity from the center toward large line shifts. It is remarkable that a such rogue wave had a rather small effect on the profile. This is probably due to the fact that we are in impact conditions for this line. In impact regimes, decorrelation is very effective, leaving only a small broadening contribution to the type of rogue waves that we considered. A larger broadening effect would be observed by considering wave collapse, a phenomenon occurring as $W$ takes larger values of the order of 1 or more for such plasma conditions. The emitters then experience the effect of a sequence of solitons which can significantly increase the broadening [37].

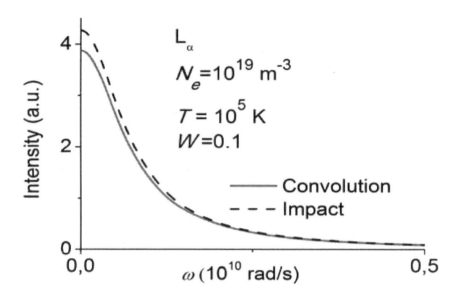

**Figure 4.** $L_\alpha$ in a plasma with a density $N_e = 10^{19}$ m$^{-3}$, and a temperature $T = 10^5$ K, calculated with an impact approximation (dashed line), and with a Fourier transform of the DAF in Figure 3 (solid line).

## 5. Conclusions

Impact approximation mainly consists of saying that, on an average, it takes many collisions to change the quantum state of an atom. When this approximation is valid, the effect of the numerous fluctuating interactions of the emitter with the perturbers can be expressed with a constant collision operator. We have briefly described several models using impact approximation. A wide variety of impact models have been proposed, ranging from full quantum calculations to semiclassical approaches. Impact calculations allow expression of the width and shift of a line in terms of quantum scattering cross-sections. Such calculations have enabled many improvements in the application of quantum theory for obtaining observable quantities such as a line shape. The comparison between experimental and theoretical spectra is of great benefit for the validation of such models. It is thus crucial to be able to rapidly obtain numerous spectra for the lines of many atoms and ions. This is possible using models such as the semiclassical perturbation or the semi-empirical formalism. We have also shown how a computer simulation can reproduce the results of the impact approximation for hydrogen lines. Such simulations involve several thousand particles, however, and are certainly not the most efficient technique for obtaining the impact profile. The main advantage of simulations is that they can go beyond the impact approximation, for situations with many perturbers acting simultaneously on the emitter. We have briefly recalled the problem of ion dynamics, and have proposed a simple simulation for a calculation of the effect of Langmuir rogue waves of $L_\alpha$ in an impact regime.

**Acknowledgments:** This work is supported by the funding agency Campus France (Pavle Savic PHC project 36237PE). This work has also been carried out within the framework of the EUROfusion Consortium and has received funding from the Euratom research and training program 2014–2018 under Grant agreement no 633053. The views and opinions expressed herein do not necessarily reflect those of the European Commission.

**Author Contributions:** This work is based on the numerous contributions of all the authors.

## References

1.    Dimitrijević, M.S.; Konjević, N. Stark widths of doubly- and triply-ionized atom lines. *J. Quant. Spectrosc. Radiat. Transf.* **1980**, *24*, 451–459. [CrossRef]
2.    Baranger, M. General Impact Theory of Pressure Broadening. *Phys. Rev.* **1958**, *112*, 855–865. [CrossRef]
3.    McMahon, J.M.; Morales, M.A.; Pierleoni, C.; Ceperley, D.M. The properties of hydrogen and helium under extreme conditions. *Rev. Mod. Phys.* **2012**, *84*, 1607–1653. [CrossRef]

4. Barnes, K.S.; Peach, G. The shape and shift of the resonance line of Ca+ perturbed by electron collisions. *J. Phys. B* **1970**, *3*, 350–362. [CrossRef]
5. Bely, O.; Griem, H.R. Quantum-mechanical calculation for the electron-impact broadening of the resonance line of singly ionized magnesium. *Phys. Rev. A* **1970**, *1*, 97–105. [CrossRef]
6. Elabidi, H.; Ben Nessib, N.; Sahal-Bréchot, S. Quantum-mechanical calculations of the electron-impact broadening of spectral lines for intermediate coupling. *J. Phys. B* **2004**, *37*, 63–71. [CrossRef]
7. Elabidi, H.; Sahal-Bréchot, S.; Dimitrijević, M.S. Quantum Stark broadening of Ar XV lines. Strong collision and quadrupolar potential contributions. *Adv. Space Res.* **2014**, *54*, 1184–1189. [CrossRef]
8. Griem, H.R. Semi-empirical formulas for the electron-impact widths and shifts of isolated ion lines in plasmas. *Phys. Rev.* **1968**, *165*, 258–266. [CrossRef]
9. Van Regemorter, H. Rate of collisionnal excitation in stellar atmospheres. *Astrophys. J.* **1962**, *136*, 906–915. [CrossRef]
10. Griem, H.R.; Kolb, A.C.; Shen, K.Y. Stark broadening of hydrogen lines in a plasma. *Phys. Rev.* **1959**, *116*, 4–16. [CrossRef]
11. Sahal-Bréchot, S.; Dimitrijević, M.S.; Ben Nessib, N. Widths and shifts of isolated lines of neutral and ionized atoms perturbed by collisions with electrons and ions: An outline of the semiclassical perturbation (SCP) method and of the approximations used for the calculations. *Atoms* **2014**, *2*, 225–252. [CrossRef]
12. Sahal-Bréchot, S.; Dimitrijević, M.S.; Moreau, N. *STARK-B Database*; LERMA, Observatory of Paris, France and Astronomical Observatory: Belgrade, Serbia, 2014; Available online: http://stark-b.obspm.fr (accessed on 12 August 2017).
13. Griem, H.R. *Principles of Plasma Spectroscopy*; Cambridge University Press: Cambridge, UK, 1997.
14. Alexiou, S.; Lee, R.W. Semiclassical calculations of line broadening in plasmas: Comparison with quantal results. *J. Quant. Spectrosc. Radiat. Transf.* **2006**, *99*, 10–20. [CrossRef]
15. Elabidi, H.; Sahal-Bréchot, S.; Ben Nessib, N. Quantum Stark broadening of the 3s-3p spectral lines in Li-like ions; Z-scaling and comparison with semi-classical perturbation theory. *Eur. Phys. J. D* **2009**, *54*, 51–64. [CrossRef]
16. Wiese, W.L.; Kelleher, D.E.; Paquette, D.R. Detailed study of the Stark broadening of Balmer lines in a high density plasma. *Phys. Rev. A* **1972**, *6*, 1132–1153. [CrossRef]
17. Voslamber, D. Unified model for Stark broadening. *Z. Naturforsch.* **1969**, *24*, 1458–1472.
18. Smith, E.W.; Cooper, J.; Vidal, C.R. Unified classical-path treatment of Stark broadening in plasmas. *Phys. Rev.* **1969**, *185*, 140–151. [CrossRef]
19. Grützmacher, K.; Wende, B. Discrepancies between the Stark–broadening theories of hydrogen and measurements of Lyman-α Stark profiles in a dense equilibrium plasma. *Phys. Rev. A* **1977**, *16*, 243–246. [CrossRef]
20. Stamm, R.; Voslamber, D. On the role of ion dynamics in the Stark broadening of hydrogen lines. *J. Quant. Spectrosc. Radiat. Transf.* **1979**, *22*, 599–609. [CrossRef]
21. Stamm, R.; Smith, E.W.; Talin, B. Study of hydrogen Stark profiles by means of computer simulation. *Phys. Rev. A* **1984**, *30*, 2039–2046. [CrossRef]
22. Stamm, R.; Talin, B.; Pollock, E.L.; Iglesias, C.A. Ion-dynamics effects on the line shapes of hydrogenic emitters in plasmas. *Phys. Rev. A* **1986**, *34*, 4144–4152. [CrossRef]
23. Rosato, J.; Capes, H.; Stamm, R. Ideal Coulomb plasma approximation in line shapes models: Problematic issues. *Atoms* **2014**, *2*, 253–258. [CrossRef]
24. Vesely, F. *Computational Physics, an Introduction*; Plenum Press: New York, NY, USA, 1994.
25. Brissaud, A.; Frisch, U. Theory of Stark broadening-II exact line profile with model microfield. *J. Quant. Spectrosc. Radiat. Transf.* **1971**, *11*, 1767–1783. [CrossRef]
26. Stehlé, C. Stark broadening of hydrogen Lyman and Balmer in the conditions of stellar envelopes. *Astron. Astrophys. Suppl. Ser.* **1994**, *104*, 509–527.
27. Talin, B.; Calisti, A.; Godbert, L.; Stamm, R.; Lee, R.W.; Klein, L. Frequency-fluctuation model for line-shape calculations in plasma spectroscopy. *Phys. Rev. A* **1995**, *51*, 1918–1928. [CrossRef] [PubMed]
28. Spectral Line Shapes in Plasmas (SLSP) Code Comparison Workshop. Available online: http://plasma-gate. weizmann.ac.il/slsp/ (accessed on 20 July 2017).
29. Moslem, W.M.; Shukla, P.K.; Eliasson, B. Surface plasma rogue waves. *EPL* **2011**, *96*, 25002. [CrossRef]
30. Ahmed, S.M.; Metwally, M.S.; El-Hafeez, S.A.; Moslem, W.M. On the generation of rogue waves in dusty

plasmas due to modulation instability of nonlinear Schrödinger equation. *Appl. Math Inf. Sci.* **2016**, *10*, 317–323. [CrossRef]

31.    Mc Kerr, M.; Kourakis, I.; Haas, F. Freak waves and electrostatic wavepacket modulation in a quantum electron–positron–ion plasma. *Plasma Phys. Control. Fusion* **2014**, *56*, 035007. [CrossRef]

32.    Erkintalo, M.; Genty, G.; Dudley, J.M. Rogue-wave-like characteristics in femtosecond supercontinuum generation. *Opt. Lett.* **2009**, *34*, 2468–2470. [CrossRef] [PubMed]

33.    Bludov, Y.V.; Konotop, V.V.; Akhmediev, N. Matter rogue waves. *Phys. Rev. A* **2009**, *80*, 033610. [CrossRef]

34.    Zakharov, V.E. Collapse of Langmuir waves. *Sov. Phys. JETP* **1972**, *35*, 908–914.

35.    Robinson, P.A. Nonlinear wave collapse and strong turbulence. *Rev. Mod. Phys.* **1997**, *69*, 507–573. [CrossRef]

36.    Bailung, H.; Sharma, S.K.; Nakamura, Y. Observation of Peregrine solitons in a multicomponent plasma with negative ions. *Phys. Rev. Lett.* **2011**, *107*, 255005. [CrossRef] [PubMed]

37.    Hannachi, I.; Stamm, R.; Rosato, J.; Marandet, Y. Effect of nonlinear wave collapse on line shapes in a plasma. *EPL* **2016**, *114*, 23002. [CrossRef]

# 9

# Radiative and Collisional Molecular Data and Virtual Laboratory Astrophysics

Vladimir A. Srećković [1,*], Ljubinko M. Ignjatović [1], Darko Jevremović [2], Veljko Vujčić [2,3] and Milan S. Dimitrijević [2,4]

[1]  Institute of Physics, Belgrade University, Pregrevica 118, Zemun, 11080 Belgrade, Serbia; ljuba@ipb.ac.rs
[2]  Astronomical Observatory, Volgina 7, 11060 Belgrade, Serbia; darko@aob.rs (D.J.); veljko@aob.rs (V.V.); mdimitrijevic@aob.rs (M.S.D.)
[3]  Faculty of Organizational Sciences, University of Belgrade, Jove Ilica 154, 11000 Belgrade, Serbia
[4]  LERMA, Observatoire de Paris, UMR CNRS 8112, UPMC, 92195 Meudon CEDEX, France
*   Correspondence: vlada@ipb.ac.rs

Academic Editor: Elmar Träbert

**Abstract:** Spectroscopy has been crucial for our understanding of physical and chemical phenomena. The interpretation of interstellar line spectra with radiative transfer calculations usually requires two kinds of molecular input data: spectroscopic data (such as energy levels, statistical weights, transition probabilities, etc.) and collision data. This contribution describes how such data are collected, stored, and which limitations exist. Also, here we summarize challenges of atomic/molecular databases and point out our experiences, problems, etc., which we are faced with. We present overview of future developments and needs in the areas of radiative transfer and molecular data.

**Keywords:** atomic/molecular data; radiative and collisional processes; stars

## 1. Introduction

Many fields in astronomy such as astrophysics, astrochemistry and astrobiology, depend on data for atomic and molecular (A + M) collision and radiative processes. Among these amount of data collections there are atomic and molecular processes and spectral regions that even today are poorly represented. Therefore, there is an urgent need to collect these data in the databases as well as to develop methods for improving the existing ones. Also, this require a joint effort both of scientists and IT software specialists to develop state-of-the-art infrastructures satisfying their needs, such as Virtual Laboratory [1–3].

*The Base Astrophysical Targets*

Nowadays, the data in in the field of astrophysics modeling are especially important and needed for simulations/calculations. For example the A + M data for hydrogen are important for development of atmosphere models of solar and near solar type stars and for radiative transport investigations as well as an understanding of the kinetics of stellar and other astrophysical plasmas [4,5]. Modern codes for stellar atmosphere modelling, like e.g., PHOENIX (see e.g., [6–8]) require the knowledge of atomic data, so that the access to such atomic data, via online databases become very important.

The helium A + M data are of interest particularly for helium-rich white dwarf atmospheres investigations [9,10]. Such data are also important in modelling early Universe chemistry (see the paper of Coppola et al., 2013 [11]). The data for H and some metal atoms like Li, Na, Si are important for the exploring of the geo-cosmical plasmas, the interstellar medium as well as for studies of the early Universe chemistry and for the modelling of stellar and solar atmospheres (see, e.g., [12,13]).

Recently, in the papers [14–18] it has been pointed out that the photodissociation of the diatomic molecular ion in the symmetric and non-symmetric cases, are of astrophysical relevance and could be important in modeling of specific stellar atmosphere layers and they should be included in some chemical models. In the symmetric case, it was considered the processes of molecular ion photodissociation (bound-free) and ion-atom photoassociation (free-bound):

$$hv + A_2^+ \Longleftrightarrow A + A^+, \tag{1}$$

where $A$ and $A^+$ are atom and ion in their ground states, and $A_2^+$ is molecular-ion in the ground electronic state.

In the non-symmetric case, the similar processes of photodissociation/photoassociation are:

$$hv + AM^+ \Longleftrightarrow A^+ + M, \tag{2}$$

where $M$ is an atom whose ionization potential is less than the corresponding value for atom $A$. $AM^+$ is also molecular-ion in the ground electronic state.

In the general case molecular ion $A_2^+$ or $AM^+$ can be in one of the states from the group which contains the ground electronic state. For the solar atmosphere, $A$ usually denotes atom H(1s) and $M$ one of the relevant metal atoms (Mg, Si, Ca, Na) [14–16], but there are cases where $A$ = He, and $M$ = H, Mg, Si, Ca, Na. For the helium-rich white dwarf atmospheres $A$ denotes He(1s$^2$) and $M$ denotes, H(1s), and eventually carbon or oxygen [19,20].

Recently, the results from [16] show the importance of including the symmetric processes with $A$ = H(1s) in the stellar atmosphere models like [5]. Also, for modeling the DB white dwarf atmospheres results for case $A$ = He(1s$^2$) have been used (Koester 2016, private communication). The photodissociation of HeH$^+$ has been extensively studied both from a theoretical and experimentalist point of view and inserted in chemical networks describing the formation and destruction of primordial molecules.

It is well known [21] that the chemical composition of the primordial gas consists of electrons and species such as: helium- He, He$^+$, He$^{2+}$ and HeH$^+$; hydrogen- H, H$^-$, H$^+$, H$_2^+$ and H$_2$; deuterium- $D$, $D^+$, $HD$, $HD^+$ and $HD^-$; lithium- Li, Li$^+$, Li$^-$, LiH$^-$ and LiH$^+$. Evaluation of chemical abundances in the standard BB model are calculated from a set of chemical reactions for the early universe [21] and is presented at Figure 1 from [21]. One can see that among them are species like molecular ions H$_2^+$, $HD^+$, HeH$^+$, etc., whose role in the primordial star formation is important.

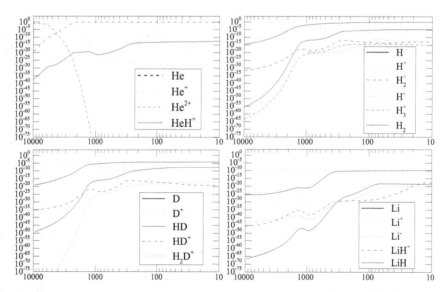

**Figure 1.** Evaluation of chemical abundances in the standard Big Bang (BB) model. Vertical axes are the relative abundances and the horizontal axes are relative to the redshift (taken from [21]).

## 2. MolD

### 2.1. Database Description

MolD database contains cross-sections for photodissociation processes [22], as well as corresponding data on molecular species and molecular state characterisations. MolD project is part of Serbian Virtual Observatory (SerVO)[1] and Virtual Atomic and Molecular Data Center (VAMDC)[2] (see Figure 2).

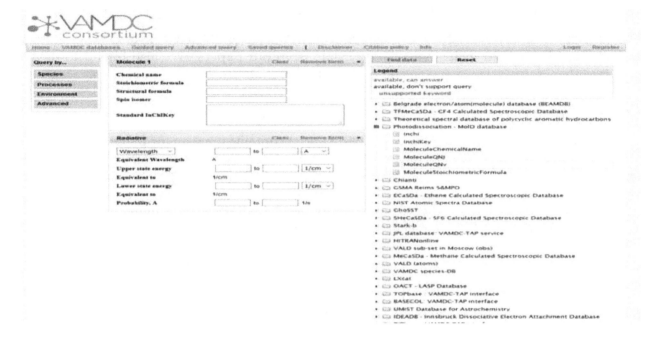

**Figure 2.** Snapshot from the query page from the Virtual Atomic and Molecular Data Center (VAMDC) portal.

MolD application is implemented as a customisation and extension of NodeSoftware provided by VAMDC. It complies to VAMDC interoperability standards and protocols for distributed remote queries. The underlying technology is Python-based, with Django as a Web framework [23], MySQL as a relational database system [24]. The web application runs on Apache web server.

The data model of MolD application is tailored to specifically suit the needs of the theoretical photodissociation data, and yet to easily map onto VAMDC's standardized XSAMS[3] (XML Schema for Atoms, Molecules and Solids format) schema for representation and exchange of atomic and molecular data[4].

### 2.2. Accessing MolD Data

MolD data can be accessed in several ways:

- Via MolD homepage (http://servo.aob.rs/mold). There is an AJAX-enabled (Asynchronous JavaScript and XML) web form for data querying as well as calculating and plotting average thermal cross sections along available wavelengths for a given temperatiure.

---

[1]  http://servo.aob.rs
[2]  http://vamdc.eu
[3]  http://vamdc-standards.readthedocs.io/en/latest/dataModel/vamdcxsams/structure.html
[4]  http://standards.vamdc.eu

- Via VAMDC portal (http://portal.vamdc.eu), where one can pose a distributed query to 30 databases across the European scientific institutes.
- Via standalone applications which support VAMDC-TAP (Table Access Protoco) for data access and tranformation to VAMDC-XSAMS.

During 2017, MolD entered *stage 3* of development. Currently, the database includes cross-section data for processes which involve species such as $He_2^+$, $H_2^+$, $MgH^+$, $HeH^+$, $LiH^+$, $NaH^+$. These processes are important for exploring of the interstellar medium, the early Universe chemistry as well as the modeling of different stellar and solar atmospheres (see papers [11,15,16,20,22]).

Our plans include transition to new versions of Django framework and NodeSoftware, with ongoing incremental inclusion of A + M data from our papers. We also intend to develop a more intuitive interface for querying and presentation of multidimensional data on our website.

### 2.3. Example: $H_2^+$ Molecular Ion

MolD is available online from the end of 2014 and it contains the data for the photodissociation processes Equation (1) with $A = H(1s)$ and $A = He(1s^2)$. Also it contains the relevant data for some other non-symmetric photodissociation processes Equation (2) where $M$ = Li, Na, Mg or He.

**The methods of calculation.** The cross-sections for the photodissociation of individual ro-vibrational state of the considered molecular ion $H_2^+$ is determined in the dipole approximation [14]:

$$\sigma_{J,v}(\lambda) = \frac{8\pi^3}{3\lambda}\left[\frac{(J+1)|D_{E,J+1;v,J}|^2 + J|D_{E,J-1;v,J}|^2}{2J+1}\right],\tag{3}$$

and the corresponding averaged thermal cross sections are given by:

$$\sigma_{phd}(\lambda, T) = \frac{1}{Z}\sum_{J}\sum_{v}g_{J;v}(2J+1)e^{-\frac{E_{Jv}-E_{00}}{k_B T}}\sigma_{J,v}(\lambda).\tag{4}$$

where $T$ is temperature, $\lambda$-wavelength, $D_{E,J+1;v,J}$ is the radial matrix element [25], $E_{Jv}$ is the energy of the individual states with the angular and vibrational quantum numbers $J$ and $v$ respectively, and $Z$ is the partition function

$$Z = \sum_{J}\sum_{v}g_{J;v}(2J+1)e^{-\frac{E_{Jv}-E_{0,0}}{k_B T}}.\tag{5}$$

In this expression the product $g_{J;v}\times(2J+1)$ is the statistical weight of the considered state and the coefficient $g_{J;v}$ depends on the "the spin of the nuclei".

The photo-dissociation crosssection $\sigma_{phd}(\lambda, T)$ given by Equation (4), as well as the coefficients $K_{ia}(\lambda, T)$, are determined within the approximation where the processes are treated as the result of the radiative transitions between the ground and the first excited adiabatic electronic state of the molecular ion $H_2^+$ which are caused by the interaction of the electron component of the ion-atom system with the free electromagnetic field taken in the dipole approximation.

For determination of $\sigma_{phd}(\lambda, T)$, as well as the coefficients $K_{ia}(\lambda, T)$ it is important to know the dipole matrix element $D_{12}(R)$ defined by relations

$$D_{12}(R) = \mathbf{D}_{12}(\mathbf{R}), \quad \mathbf{D}_{12}(\mathbf{R}) = \langle 1|\mathbf{D}(\mathbf{R})|2\rangle\tag{6}$$

where $R = |\mathbf{R}|$ and $\mathbf{D}(\mathbf{R})$ is the operator of electron dipole momentum. The mentioned adiabatic electronic states, $X^2\sum_g^+$ and $A^2\sum_u^-$, are denoted here with $|1\rangle$ and $|2\rangle$ and $R$ is the internuclear distance in the considered ion-atom system.

The described mechanism of the processes causes absorption of the photon with energy $\epsilon_\lambda$ near the resonant point $R = R_\lambda$, where $R_\lambda$ is the root of the equation

$$U_{12}(R) \equiv U_1(R) - U_2(R) = \epsilon_\lambda. \tag{7}$$

where $U_1(R)$ corresponds to the ground electronic state, and $U_2(R)$—to the first excited electronic state.

In Figure 3 are presented the data for $\sigma_{J,v}(\lambda)$ Equation (3) for the case $J = 0$ and $v = 10$, in the wavelength region $50\,\text{nm} \leq \lambda \leq 1500\,\text{nm}$.

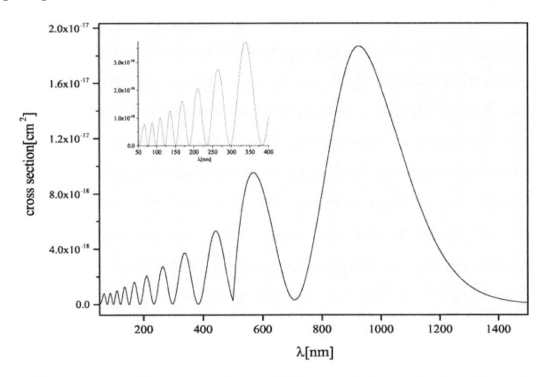

**Figure 3.** The behaviour of the cross-section $\sigma_{J,v}(\lambda)$ Equation (3) for $J = 0$ and $v = 10$, as a function of $\lambda$.

**The spectral coefficients.** The absorption process (1) i.e., processes of molecular ion photodissociation (bound-free) for the case $A = H$ is characterized by partial spectral absorption coefficients $\kappa_{ia}(\lambda)$ (see e.g., [25]) taken in the form

$$\kappa_{ia}(\lambda) = \sigma_{phd}(\lambda, T) N(H_2^+) \tag{8}$$

where $N(H_2^+)$ is density of the $H_2^+$ and $\sigma_{phd}(\lambda, T)$ is average cross-section for photo-dissociation of this molecular ion given by Equation (4). As in previous papers [14,25] the partial spectral absorption coefficient $\kappa_{ia}(\lambda)$ is also used in the form

$$\kappa_{ia}(\lambda) = K_{ia}(\lambda, T) N(H) N(H^+) \tag{9}$$

where the coefficient $K_{ia}(\lambda, T)$ is connected with $\sigma_{phd}(\lambda, T)$ by the relations

$$K_{ia}(\lambda, T) = \sigma_{phd}(\lambda, T) \chi^{-1}, \quad \chi = \frac{N(H)N(H^+)}{N(H_2^+)}. \tag{10}$$

In accordance with the definition of the absorption coefficient $\kappa_{ia}(\lambda)$, the coefficient $K_{ia}(\lambda, T)$ is given in units $(\text{cm}^5)$. The results for the average photodissociation cross-section $\sigma_{phd}(\lambda, T)$ for $H_2^+$ molecular ion are illustrated by Figure 4. The curves in this figure show the behavior of $\sigma_{phd}(\lambda, T)$ as a function of $\lambda$ for a wide range of temperatures $T$, which are relevant for the stellar atmosphere (e.g., solar photosphere). The values of the coefficient $K_{ia}(\lambda, T)$, defined by Equation (10), are presented in the Table 1 for the regions $90\,\mathrm{nm} \leq \lambda \leq 370\,\mathrm{nm}$ with small wavelength steps and for $3000\,\mathrm{K} \leq T \leq 10,000\,\mathrm{K}$ in order to enable easier use (interpolation) of this results. This allows direct calculation of the spectral absorption coefficients during the process of applying a any atmosphere model with the given parameters of plasma and composition.

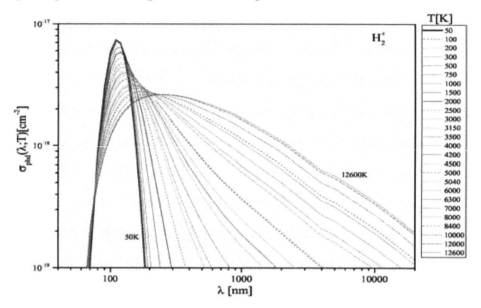

**Figure 4.** The behaviour of the averaged cross-section $\sigma_{phd}(\lambda, T)$ for photodissociation of the $H_2^+$ molecular ion, as a function of $\lambda$ and $T$.

**Table 1.** The coefficient $K_{ia}(\mathrm{cm}^5)$ Equation (9) as a function of $\lambda$ and $T$.

| $\lambda$ (nm) | 3000 K | 4000 K | 5000 K | 6000 K | 7000 K | 8000 K | 9000 K | 10,000 K |
|---|---|---|---|---|---|---|---|---|
| 90 | 3.50E−37 | 2.62E−38 | 5.50E−39 | 1.93E−39 | 9.04E−40 | 5.07E−40 | 3.20E−40 | 2.20E−40 |
| 91 | 3.75E−37 | 2.78E−38 | 5.80E−39 | 2.03E−39 | 9.47E−40 | 5.30E−40 | 3.34E−40 | 2.29E−40 |
| 92 | 3.99E−37 | 2.94E−38 | 6.10E−39 | 2.12E−39 | 9.90E−40 | 5.53E−40 | 3.48E−40 | 2.39E−40 |
| 93 | 4.23E−37 | 3.10E−38 | 6.40E−39 | 2.22E−39 | 1.03E−39 | 5.76E−40 | 3.62E−40 | 2.48E−40 |
| 94 | 4.47E−37 | 3.25E−38 | 6.69E−39 | 2.31E−39 | 1.07E−39 | 5.98E−40 | 3.76E−40 | 2.57E−40 |
| 95 | 4.71E−37 | 3.41E−38 | 6.98E−39 | 2.41E−39 | 1.12E−39 | 6.20E−40 | 3.89E−40 | 2.66E−40 |
| 96 | 4.95E−37 | 3.56E−38 | 7.27E−39 | 2.50E−39 | 1.16E−39 | 6.42E−40 | 4.03E−40 | 2.75E−40 |
| 97 | 5.18E−37 | 3.70E−38 | 7.54E−39 | 2.59E−39 | 1.20E−39 | 6.63E−40 | 4.16E−40 | 2.84E−40 |
| 98 | 5.40E−37 | 3.85E−38 | 7.82E−39 | 2.68E−39 | 1.24E−39 | 6.85E−40 | 4.29E−40 | 2.92E−40 |
| 99 | 5.62E−37 | 3.99E−38 | 8.08E−39 | 2.76E−39 | 1.27E−39 | 7.05E−40 | 4.41E−40 | 3.01E−40 |
| 100 | 5.82E−37 | 4.12E−38 | 8.34E−39 | 2.85E−39 | 1.31E−39 | 7.26E−40 | 4.54E−40 | 3.09E−40 |
| 101 | 6.03E−37 | 4.25E−38 | 8.59E−39 | 2.93E−39 | 1.35E−39 | 7.46E−40 | 4.66E−40 | 3.17E−40 |
| 102 | 6.22E−37 | 4.38E−38 | 8.83E−39 | 3.01E−39 | 1.38E−39 | 7.65E−40 | 4.78E−40 | 3.26E−40 |
| 103 | 6.40E−37 | 4.50E−38 | 9.06E−39 | 3.09E−39 | 1.42E−39 | 7.84E−40 | 4.90E−40 | 3.34E−40 |
| 104 | 6.58E−37 | 4.62E−38 | 9.29E−39 | 3.16E−39 | 1.45E−39 | 8.03E−40 | 5.02E−40 | 3.41E−40 |
| 105 | 6.74E−37 | 4.73E−38 | 9.51E−39 | 3.24E−39 | 1.49E−39 | 8.21E−40 | 5.13E−40 | 3.49E−40 |
| 106 | 6.90E−37 | 4.83E−38 | 9.72E−39 | 3.31E−39 | 1.52E−39 | 8.39E−40 | 5.24E−40 | 3.57E−40 |
| 107 | 7.04E−37 | 4.93E−38 | 9.92E−39 | 3.38E−39 | 1.55E−39 | 8.57E−40 | 5.35E−40 | 3.64E−40 |
| 108 | 7.18E−37 | 5.03E−38 | 1.01E−38 | 3.44E−39 | 1.58E−39 | 8.74E−40 | 5.46E−40 | 3.72E−40 |
| 109 | 7.30E−37 | 5.11E−38 | 1.03E−38 | 3.51E−39 | 1.61E−39 | 8.90E−40 | 5.56E−40 | 3.79E−40 |
| 110 | 7.41E−37 | 5.20E−38 | 1.05E−38 | 3.57E−39 | 1.64E−39 | 9.06E−40 | 5.67E−40 | 3.86E−40 |
| 111 | 7.52E−37 | 5.28E−38 | 1.06E−38 | 3.62E−39 | 1.67E−39 | 9.22E−40 | 5.77E−40 | 3.93E−40 |
| 112 | 7.61E−37 | 5.35E−38 | 1.08E−38 | 3.68E−39 | 1.69E−39 | 9.37E−40 | 5.86E−40 | 3.99E−40 |
| 113 | 7.69E−37 | 5.42E−38 | 1.09E−38 | 3.73E−39 | 1.72E−39 | 9.52E−40 | 5.96E−40 | 4.06E−40 |
| 114 | 7.77E−37 | 5.48E−38 | 1.11E−38 | 3.79E−39 | 1.74E−39 | 9.66E−40 | 6.05E−40 | 4.12E−40 |
| 115 | 7.83E−37 | 5.54E−38 | 1.12E−38 | 3.83E−39 | 1.77E−39 | 9.80E−40 | 6.14E−40 | 4.18E−40 |
| 116 | 7.89E−37 | 5.59E−38 | 1.13E−38 | 3.88E−39 | 1.79E−39 | 9.93E−40 | 6.22E−40 | 4.24E−40 |

**Table 1.** *Cont.*

| λ (nm) | 3000 K | 4000 K | 5000 K | 6000 K | 7000 K | 8000 K | 9000 K | 10,000 K |
|---|---|---|---|---|---|---|---|---|
| 117 | 7.93E−37 | 5.63E−38 | 1.15E−38 | 3.93E−39 | 1.81E−39 | 1.01E−39 | 6.31E−40 | 4.30E−40 |
| 118 | 7.97E−37 | 5.68E−38 | 1.16E−38 | 3.97E−39 | 1.83E−39 | 1.02E−39 | 6.39E−40 | 4.36E−40 |
| 119 | 8.00E−37 | 5.71E−38 | 1.17E−38 | 4.01E−39 | 1.85E−39 | 1.03E−39 | 6.46E−40 | 4.42E−40 |
| 120 | 8.02E−37 | 5.75E−38 | 1.18E−38 | 4.05E−39 | 1.87E−39 | 1.04E−39 | 6.54E−40 | 4.47E−40 |
| 121 | 8.03E−37 | 5.78E−38 | 1.18E−38 | 4.08E−39 | 1.89E−39 | 1.05E−39 | 6.62E−40 | 4.52E−40 |
| 122 | 8.04E−37 | 5.80E−38 | 1.19E−38 | 4.12E−39 | 1.91E−39 | 1.06E−39 | 6.69E−40 | 4.57E−40 |
| 123 | 8.04E−37 | 5.82E−38 | 1.20E−38 | 4.15E−39 | 1.93E−39 | 1.07E−39 | 6.76E−40 | 4.62E−40 |
| 124 | 8.03E−37 | 5.84E−38 | 1.21E−38 | 4.18E−39 | 1.94E−39 | 1.08E−39 | 6.83E−40 | 4.67E−40 |
| 125 | 8.01E−37 | 5.85E−38 | 1.21E−38 | 4.21E−39 | 1.96E−39 | 1.09E−39 | 6.89E−40 | 4.72E−40 |
| 126 | 7.99E−37 | 5.86E−38 | 1.22E−38 | 4.23E−39 | 1.97E−39 | 1.10E−39 | 6.96E−40 | 4.77E−40 |
| 127 | 7.97E−37 | 5.87E−38 | 1.22E−38 | 4.26E−39 | 1.99E−39 | 1.11E−39 | 7.02E−40 | 4.81E−40 |
| 128 | 7.94E−37 | 5.87E−38 | 1.23E−38 | 4.28E−39 | 2.00E−39 | 1.12E−39 | 7.08E−40 | 4.86E−40 |
| 129 | 7.90E−37 | 5.87E−38 | 1.23E−38 | 4.30E−39 | 2.01E−39 | 1.13E−39 | 7.14E−40 | 4.90E−40 |
| 130 | 7.86E−37 | 5.87E−38 | 1.23E−38 | 4.32E−39 | 2.03E−39 | 1.14E−39 | 7.19E−40 | 4.94E−40 |
| 131 | 7.81E−37 | 5.86E−38 | 1.24E−38 | 4.34E−39 | 2.04E−39 | 1.15E−39 | 7.25E−40 | 4.98E−40 |
| 132 | 7.76E−37 | 5.86E−38 | 1.24E−38 | 4.36E−39 | 2.05E−39 | 1.15E−39 | 7.30E−40 | 5.02E−40 |
| 133 | 7.70E−37 | 5.84E−38 | 1.24E−38 | 4.37E−39 | 2.06E−39 | 1.16E−39 | 7.35E−40 | 5.06E−40 |
| 134 | 7.65E−37 | 5.83E−38 | 1.24E−38 | 4.39E−39 | 2.07E−39 | 1.17E−39 | 7.40E−40 | 5.10E−40 |
| 135 | 7.58E−37 | 5.82E−38 | 1.24E−38 | 4.40E−39 | 2.08E−39 | 1.17E−39 | 7.45E−40 | 5.13E−40 |
| 136 | 7.52E−37 | 5.80E−38 | 1.24E−38 | 4.41E−39 | 2.09E−39 | 1.18E−39 | 7.49E−40 | 5.17E−40 |
| 137 | 7.45E−37 | 5.78E−38 | 1.24E−38 | 4.42E−39 | 2.09E−39 | 1.19E−39 | 7.54E−40 | 5.20E−40 |
| 138 | 7.38E−37 | 5.76E−38 | 1.24E−38 | 4.43E−39 | 2.10E−39 | 1.19E−39 | 7.58E−40 | 5.24E−40 |
| 139 | 7.31E−37 | 5.73E−38 | 1.24E−38 | 4.43E−39 | 2.11E−39 | 1.20E−39 | 7.62E−40 | 5.27E−40 |
| 140 | 7.24E−37 | 5.71E−38 | 1.24E−38 | 4.44E−39 | 2.12E−39 | 1.20E−39 | 7.66E−40 | 5.30E−40 |
| 141 | 7.16E−37 | 5.68E−38 | 1.24E−38 | 4.45E−39 | 2.12E−39 | 1.21E−39 | 7.70E−40 | 5.33E−40 |
| 142 | 7.08E−37 | 5.65E−38 | 1.24E−38 | 4.45E−39 | 2.13E−39 | 1.21E−39 | 7.74E−40 | 5.36E−40 |
| 143 | 7.00E−37 | 5.62E−38 | 1.23E−38 | 4.45E−39 | 2.13E−39 | 1.22E−39 | 7.77E−40 | 5.39E−40 |
| 144 | 6.92E−37 | 5.59E−38 | 1.23E−38 | 4.46E−39 | 2.14E−39 | 1.22E−39 | 7.80E−40 | 5.41E−40 |
| 145 | 6.84E−37 | 5.56E−38 | 1.23E−38 | 4.46E−39 | 2.14E−39 | 1.22E−39 | 7.84E−40 | 5.44E−40 |
| 146 | 6.76E−37 | 5.52E−38 | 1.23E−38 | 4.46E−39 | 2.15E−39 | 1.23E−39 | 7.87E−40 | 5.46E−40 |
| 147 | 6.68E−37 | 5.49E−38 | 1.22E−38 | 4.46E−39 | 2.15E−39 | 1.23E−39 | 7.90E−40 | 5.49E−40 |
| 148 | 6.59E−37 | 5.45E−38 | 1.22E−38 | 4.46E−39 | 2.15E−39 | 1.23E−39 | 7.93E−40 | 5.51E−40 |
| 149 | 6.51E−37 | 5.42E−38 | 1.22E−38 | 4.45E−39 | 2.15E−39 | 1.24E−39 | 7.95E−40 | 5.54E−40 |
| 150 | 6.42E−37 | 5.38E−38 | 1.21E−38 | 4.45E−39 | 2.16E−39 | 1.24E−39 | 7.98E−40 | 5.56E−40 |
| 151 | 6.34E−37 | 5.34E−38 | 1.21E−38 | 4.45E−39 | 2.16E−39 | 1.24E−39 | 8.01E−40 | 5.58E−40 |
| 152 | 6.25E−37 | 5.30E−38 | 1.20E−38 | 4.44E−39 | 2.16E−39 | 1.25E−39 | 8.03E−40 | 5.60E−40 |
| 153 | 6.17E−37 | 5.26E−38 | 1.20E−38 | 4.44E−39 | 2.16E−39 | 1.25E−39 | 8.05E−40 | 5.62E−40 |
| 154 | 6.08E−37 | 5.22E−38 | 1.20E−38 | 4.43E−39 | 2.16E−39 | 1.25E−39 | 8.08E−40 | 5.64E−40 |
| 155 | 6.00E−37 | 5.18E−38 | 1.19E−38 | 4.43E−39 | 2.16E−39 | 1.25E−39 | 8.10E−40 | 5.66E−40 |
| 156 | 5.91E−37 | 5.14E−38 | 1.19E−38 | 4.42E−39 | 2.17E−39 | 1.25E−39 | 8.12E−40 | 5.68E−40 |
| 157 | 5.83E−37 | 5.10E−38 | 1.18E−38 | 4.42E−39 | 2.17E−39 | 1.26E−39 | 8.14E−40 | 5.70E−40 |
| 158 | 5.74E−37 | 5.06E−38 | 1.18E−38 | 4.41E−39 | 2.17E−39 | 1.26E−39 | 8.16E−40 | 5.72E−40 |
| 159 | 5.66E−37 | 5.02E−38 | 1.17E−38 | 4.40E−39 | 2.17E−39 | 1.26E−39 | 8.18E−40 | 5.73E−40 |
| 160 | 5.58E−37 | 4.98E−38 | 1.17E−38 | 4.39E−39 | 2.17E−39 | 1.26E−39 | 8.19E−40 | 5.75E−40 |
| 162 | 5.41E−37 | 4.90E−38 | 1.16E−38 | 4.37E−39 | 2.16E−39 | 1.26E−39 | 8.22E−40 | 5.78E−40 |
| 163 | 5.33E−37 | 4.85E−38 | 1.15E−38 | 4.36E−39 | 2.16E−39 | 1.26E−39 | 8.24E−40 | 5.79E−40 |
| 164 | 5.25E−37 | 4.81E−38 | 1.14E−38 | 4.35E−39 | 2.16E−39 | 1.27E−39 | 8.25E−40 | 5.81E−40 |
| 165 | 5.17E−37 | 4.77E−38 | 1.14E−38 | 4.34E−39 | 2.16E−39 | 1.27E−39 | 8.27E−40 | 5.82E−40 |
| 166 | 5.09E−37 | 4.73E−38 | 1.13E−38 | 4.33E−39 | 2.16E−39 | 1.27E−39 | 8.28E−40 | 5.83E−40 |
| 167 | 5.02E−37 | 4.68E−38 | 1.13E−38 | 4.32E−39 | 2.16E−39 | 1.27E−39 | 8.29E−40 | 5.85E−40 |
| 168 | 4.94E−37 | 4.64E−38 | 1.12E−38 | 4.31E−39 | 2.16E−39 | 1.27E−39 | 8.30E−40 | 5.86E−40 |
| 169 | 4.86E−37 | 4.60E−38 | 1.12E−38 | 4.30E−39 | 2.15E−39 | 1.27E−39 | 8.31E−40 | 5.87E−40 |
| 170 | 4.79E−37 | 4.56E−38 | 1.11E−38 | 4.29E−39 | 2.15E−39 | 1.27E−39 | 8.32E−40 | 5.88E−40 |
| 171 | 4.71E−37 | 4.52E−38 | 1.10E−38 | 4.27E−39 | 2.15E−39 | 1.27E−39 | 8.33E−40 | 5.89E−40 |
| 172 | 4.64E−37 | 4.48E−38 | 1.10E−38 | 4.26E−39 | 2.15E−39 | 1.27E−39 | 8.34E−40 | 5.90E−40 |
| 173 | 4.57E−37 | 4.43E−38 | 1.09E−38 | 4.25E−39 | 2.14E−39 | 1.27E−39 | 8.34E−40 | 5.91E−40 |
| 174 | 4.50E−37 | 4.39E−38 | 1.09E−38 | 4.24E−39 | 2.14E−39 | 1.27E−39 | 8.35E−40 | 5.92E−40 |
| 175 | 4.43E−37 | 4.35E−38 | 1.08E−38 | 4.22E−39 | 2.14E−39 | 1.27E−39 | 8.36E−40 | 5.93E−40 |

**Table 1.** *Cont.*

| $\lambda$ (nm) | 3000 K | 4000 K | 5000 K | 6000 K | 7000 K | 8000 K | 9000 K | 10,000 K |
|---|---|---|---|---|---|---|---|---|
| 176 | 4.36E−37 | 4.31E−38 | 1.07E−38 | 4.21E−39 | 2.13E−39 | 1.27E−39 | 8.36E−40 | 5.94E−40 |
| 177 | 4.29E−37 | 4.27E−38 | 1.07E−38 | 4.20E−39 | 2.13E−39 | 1.27E−39 | 8.37E−40 | 5.94E−40 |
| 178 | 4.22E−37 | 4.23E−38 | 1.06E−38 | 4.18E−39 | 2.13E−39 | 1.27E−39 | 8.37E−40 | 5.95E−40 |
| 179 | 4.15E−37 | 4.19E−38 | 1.06E−38 | 4.17E−39 | 2.12E−39 | 1.27E−39 | 8.38E−40 | 5.96E−40 |
| 180 | 4.09E−37 | 4.15E−38 | 1.05E−38 | 4.16E−39 | 2.12E−39 | 1.27E−39 | 8.38E−40 | 5.96E−40 |
| 181 | 4.02E−37 | 4.11E−38 | 1.04E−38 | 4.14E−39 | 2.12E−39 | 1.27E−39 | 8.38E−40 | 5.97E−40 |
| 182 | 3.96E−37 | 4.07E−38 | 1.04E−38 | 4.13E−39 | 2.11E−39 | 1.26E−39 | 8.39E−40 | 5.98E−40 |
| 183 | 3.90E−37 | 4.03E−38 | 1.03E−38 | 4.11E−39 | 2.11E−39 | 1.26E−39 | 8.39E−40 | 5.98E−40 |
| 184 | 3.84E−37 | 4.00E−38 | 1.03E−38 | 4.10E−39 | 2.11E−39 | 1.26E−39 | 8.39E−40 | 5.99E−40 |
| 185 | 3.78E−37 | 3.96E−38 | 1.02E−38 | 4.09E−39 | 2.10E−39 | 1.26E−39 | 8.39E−40 | 5.99E−40 |
| 186 | 3.72E−37 | 3.92E−38 | 1.01E−38 | 4.07E−39 | 2.10E−39 | 1.26E−39 | 8.39E−40 | 6.00E−40 |
| 187 | 3.66E−37 | 3.88E−38 | 1.01E−38 | 4.06E−39 | 2.09E−39 | 1.26E−39 | 8.39E−40 | 6.00E−40 |
| 188 | 3.60E−37 | 3.84E−38 | 1.00E−38 | 4.04E−39 | 2.09E−39 | 1.26E−39 | 8.40E−40 | 6.01E−40 |
| 189 | 3.54E−37 | 3.81E−38 | 9.95E−39 | 4.03E−39 | 2.09E−39 | 1.26E−39 | 8.40E−40 | 6.01E−40 |
| 190 | 3.49E−37 | 3.77E−38 | 9.89E−39 | 4.01E−39 | 2.08E−39 | 1.26E−39 | 8.40E−40 | 6.02E−40 |
| 191 | 3.43E−37 | 3.73E−38 | 9.83E−39 | 4.00E−39 | 2.08E−39 | 1.26E−39 | 8.39E−40 | 6.02E−40 |
| 192 | 3.38E−37 | 3.70E−38 | 9.77E−39 | 3.98E−39 | 2.07E−39 | 1.26E−39 | 8.39E−40 | 6.02E−40 |
| 193 | 3.33E−37 | 3.66E−38 | 9.71E−39 | 3.97E−39 | 2.07E−39 | 1.25E−39 | 8.39E−40 | 6.03E−40 |
| 194 | 3.27E−37 | 3.63E−38 | 9.65E−39 | 3.95E−39 | 2.06E−39 | 1.25E−39 | 8.39E−40 | 6.03E−40 |
| 195 | 3.22E−37 | 3.59E−38 | 9.59E−39 | 3.94E−39 | 2.06E−39 | 1.25E−39 | 8.39E−40 | 6.03E−40 |
| 196 | 3.17E−37 | 3.56E−38 | 9.54E−39 | 3.92E−39 | 2.06E−39 | 1.25E−39 | 8.39E−40 | 6.03E−40 |
| 197 | 3.12E−37 | 3.52E−38 | 9.48E−39 | 3.91E−39 | 2.05E−39 | 1.25E−39 | 8.38E−40 | 6.04E−40 |
| 198 | 3.08E−37 | 3.49E−38 | 9.42E−39 | 3.89E−39 | 2.05E−39 | 1.25E−39 | 8.38E−40 | 6.04E−40 |
| 199 | 3.03E−37 | 3.46E−38 | 9.36E−39 | 3.88E−39 | 2.04E−39 | 1.25E−39 | 8.38E−40 | 6.04E−40 |
| 200 | 2.98E−37 | 3.42E−38 | 9.31E−39 | 3.86E−39 | 2.04E−39 | 1.24E−39 | 8.38E−40 | 6.04E−40 |
| 205 | 2.76E−37 | 3.26E−38 | 9.03E−39 | 3.79E−39 | 2.01E−39 | 1.24E−39 | 8.35E−40 | 6.04E−40 |
| 210 | 2.56E−37 | 3.11E−38 | 8.75E−39 | 3.71E−39 | 1.99E−39 | 1.23E−39 | 8.33E−40 | 6.04E−40 |
| 215 | 2.37E−37 | 2.97E−38 | 8.49E−39 | 3.64E−39 | 1.96E−39 | 1.22E−39 | 8.29E−40 | 6.03E−40 |
| 220 | 2.20E−37 | 2.83E−38 | 8.24E−39 | 3.57E−39 | 1.94E−39 | 1.21E−39 | 8.25E−40 | 6.02E−40 |
| 225 | 2.04E−37 | 2.71E−38 | 7.99E−39 | 3.50E−39 | 1.91E−39 | 1.20E−39 | 8.21E−40 | 6.01E−40 |
| 230 | 1.90E−37 | 2.59E−38 | 7.76E−39 | 3.43E−39 | 1.89E−39 | 1.19E−39 | 8.17E−40 | 5.99E−40 |
| 235 | 1.77E−37 | 2.48E−38 | 7.53E−39 | 3.36E−39 | 1.86E−39 | 1.18E−39 | 8.13E−40 | 5.98E−40 |
| 240 | 1.66E−37 | 2.37E−38 | 7.32E−39 | 3.30E−39 | 1.84E−39 | 1.17E−39 | 8.08E−40 | 5.96E−40 |
| 245 | 1.55E−37 | 2.27E−38 | 7.11E−39 | 3.23E−39 | 1.81E−39 | 1.16E−39 | 8.04E−40 | 5.94E−40 |
| 250 | 1.45E−37 | 2.18E−38 | 6.92E−39 | 3.17E−39 | 1.79E−39 | 1.15E−39 | 7.99E−40 | 5.92E−40 |
| 255 | 1.36E−37 | 2.09E−38 | 6.73E−39 | 3.11E−39 | 1.76E−39 | 1.13E−39 | 7.94E−40 | 5.90E−40 |
| 260 | 1.27E−37 | 2.01E−38 | 6.54E−39 | 3.05E−39 | 1.74E−39 | 1.12E−39 | 7.89E−40 | 5.87E−40 |
| 265 | 1.20E−37 | 1.93E−38 | 6.37E−39 | 2.99E−39 | 1.72E−39 | 1.11E−39 | 7.83E−40 | 5.84E−40 |
| 270 | 1.12E−37 | 1.85E−38 | 6.20E−39 | 2.94E−39 | 1.69E−39 | 1.10E−39 | 7.78E−40 | 5.82E−40 |
| 275 | 1.06E−37 | 1.78E−38 | 6.04E−39 | 2.89E−39 | 1.67E−39 | 1.09E−39 | 7.72E−40 | 5.79E−40 |
| 280 | 9.99E−38 | 1.72E−38 | 5.89E−39 | 2.83E−39 | 1.65E−39 | 1.08E−39 | 7.67E−40 | 5.76E−40 |
| 285 | 9.43E−38 | 1.66E−38 | 5.74E−39 | 2.78E−39 | 1.63E−39 | 1.07E−39 | 7.62E−40 | 5.73E−40 |
| 290 | 8.92E−38 | 1.60E−38 | 5.60E−39 | 2.73E−39 | 1.61E−39 | 1.06E−39 | 7.56E−40 | 5.70E−40 |
| 295 | 8.45E−38 | 1.54E−38 | 5.47E−39 | 2.69E−39 | 1.59E−39 | 1.05E−39 | 7.51E−40 | 5.67E−40 |
| 300 | 8.00E−38 | 1.49E−38 | 5.34E−39 | 2.64E−39 | 1.57E−39 | 1.04E−39 | 7.46E−40 | 5.64E−40 |
| 305 | 7.60E−38 | 1.44E−38 | 5.22E−39 | 2.60E−39 | 1.55E−39 | 1.03E−39 | 7.41E−40 | 5.61E−40 |
| 310 | 7.22E−38 | 1.39E−38 | 5.10E−39 | 2.56E−39 | 1.53E−39 | 1.02E−39 | 7.36E−40 | 5.59E−40 |
| 315 | 6.86E−38 | 1.35E−38 | 4.99E−39 | 2.52E−39 | 1.51E−39 | 1.01E−39 | 7.32E−40 | 5.56E−40 |
| 320 | 6.53E−38 | 1.31E−38 | 4.88E−39 | 2.48E−39 | 1.49E−39 | 1.01E−39 | 7.27E−40 | 5.53E−40 |
| 325 | 6.23E−38 | 1.27E−38 | 4.78E−39 | 2.44E−39 | 1.48E−39 | 9.96E−40 | 7.22E−40 | 5.51E−40 |
| 330 | 5.94E−38 | 1.23E−38 | 4.68E−39 | 2.40E−39 | 1.46E−39 | 9.88E−40 | 7.17E−40 | 5.48E−40 |
| 335 | 5.67E−38 | 1.19E−38 | 4.58E−39 | 2.37E−39 | 1.44E−39 | 9.79E−40 | 7.13E−40 | 5.45E−40 |
| 340 | 5.42E−38 | 1.16E−38 | 4.49E−39 | 2.33E−39 | 1.43E−39 | 9.71E−40 | 7.08E−40 | 5.43E−40 |
| 345 | 5.19E−38 | 1.13E−38 | 4.40E−39 | 2.30E−39 | 1.41E−39 | 9.63E−40 | 7.04E−40 | 5.40E−40 |

**Table 1.** *Cont.*

| λ (nm) | 3000 K | 4000 K | 5000 K | 6000 K | 7000 K | 8000 K | 9000 K | 10,000 K |
|--------|--------|--------|--------|--------|--------|--------|--------|----------|
| 350 | 4.97E−38 | 1.09E−38 | 4.31E−39 | 2.26E−39 | 1.40E−39 | 9.55E−40 | 6.99E−40 | 5.37E−40 |
| 355 | 4.76E−38 | 1.06E−38 | 4.23E−39 | 2.23E−39 | 1.38E−39 | 9.47E−40 | 6.95E−40 | 5.35E−40 |
| 360 | 4.57E−38 | 1.04E−38 | 4.15E−39 | 2.20E−39 | 1.37E−39 | 9.39E−40 | 6.90E−40 | 5.32E−40 |
| 365 | 4.38E−38 | 1.01E−38 | 4.07E−39 | 2.17E−39 | 1.35E−39 | 9.32E−40 | 6.86E−40 | 5.29E−40 |
| 370 | 4.21E−38 | 9.82E−39 | 4.00E−39 | 2.14E−39 | 1.34E−39 | 9.24E−40 | 6.81E−40 | 5.26E−40 |

**Solar atmosphere: absorption processes.** The influence of radiation processes (1) can bee estimated by comparing their intensities with the intensities of known concurrent radiation processes, namely:

$$\varepsilon_\lambda + \left\{ \begin{array}{c} H^- \\ H + e' \end{array} \right. \Rightarrow H + e'' \tag{11}$$

$$\varepsilon_\lambda + \left\{ \begin{array}{c} H^* \\ H^+ + e' \end{array} \right. \Rightarrow H^+ + e'' \tag{12}$$

The relative contributions of the processes (1), with respect to processes (11) and (12), is described by the quantities $F_\kappa$ defined by relation

$$F_\kappa = \frac{\kappa_{ia}}{\kappa_{ea} + \kappa_{ei}} \tag{13}$$

where $\kappa_{ea}$ is absorption spectral coefficient of processes (11) and $\kappa_{ei}$ is absorption spectral coefficient of processes (12) (see papers [14,16]).

Similarly to the He case in DB white dwarf atmosphere [25] calculations of the absorption coefficient were performed for the solar photosphere and lower chromosphere by means of a standard Solar atmosphere model [5], and the total contribution of the processes (1) to the solar opacity was estimated [16]. The results of the calculations of the parameter $F_\kappa$ for $92\,\text{nm} \le \lambda \le 350\,\text{nm}$ are presented in Figure 5. The figure show that in the significant part of the considered region of altitudes ($-75\,\text{km} \le h \le 1065\,\text{km}$) the absorption process (1) together give the contribution which varies from about 10% to about 90% of the contribution of the absorption process (11) and (12), which are considered as the main absorption processes [16].

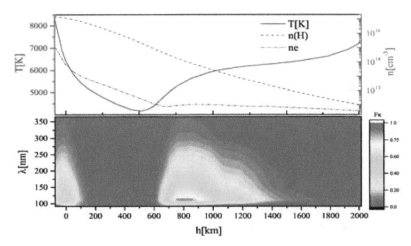

**Figure 5.** Upper panel: The behavior of the temperature $T$, $N_H$ and $Ne$ as a function of height $h$ within the considered part of the solar atmosphere model; lower panel: A surface plot of the quantity $F_\kappa = \kappa_{ia}/(\kappa_{ea} + \kappa_{ei})$ (data taken from [14]) as a function of $\lambda$ and height $h$ for a model of the solar photosphere [5].

On the basis of the above, it can be concluded that photodissociation processes represent important channels for destruction of molecules in lot of astrophysical environments and features of the interacting radiation are important in their spectral analyses.

## 3. Future Developments and Concluding Remarks

Exploitation the full potential of A + M data and database services is an ongoing challenge in virtual data centers. There are still many limitations and problems that users are facing with such as poor documentation, missing of data evaluation, no open access, etc. The aim of MolD database is to be accessible, and be used by the wider scientific community, through VAMDC and to follow certain protocols and defined rules in order to eliminate such limitations and problems.

The next step of development i.e., the stage three of MolD development will be the implementation of possibility to fit the tabulated data. We plan to develop fitting formulas for photodissociation cross section as the function of the temperature and wavelength. Also, we intend to update the current database with newly calculated/measured data.

The continuation of such developments and services such as constantly updated online A + M database, is crucial in the field of astrophysics and modern physics due to its rapid development and make an immense impact on the way science is done in the developing world.

**Acknowledgments:** The authors are thankful to the Ministry of Education, Science and Technological Development of the Republic of Serbia for the support of this work within the projects 176002 and III44002. This work has also been supported by the VAMDC (Virtual Atomic and Molecular Data Centre). VAMDC is funded under the Combination of Collaborative Projects and Coordination and Support Actions Funding Scheme of The Seventh Framework Program. We acknowledge the contribution of late A.A. Mihajlov who collected and calculated much of the data that are presented within the MolD database.

**Author Contributions:** This work is based on the numerous contributions of all the authors.

## References

1. Marinković, B.P.; Vujčić, V.; Sushko, G.; Vudragović, D.; Marinković, D.B.; Đorđević, S.; Ivanović, S.; Nešić, M.; Jevremović, D.; Solov'yov, A.V.; et al. Development of collisional data base for elementary processes of electron scattering by atoms and molecules. *Nucl. Instr. Meth. Phys. Res. B* **2015**, *354*, 90–95.
2. Dubernet, M.L.; Antony, B.; Ba, Y.A.; Babikov, Y.L.; Bartschat, K.; Boudon, V.; Braams, B.; Chung, H.K.; Daniel, F.; Delahaye, F.; et al. The virtual atomic and molecular data centre (VAMDC) consortium. *J. Phys. B* **2016**, *49*, 074003.
3. Marinković, B.P.; Jevremović, D.; Srećković, V.A.; Vujčić, V.; Ignjatović, L.M.; Dimitrijević, M.S.; Mason, N.J. BEAMDB and MolD—Databases for atomic and molecular collisional and radiative processes: Belgrade nodes of VAMDC. *Eur. Phys. J. D* **2017**, *71*, 158.
4. Christensen-Dalsgaard, J.; Dappen, W.; Ajukov, S.; Anderson, E.; Antia, H.; Basu, S.; Baturin, V.; Berthomieu, G.; Chaboyer, B.; Chitre, S.; et al. The current state of solar modeling. *Science* **1996**, *272*, 1286.
5. Fontenla, J.; Curdt, W.; Haberreiter, M.; Harder, J.; Tian, H. Semiempirical models of the solar atmosphere. III. Set of non-LTE models for far-ultraviolet/extreme-ultraviolet irradiance computation. *Astrophys. J.* **2009**, *707*, 482.
6. Hauschildt, P.; Baron, E. Cool stellar atmospheres with PHOENIX. *Mem. Soc. Astron. Ital. Suppl.* **2005**, *7*, 140.
7. Hauschildt, P.H.; Baron, E. A 3D radiative transfer framework-VI. PHOENIX/3D example applications. *Astron. Astrophys.* **2010**, *509*, A36.
8. Husser, T.O.; Wende-von Berg, S.; Dreizler, S.; Homeier, D.; Reiners, A.; Barman, T.; Hauschildt, P.H. A new extensive library of PHOENIX stellar atmospheres and synthetic spectra. *Astron. Astrophys.* **2013**, *553*, A6.
9. Augustovičová, L.; Špirko, V.; Kraemer, W.; Soldán, P. Radiative association of $He_2^+$ revisited. *Astron. Astrophys.* **2013**, *553*, A42.
10. Koester, D. Model atmospheres for DB white dwarfs. *Astron. Astrophys. Suppl. Ser.* **1980**, *39*, 401–409.
11. Coppola, C.M.; Galli, D.; Palla, F.; Longo, S.; Chluba, J. Non-thermal photons and $H_2$ formation in the early Universe. *Mon. Not. R. Astron. Soc.* **2013**, *434*, 114–122.

12. Klyucharev, A.; Bezuglov, N.; Matveev, A.; Mihajlov, A.; Ignjatović, L.M.; Dimitrijević, M. Rate coefficients for the chemi-ionization processes in sodium-and other alkali-metal geocosmical plasmas. *New Astron. Rev.* **2007**, *51*, 547–562.

13. Sugimura, K.; Coppola, C.M.; Omukai, K.; Galli, D.; Palla, F. Role of the channel in the primordial star formation under strong radiation field and the critical intensity for the supermassive star formation. *Mon. Not. R. Astron. Soc.* **2015**, *456*, 270–277.

14. Mihajlov, A.; Ignjatovic, L.M.; Sakan, N.; Dimitrijevic, M. The influence of $H_2^+$-photo-dissociation and $(H + H^+)$-radiative collisions on the solar atmosphere opacity in UV and VUV regions. *Astron. Astrophys.* **2007**, *469*, 749–754.

15. Ignjatović, L.M.; Mihajlov, A.; Srećković, V.; Dimitrijević, M. Absorption non-symmetric ion-atom processes in helium-rich white dwarf atmospheres. *Mon. Not. R. Astron. Soc.* **2014**, *439*, 2342–2350.

16. Srećković, V.; Mihajlov, A.; Ignjatović, L.M.; Dimitrijević, M. Ion-atom radiative processes in the solar atmosphere: Quiet Sun and sunspots. *Adv. Space Res.* **2014**, *54*, 1264–1271.

17. Babb, J.F. State resolved data for radiative association of H and $H^+$ and for Photodissociation of $H_2^+$. *Astrophys. J. Suppl. Ser.* **2015**, *216*, 21.

18. Heays, A.; Bosman, A.; van Dishoeck, E. Photodissociation and photoionisation of atoms and molecules of astrophysical interest. *Astron. Astrophys.* **2017**, *602*, A105.

19. Mihajlov, A.; Ignjatović, L.M.; Srećković, V.; Dimitrijević, M.; Metropoulos, A. The non-symmetric ion-atom radiative processes in the stellar atmospheres. *Mon. Not. R. Astron. Soc.* **2013**, *431*, 589–599.

20. Ignjatović, L.M.; Mihajlov, A.; Srećković, V.; Dimitrijević, M. The ion-atom absorption processes as one of the factors of the influence on the sunspot opacity. *Mon. Not. R. Astron. Soc.* **2014**, *441*, 1504–1512.

21. Puy, D.; Dubrovich, V.; Lipovka, A.; Talbi, D.; Vonlanthen, P. Molecular fluorine chemistry in the early Universe. *Astron. Astrophys.* **2007**, *476*, 685–689.

22. Vujčić, V.; Jevremović, D.; Mihajlov, A.; Ignjatović, L.M.; Srećković, V.; Dimitrijević, M.; Malović, M. MOL-D: A Collisional Database and Web Service within the Virtual Atomic and Molecular Data Center. *J. Astrophys. Astron.* **2015**, *36*, 693–703.

23. Forcier, J.; Bissex, P.; Chun, W.J. *Python Web Development with Django*; Addison-Wesley Professional: Indianapolis, IN, USA, 2008.

24. Widenius, M.; Axmark, D. *MySQL Reference Manual: Documentation from the Source*; O'Reilly Media, Inc.: Sebastopol, CA, USA, 2002.

25. Ignjatović, L.M.; Mihajlov, A.; Sakan, N.; Dimitrijević, M.; Metropoulos, A. The total and relative contribution of the relevant absorption processes to the opacity of DB white dwarf atmospheres in the UV and VUV regions. *Mon. Not. R. Astron. Soc.* **2009**, *396*, 2201–2210.

# 10

# The Application of the Cut-Off Coulomb Model Potential for the Calculation of Bound-Bound State Transitions

Nenad M. Sakan [1,*], Vladimir A. Srećković [1], Zoran J. Simić [2] and Milan S. Dimitrijević [2,3]

[1] Institute of Physics, Belgrade University, Pregrevica 118, 11080 Zemun, Belgrade, Serbia; vlada@ipb.ac.rs
[2] Astronomical Observatory, Volgina 7, 11060 Belgrade, Serbia; zsimic@aob.rs (Z.J.S.); mdimitrijevic@aob.rs (M.S.D.)
[3] LERMA (Laboratoire d'Etudes du Rayonnement et de la Matière en Astrophysique et Atmosphères), Observatoire de Paris, UMR CNRS 8112, UPMC, 92195 Meudon CEDEX, France
* Correspondence: nsakan@ipb.ac.rs

**Abstract:** In this contribution, we present results of bound state transition modeling using the cut-off Coulomb model potential. The cut-off Coulomb potential has proven itself as a model potential for the dense hydrogen plasma. The main aim of our investigation include further steps of improvement of the usage of model potential. The results deal with partially ionized dense hydrogen plasma. The presented results covers $N_e = 6.5 \times 10^{18}$ cm$^{-3}$, $T = 18,000$ K and $N_e = 1.5 \times 10^{19}$ cm$^{-3}$, $T = 23,000$ K, where the comparison with the experimental data should take place, and the theoretical values for comparison. Since the model was successfully applied on continuous photoabsorption of dense hydrogen plasma in the broad area of temperatures and densities, it is expected to combine both continuous and bound-bound photoabsorption within single quantum mechanical model with the same success.

**Keywords:** atomic processes; radiative transfer; Sun: atmosphere; Sun: photosphere; stars: atmospheres; white dwarfs

## 1. Introduction

The problems of plasma opacity, energy transport and radiative transfer under moderate and strong non-ideality are of interest in theoretical and experimental research [1–4]. The strong coupling and density effects in plasma radiation were the subject of numerous experimental and theoretical studies in the last decades. Here, we keep in mind the plasma of the inner layers of the solar atmosphere, as well as of partially ionized layers of other stellar atmospheres—for example, the atmospheres of DA white dwarfs with effective temperatures between 4500 K and 30,000 K.

In this paper, we presented a new model way of describing atomic photo-absorption processes in dense, strongly ionized hydrogen plasmas, which is based on the approximation of the cut-off Coulomb potential. By now, this approximation has been used in order to describe transport properties of dense plasmas (e.g., [1,5,6]), but it was clear that it could be applied to some absorption processes in non-ideal plasmas too [3,7–9]. This topic itself, and the search for more consistent models of screening and more realistic potentials in plasmas is still continuing and is very real (e.g., [10–13] and references therein).

The neutral hydrogen acts as an absorber in plasma, and its concentration determination is essential in order to calculate the absorption coefficients. The neutral hydrogen concentration from experimental data from [14] is used, and, since the result presented here is in the area where the Saha equation is valid, the neutral concentrations using the Saha equation with the ionization potential correction is possible. Since this is related to the further step, bound level broadening mechanism, the procedure was not elaborated here. In addition, the emitter, neutral hydrogen, interacts with other plasma species such as electrons and ions. In such a way, the theoretical hydrogen plasma model results as well as NIST database values act as a limiting case for verification of solutions, in the case when plasma influence is small or diminishes, e.g., $r_c \rightarrow \infty$ in later case.

In order to calculate the total absorption coefficient, the same broadening mechanism for the bound state levels should be applied. Up until now, the continuous absorption processes were calculated with the help of parametric broadening for the bound-free absorption [7]. The broadening processes play a more important role in bound-bound processes, and, as such, the more realistic theoretical model of broadening of bound state levels or a result of adequate molecular dynamics (MD) simulations should be used. Only in such a case could the total absorption coefficients, obtained within the same broadening model, be compared with real experimental data. This work is in progress.

A first step in extending of the model with additional processes is the bound state transition processes inclusion. The bound state transition processes are stated as the most important goal in the development of the self containing model, capable of describing optical as well as dynamical characteristics of dense hydrogen plasma. The characteristics of the bound state transitions in plasma diagnostics are well known, an almost mandatory method [15]. The usage of the hydrogen lines as a probing method for the plasma characteristics is also well known and widely used in plasmas of moderate and small non-ideality [16]. In accordance with that previously mentioned, the continuation of research of the presented approach is the inclusion of the bound-bound photo absorption process using the same model potential. The investigated process is the bound-bound i.e., photo absorption processes:

$$\varepsilon_\lambda + \mathrm{H}^*(n_i, l_i) \rightarrow \mathrm{H}^*(n_f, l_f), \tag{1}$$

where $n$ and $l$ are the principal and the orbital quantum number of hydrogen-atom excited states, hydrogen atom in its initial state $|n_i, l_i >$ is presented by $\mathrm{H}^*(n_i, l_i)$, its final state $|n_f, l_f >$ by $\mathrm{H}^*(n_f, l_f)$, and $\varepsilon_\lambda$ presents absorbed photon energy.

## 2. Theory

### 2.1. The Approximation of the Cut-Off Coulomb Potential

The absorption processes (1) in astrophysical plasma are considered here as a result of radiative transition in the whole system "electron–ion pair (atom) + the neighborhood", namely: $\varepsilon_\lambda + (\mathrm{H}^+ + e)_i + S_{rest} \rightarrow (\mathrm{H}^+ + e)_f + S'_{rest}$, where $S_{rest}$ and $S'_{rest}$ denote the rest of the considered plasma. However, as it is well known, many-body processes can sometimes be simplified by their transformation to the corresponding single-particle processes in an adequately chosen model potential.

As an adequate model potential for hydrogen plasma with such density, we choose, as in [3,5], the screening cut-off Coulomb potential, which satisfies above conditions, and can be presented in the form:

$$U(r; r_c) = \begin{cases} -\dfrac{e^2}{r}, & 0 < r \leq r_c, \\ U_{p;c}, & r_c < r < \infty, \end{cases} \tag{2}$$

where the cut-off radius $r_c$ is defined by relation (3), as it is illustrated by Figure 1a. Here, $e$ is the modulus of the electron charge, $r$—distance from the ion, and cut-off radius $r_c$—the characteristic screening length of the considered plasma. Namely, within this model, it is assumed that quantity

$$U_{p;c} = -\frac{e^2}{r_c} \tag{3}$$

is the mean potential energy of an electron in the considered hydrogen plasma.

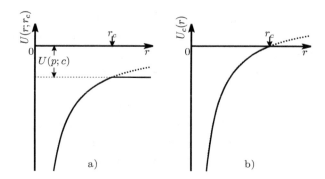

**Figure 1.** Behaviour of the potentials $U(r; r_c)$ and $U_c(r)$, where $r_c$ is the cut-off parameter. In (**a**) the potetntial goes to the mean potential energy of an electron in the considered hydrogen plasma $U_{p;c}$, and in (**b**), the value of $U_{p;c}$ is taken as zero energy, i.e., $U_c = U(r; r_c) - U_{p;c}$, in the more applicable form used here.

As in [3,5,8], we will take the value $U_{p;c}$ as the zero of energy. After that, the potential Equation (2) is transformed to the form

$$U_c(r) = \begin{cases} -\dfrac{e^2}{r} + \dfrac{e^2}{r_c}, & 0 < r \leq r_c, \\ 0, & r_c < r < \infty, \end{cases} \qquad (4)$$

which is illustrated by Figure 1b. Here, $e$ is the modulus of the electron charge, $r$—distance from the ion, and cut-off radius $r_c$—the characteristic screening length of the considered plasma.

It is important that the cut-off radius $r_c$ can be determined as a given function of $N_e$ and $T$, using two characteristic lengths:

$$r_i = \left( \frac{k_B T}{4\pi N_i e^2} \right)^{1/2} \qquad r_{s;i} = \left( \frac{3}{4\pi N_i} \right)^{\frac{1}{3}}, \qquad (5)$$

namely, taking that $N_i = N_e$ and $r_c = a_{c;i} \cdot r_i$ we can directly determine the factor $a_{c;i}$ as a function of ratio $r_{s;i}/r_i$, on the basis of the data about the mean potential energy of the electron in the single ionized plasma from [12]. The behavior of $a_{c;i}$ in a wide region of values of $r_{s;i}/r_i$ is presented in Figure 2.

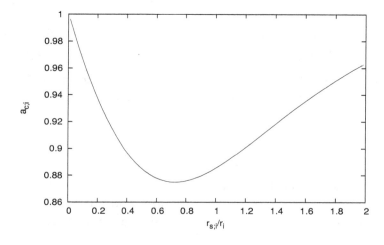

**Figure 2.** Behavior of the parameters $a_{c;i} \equiv r_c/r_i$ as the function of the ratio $r_{s;i}/r_i$, where $r_i$ is given by Equation (5) and $r_{s;i}$ is the ion Wigner–Seitz radius for the considered electron-ion plasma (5). The presented curve is obtained on the basis of data presented in [12].

## 2.2. The Calculated Quantities

In accordance with that, the behavior of the dipole matrix element is investigated here. It is given by

$$\hat{D}(r; r_c; n_i, l_i; n_f, l_f) = < n_f, l_f |\mathbf{r}| n_i, l_i >, \tag{6}$$

where the wave functions $|n_i, l_i >$ and $|n_f, l_f >$ are initial and final state wave functions obtained within the model of cut-off Coulomb potential, for the calculations of the plasma characteristics, or the theoretical hydrogen ones in order to additionally check the model.

For the calculation of oscillator strength, we use expressions from [17–19]

$$f(n_f, l_f; n_i, l_i; r_c) = \frac{1}{3} \frac{\nu}{Ry} \left[\frac{max(l_f, l_i)}{2l_f + 1}\right] \hat{D}(r; r_c; n_i, l_i; n_f, l_f)^2, \tag{7}$$

where $Ry$ is the Rydberg constant, in the same units as the frequency $\nu$ of the transition $(n, l) \to (n', l')$. The calculated data are presented in Table 1.

Our future plan is to include broadening processes and make it possible to directly compare data with the experimental ones (e.g., the data from [14]). In continuation, some of the bound-bound theoretical aspects are shown.

The total absorption cross section of the line could be linked with the dipole moment directly with the help of relation

$$\sigma_0(\omega = \omega_{fi}) = \frac{1}{3} \frac{g_2}{g_1} \frac{\pi \omega_{fi}}{\varepsilon_0 \hbar c} \hat{D}(r; r_c; n_i, l_i; n_f, l_f)^2. \tag{8}$$

It should be noted that the absorption line profile is needed in order to apply the obtained results to the absorption calculations:

$$\sigma(\omega) = \sigma_0(\omega = \omega_{fi}) g(\omega, \omega_{fi}). \tag{9}$$

Here, $g(\omega, \omega_{fi})$ is the spectral line profile and could be obtained, for instance, as a result of a molecular dynamics (MD) simulations [20–22] or as a result of theoretical modeling and spectral line data [23–27]. The lineshape is normalized to unity area, e.g.,

$$\int_{-\infty}^{\infty} g(\omega)d\omega = 1. \tag{10}$$

The line profile is a convolution of the broadening of the initial and final state energy levels, e.g.,

$$g(\omega) = g_i(\omega) * g_f(\omega). \tag{11}$$

Here, $g_i$ and $g_f$ are energy level broadening shapes of the initial and final states.

From here, the absorption coefficient could be calculated

$$k(\omega) = N_i \int_f \int_i g_i(\omega) g_f(\omega) \sigma_0. \tag{12}$$

Here, $N_i$ is the concentration of hydrogen in initial level $|n_i, l_i >$. The broadening of the levels should be considered separately in order to use the same broadening profiles for the calculation of the bound-free transition continuous photo absorption.

There are a few steps to be carried out, and the first to be taken into account is to use a parametric level and line broadening model. The second one is to use a theoretical model for spectral line broadening and try to deconvolve level broadening values, and the final step should be the usage of the MD code [20–22]. A candidate for the coupling of the MD code is LAMMPS, which is capable of simulating a large scale molecular dynamics simulations, and it also could include particular interaction potential as well as all relevant processes of interest not included in the initial code as an additional module. The broadening parameters could be calculated from average energy and particle distributions and their temporal distributions, or, as an iterative procedure, a more real plasma potential could be obtained [20,21,28,29]. In the former case, the solutions would not necessary to be described as a set of analytical functions. Since there are more steps, we did not elaborate on the ideas until we obtained some initial results.

**Table 1.** The oscillator strength values for the "short", $Ne = 1.5 \times 10^{19}$ cm$^{-3}$, $T = 23,000$ K, and "long" pulse, $Ne = 6.5 \times 10^{18}$ cm$^{-3}$, $T = 18,000$ K, from [14], as well as a theoretical hydrogen case calculated by code from [18], and NIST spectral database values [30].

| $\|n_i, l_i>$ | $\|n_f, l_f>$ | Short Pulse $f(n_f, l_f; n_i, l_i; r_c)$ | Long Pulse $f(n_f, l_f; n_i, l_i; r_c)$ | Theory $f(n_f, l_f; n_i, l_i)$ | NIST $f(n_f, l_f; n_i, l_i)$ |
|---|---|---|---|---|---|
| $\|1,0>$ | $\|2,1>$ | 0.416197 | 0.416197 | 0.416200 | 0.416400 |
| $\|1,0>$ | $\|3,1>$ | 0.079102 | 0.079102 | 0.079101 | 0.079120 |
| $\|1,0>$ | $\|4,1>$ | 0.028923 | 0.028991 | 0.028991 | 0.029010 |
| $\|1,0>$ | $\|5,1>$ | | 0.013728 | 0.013938 | 0.013950 |
| $\|2,0>$ | $\|3,1>$ | 0.434865 | 0.434865 | 0.434870 | 0.435100 |
| $\|2,0>$ | $\|4,1>$ | 0.102533 | 0.102756 | 0.102760 | 0.102800 |
| $\|2,0>$ | $\|5,1>$ | | 0.041425 | 0.041930 | 0.041950 |
| $\|2,1>$ | $\|3,0>$ | 0.013589 | 0.013589 | 0.013590 | 0.013600 |
| $\|2,1>$ | $\|3,2>$ | 0.695785 | 0.695785 | 0.695780 | 0.696100 |
| $\|2,1>$ | $\|4,0>$ | 0.003035 | 0.003045 | 0.003045 | 0.003046 |
| $\|2,1>$ | $\|4,2>$ | 0.121659 | 0.121795 | 0.121800 | 0.102800 |
| $\|2,1>$ | $\|5,0>$ | | 0.001191 | 0.001213 | 0.001214 |
| $\|2,1>$ | $\|5,2>$ | | 0.043962 | 0.044371 | 0.044400 |
| $\|3,0>$ | $\|4,1>$ | 0.483750 | 0.484708 | 0.484710 | 0.484900 |
| $\|3,0>$ | $\|5,1>$ | | 0.119310 | 0.121020 | 0.121100 |
| $\|3,1>$ | $\|4,0>$ | 0.032165 | 0.032250 | 0.032250 | 0.032280 |
| $\|3,1>$ | $\|4,2>$ | 0.617675 | 0.618282 | 0.618290 | 0.618600 |
| $\|3,1>$ | $\|5,0>$ | | 0.007299 | 0.007428 | 0.007433 |
| $\|3,1>$ | $\|5,2>$ | | 0.138013 | 0.139230 | 0.139300 |
| $\|3,2>$ | $\|4,1>$ | 0.010971 | 0.010992 | 0.010992 | 0.011000 |
| $\|3,2>$ | $\|4,3>$ | 1.017260 | 1.017520 | 1.017500 | 1.018000 |
| $\|3,2>$ | $\|5,1>$ | | 0.002180 | 0.002210 | 0.002211 |
| $\|3,2>$ | $\|5,3>$ | | 0.156046 | 0.156640 | 0.156700 |
| $\|4,0>$ | $\|5,1>$ | | 0.537527 | 0.544150 | 0.544400 |
| $\|4,1>$ | $\|5,0>$ | | 0.052123 | 0.052907 | 0.052940 |
| $\|4,1>$ | $\|5,2>$ | | 0.604678 | 0.609290 | 0.609700 |
| $\|4,2>$ | $\|5,1>$ | | 0.027498 | 0.027822 | 0.027840 |
| $\|4,2>$ | $\|5,3>$ | | 0.887328 | 0.890250 | 0.890600 |
| $\|4,3>$ | $\|5,2>$ | | 0.008809 | 0.008871 | 0.008877 |
| $\|4,3>$ | $\|5,4>$ | | 1.344790 | 1.345800 | 1.346000 |

## 3. Results and Discussion

The convergence toward theoretical values for hydrogen is examined and proven for the bound-bound transitions that could appear in the investigated area of electron concentrations and temperatures, [31]. Since the initial results have proven that the model potential could be used for the bound-bound state transition calculations, further investigation was completed.

The next step towards the application of the model results is a calculation of the oscillator strength values for the plasma parameters denoted as "short", with $N_e = 1.5 \times 10^{19}$ cm$^{-3}$, $T = 23,000$ K and "long" pulse, with $N_e = 6.5 \times 10^{18}$ cm$^{-3}$, $T = 18,000$ K from [14]. The radius $r_i$, given by Equation (5), Wigner–Seitz and cut-off radius as well as atom concentrations given by the Saha equation are $r_i = 51.0655$ a.u., $r_{s;i} = 47.5339$ a.u., $r_c = 44.9907$ a.u., $N_a = 1.9 \times 10^{19}$ cm$^{-3}$ for short and $r_i = 68.6260$ a.u., $r_{s;i} = 62.8149$ a.u., $r_c = 60.4062$ a.u., $N_a = 3.4 \times 10^{19}$ cm$^{-3}$ for long pulse.

The essential quality of the presented approach is that the values are calculated with the help of wave functions that are obtained as a completely analytical solution in the frame of the presented model. The influence of the numerical procedure is minimized, and, as a consequence, the possibility of introducing the numerical artifacts into the solution is also minimized.

Since the convergence towards the theoretical, unperturbed Coulomb potential is investigated in [31], and the results of the presented calculations converge uniformly towards theoretical values, it is expected that the presented model could obtain good data in a wider variety of plasma parameters than shown here.

There is a need to introduce the broadening mechanism for the energy levels in the presented approach. There is a possibility to couple the presented model calculations with the MD code in order to obtain a consistent model for the broadening of the bound levels. Those results influence the bound-free and bound-bound absorption profiles and are essential for the hydrogen plasma optical properties calculation in the frame of the cut-off Coulomb potential model approach.

## 4. Conclusions

The presented results are a step forward towards inclusion of the entire photo-absorption processes for hydrogen atoms in plasma within the frame of the cut-off Coulomb potential model. One of the benefits of the presented results is a completely quantum mechanical solution for the cut-off Coulomb model potential, obtained from wave functions that are analytical and represented with the help of special functions, e.g., the influence of additional numerical source of errors is minimized as much as possible. Along with this, the solutions converge towards theoretical pure Coulomb potential ones as the cut-off radius converges to infinity, e.g., the influence of plasma diminishes. Further steps in application of the presented results, the inclusion of the bound-bound transition processes within the continuous absorption model, as well as its application in the analysis of the spectral absorption processes is to be carried out. In order to make such calculations, it is needed to include the models for the bound state level broadening. Our plan is to present the results obtained during this investigation in a database that can be accessed directly through http://servo.aob.rs as a web service, similar to the existing databases [26,27,32].

**Acknowledgments:** The authors are thankful to the Ministry of Education, Science and Technological Development of the Republic of Serbia for the support of this work within the projects 176002 and III44002. We acknowledge the contribution of the late Dr A. A. Mihajlov, who participated in the initial discussions and preparation of this paper

**Author Contributions:** N. M. S. have adopted the cut-off Coulomb potential model for the optical plasma characteistics; N. M. S. developed the code and performed the calculations; M. S. D., Z. J. S. and V. A. S. analysed the data; N. M. S., V. A. S. and M. S. D. wrote the paper; M. S. D., V. A. S. and Z. J. S. have edited the paper.

## References

1.  Fortov, V.E.; Iakubov, I.T. *The Physics of Non-Ideal Plasma*; World Scientific: Singapore, 1999.
2.  Rogers, F.J.; Iglesias, C.A. Opacity of stellar matter. *Space Sci. Rev.* **1998**, *85*, 61–70. [CrossRef]
3.  Mihajlov, A.A.; Sakan, N.M.; Srećković, V.A.; Vitel, Y. Modeling of continuous absorption of electromagnetic radiation in dense partially ionized plasmas. *J. Phys. A* **2011**, *44*, 095502. [CrossRef]
4.  Mihajlov, A.A.; Ignjatović, L.M.; Srećković, V.A.; Dimitrijević, M.S.; Metropoulos, A. The non-symmetric ion-atom radiative processes in the stellar atmospheres. *Mon. Not. R. Astron. Soc.* **2013**, *431*, 589–599. [CrossRef]
5.  Mihajlov, A.A.; Djordjević, D.; Popović, M.M.; Meyer, T.; Luft, M.; Kraeft, W.D. Determination of the Electrical Conductivity of a Plasma on the Basis of the Coulomb cut-off Potential Model. *Contrib. Plasma Phys.* **1989**, *29*, 441–446. [CrossRef]
6.  Ignjatović, L.M.; Srećković, V.A.; Dimitrijević, M.S. The Screening Characteristics of the Dense Astrophysical Plasmas: The Three-Component Systems. *Atoms* **2017**, *5*, 42. [CrossRef]

7.   Mihajlov, A.A.; Sakan, N.M.; Srećković, V.A.; Vitel, Y. Modeling of the Continuous Absorption of Electromagnetic Radiation in Dense Hydrogen Plasma. *Balt. Astron.* **2011**, *20*, 604–608. [CrossRef]

8.   Mihajlov, A.A.; Srećković, V.A.; Sakan, N.M. Inverse Bremsstrahlung in Astrophysical Plasmas: The Absorption Coefficients and Gaunt Factors. *J. Astrophys. Astron.* **2015**, *36*, 635–642. [CrossRef]

9.   Sakan, N.M.; Srećković, V.A.; Mihajlov, A.A. The application of the cut-off Coulomb potential for the calculation of a continuous spectra of dense hydrogen plasma. *Mem. Soc. Astron. Ital. Suppl.* **2005**, *7*, 221–224.

10.  Mihajlov, A.A.; Vitel, Y.; Ignjatović, L.M. The new screening characteristics of strongly non-ideal and dusty plasmas. Part 1: Single-component systems. *High Temp.* **2008**, *46*, 737–745. [CrossRef]

11.  Mihajlov, A.A.; Vitel, Y.; Ignjatović, L.M. The new screening characteristics of strongly non-ideal and dusty plasmas. Part 2: Two-component systems. *High Temp.* **2009**, *47*, 1–12. [CrossRef]

12.  Mihajlov, A.; Vitel, Y.; Ignjatović, L.M. The new screening characteristics of strongly non-ideal and dusty plasmas. Part 3: Properties and applications. *High Temp.* **2009**, *47*, 147–157. [CrossRef]

13.  Demura, A. Physical models of plasma microfield. *Int. J. Spectrosc.* **2009**, *2010*, 671073. [CrossRef]

14.  Vitel, Y.; Gavrilova, T.; D'yachkov, L.; Kurilenkov, Y.K. Spectra of dense pure hydrogen plasma in Balmer area. *J. Quant. Spectrosc. Radiat. Transf.* **2004**, *83*, 387–405. [CrossRef]

15.  Griem, H.R. *Principles of Plasma Spectroscopy*; Cambridge University Press: Cambridge, UK, 2005; Volume 2.

16.  Konjević, N.; Ivković, M.; Sakan, N. Hydrogen Balmer lines for low electron number density plasma diagnostics. *Spectrochim. Acta B* **2012**, *76*, 16–26. [CrossRef]

17.  Sobelman, I.I. Atomic spectra and radiative transitions. In *Springer Series in Chemical Physics*; Springer: Berlin, Germany, 1979.

18.  Hoang-Binh, D. A program to compute exact hydrogenic radial integrals, oscillator strengths, and Einstein coefficients, for principal quantum numbers up to n $\approx$ 1000. *Comput. Phys. Commun.* **2005**, *166*, 191–196. [CrossRef]

19.  Hilborn, R.C. Einstein coefficients, cross sections, f values, dipole moments, and all that. *Am. J. Phys.* **1982**, *50*, 982–986. [CrossRef]

20.  Stambulchik, E.; Maron, Y. Plasma line broadening and computer simulations: A mini-review. *High Energ. Dens. Phys.* **2010**, *6*, 9–14. [CrossRef]

21.  Stambulchik, E.; Fisher, D.; Maron, Y.; Griem, H.; Alexiou, S. Correlation effects and their influence on line broadening in plasmas: Application to $H_\alpha$. *High Energ. Dens. Phys.* **2007**, *3*, 272–277. [CrossRef]

22.  Talin, B.; Dufour, E.; Calisti, A.; Gigosos, M.A.; Gonzalez, M.A.; Gaztelurrutia, T.D.R.; Dufty, J.W. Molecular dynamics simulation for modeling plasma spectroscopy. *J. Phys. A* **2003**, *36*, 6049. [CrossRef]

23.  Ferri, S.; Calisti, A.; Mossé, C.; Rosato, J.; Talin, B.; Alexiou, S.; Gigosos, M.A.; González, M.A.; González-Herrero, D.; Lara, N.; et al. Ion Dynamics Effect on Stark-Broadened Line Shapes: A Cross-Comparison of Various Models. *Atoms* **2014**, *2*, 299–318. [CrossRef]

24.  Calisti, A.; Demura, A.V.; Gigosos, M.A.; González-Herrero, D.; Iglesias, C.A.; Lisitsa, V.S.; Stambulchik, E. Influence of microfield directionality on line shapes. *Atoms* **2014**, *2*, 259–276. [CrossRef]

25.  Alexiou, S.; Dimitrijević, M.S.; Sahal-Brechot, S.; Stambulchik, E.; Duan, B.; González-Herrero, D.; Gigosos, M.A. The second workshop on lineshape code comparison: Isolated lines. *Atoms* **2014**, *2*, 157–177. [CrossRef]

26.  Marinković, B.P.; Jevremović, D.; Srećković, V.A.; Vujčić, V.; Ignjatović, L.M.; Dimitrijević, M.S.; Mason, N.J. BEAMDB and MolD—Databases for atomic and molecular collisional and radiative processes: Belgrade nodes of VAMDC. *Eur. Phys. J. D* **2017**, *71*, 158. [CrossRef]

27.  Srećković, V.A.; Ignjatović, L.M.; Jevremović, D.; Vujčić, V.; Dimitrijević, M.S. Radiative and Collisional Molecular Data and Virtual Laboratory Astrophysics. *Atoms* **2017**, *5*, 31. [CrossRef]

28.  Sadykova, S.; Ebeling, W.; Valuev, I.; Sokolov, I. Electric Microfield Distributions in Li + Plasma With Account of the Ion Structure. *Contrib. Plasma Phys.* **2009**, *49*, 76–89. [CrossRef]

29.  Calisti, A.; Ferri, S.; Mossé, C.; Talin, B.; Gigosos, M.; González, M. Microfields in hot dense hydrogen plasmas. *High Energ. Dens. Phys.* **2011**, *7*, 197–202. [CrossRef]

30.  Kramida, A.; Ralchenko, Y.; Reader, J.; NIST ASD Team. *NIST Atomic Spectra Database (Ver. 5.5)*; National Institute of Standards and Technology: Gaithersburg, MD, USA, 2017. Available online: https://physics.nist.gov/asd (accessed on 30 October 2017).

31.  Sakan, N.M. The Calculation of the Photo Absorption Processes in Dense Hydrogen Plasma with the Help of Cut-Off Coulomb Potential Model. *J. Phys. Conf. Ser.* **2010**, *257*, 012036. [CrossRef]

# 11

# Semiclassical Stark Broadening Parameters of Ar VII Spectral Lines

**Milan S. Dimitrijević** [1,2,†], **Aleksandar Valjarević** [3,†] **and Sylvie Sahal-Bréchot** [2,*,†]

[1]   Astronomical Observatory, Volgina 7, 11060 Belgrade 38, Serbia; mdimitrijevic@aob.rs
[2]   LERMA, Observatoire de Paris, PSL Research University, CNRS, Sorbonne Universités,
      UPMC (Univ. Pierre & Marie Curie) Paris 06, 5 Place Jules Janssen, 92190 Meudon, France
[3]   Department of Geography, Faculty of Natural Sciences and Mathematics, University of Kosovska Mitrovica,
      Ive Lole Ribara 29, 38220 Kosovska Mitrovica, Serbia; aleksandar.valjarevic@pr.ac.rs
[*]   Correspondence: sylvie.sahal-brechot@obspm.fr
[†]   These authors contributed equally to this work.

Academic Editor: Robert C. Forrey

**Abstract:** Using the semi-classical perturbation approach in the impact approximation, full width at half maximum and shift have been determined for eight spectral lines of Ar VII, for broadening by electron-, proton-, and He III-impacts. The results are provided for temperatures from 20,000 K to 500,000 K, and for an electron density of $10^{18}$ cm$^{-3}$. The obtained results will be included in the STARK-B database, which is also in the virtual atomic and molecular data center (VAMDC).

**Keywords:** stark broadening; atomic data; atomic processes; line profiles; Ar VII

## 1. Introduction

With the development of space astronomy and satellite-born spectroscopy, trace elements—which have been without importance for astrophysics—now become increasingly important, and the corresponding data of interest for the analysis of stellar spectra. For example, spectral lines of Ar VII have been observed in the spectrum of extremely hot and massive galactic O3 If supergiant HD 93129A [1]. Additionally, Werner et al. [2] have found Ar VII lines in some of the hottest known central stars of planetary nebulae, with the effective temperatures of 95,000–110,000 K, and in (pre-) white dwarfs by analyzing high-resolution spectra from the Far Ultraviolet Spectroscopic Explorer (FUSE). We note that Stark broadening is the principal pressure broadening mechanism in such hot stars, and without the corresponding Stark broadening data, reliable analysis and modelling of high-resolution spectra are not possible.

Concerning Stark broadening parameters for Ar VII spectral lines, there is only one article where Stark broadening parameters are provided for three transitions [3]. Here, full widths at half intensity maximum (FWHM) $W$ and shifts $d$ for eight additional transitions have been calculated by using semiclassical perturbation method (SCP, [4,5]) for collisions of Ar VII ions with electrons, protons, and He III ions, since hydrogen and helium are the main constituents of stellar atmospheres.

## 2. The Impact Semiclassical Perturbation Method

The semiclassical perturbation formalism (SCP) applied here for the calculations of Stark broadening parameters, full width at half intensity maximum (FWHM-$W$), and shift of spectral line ($d$) has been formulated in [4,5], and later updates, optimisations, and innovations are presented in Sahal-Bréchot [6], Sahal-Bréchot [7], Dimitrijević et al. [8], Dimitrijević and Sahal-Bréchot [9], and Sahal-Bréchot et al. [10]. Within the frame of this method, FWHM ($W$) and shift ($d$) may be expressed by the following relation:

$$W = N \int v f(v) dv \left( \sum_{i' \neq i} \sigma_{ii'}(v) + \sum_{f' \neq f} \sigma_{ff'}(v) + \sigma_{el} \right)$$

$$d = N \int v f(v) dv \int_{R_3}^{R_D} 2\pi\rho d\rho \sin(2\varphi_p). \tag{1}$$

Here, $N$ is electron density, $f(v)$ the Maxwellian velocity distribution function for electrons, $\rho$ is the impact parameter of the incoming electron, and with $i', f'$ are denoted the perturbing levels of the initial ($i$) and final ($f$) state. The inelastic cross-section $\sigma_{jj'}(v), j = i, f$ is expressed as:

$$\sum_{i' \neq i} \sigma_{ii'}(v) = \frac{1}{2}\pi R_1^2 + \int_{R_1}^{R_D} 2\pi\rho d\rho \sum_{i' \neq i} P_{ii'}(\rho, v), \tag{2}$$

where $P_{jj'}(\rho, v), j = i, f; j' = i', f'$ is the transition probability. The elastic cross-section is

$$\sigma_{el} = 2\pi R_2^2 + \int_{R_2}^{R_D} 2\pi\rho d\rho \sin^2 \delta + \sigma_r,$$

$$\delta = (\varphi_p^2 + \varphi_q^2)^{\frac{1}{2}}. \tag{3}$$

The phase shift due to the polarization potential is $\varphi_p$ ($r^{-4}$), and due to the quadrupolar potential $\varphi_q$ ($r^{-3}$) (see Section 3 of Chapter 2 in Sahal-Bréchot [4]). $R_1$, $R_2$, $R_3$, and $R_D$ are cut-offs, defined and described in Section 1 of Chapter 3 in Sahal-Bréchot [5]. $\sigma_r$ denotes the contribution of Feshbach resonances, explained in detail in [11].

A review of the theory, all approximations and details of applications is given in Sahal-Bréchot et al. [10].

### 3. Stark Broadening Parameter Calculations

By using Equations (1)–(3) we have calculated widths (FWHM) and shifts for eight transitions (triplets) in Ar VII spectrum. The necessary atomic energy levels have been taken from Saloman [12]. The oscillator strengths—needed for calculations—have been obtained by using the method of Bates and Damgaard [13] and the tables of Oertel and Shomo [14]. When there was no corresponding data in Oertel and Shomo [14] (for some higher levels), the needed oscillator strengths have been calculated using the method of Van Regemorter et al. [15].

In Table 1 are shown widths (FWHM) and shifts of Ar VII spectral lines broadened by electron-, proton-, and He III-impacts, for a perturber density of $10^{18}$ cm$^{-3}$ and for a set of temperatures from 20,000 K to 500,000 K. The temperature range is of interest for astrophysics, laboratory plasmas, fusion research, various plasmas in technology and laser-produced plasmas. If we want to use these data for higher perturber densities, the influence of Debye screening should be checked and taken into account if needed (e.g., [16]).

The accuracy of the semiclassical perturbation method is estimated by comparison with numerous experimental data for different elements and spectral lines, and is estimated to be 20–30% (see discussion in [10]). Since Ar VII is a member of the magnesium isoelectronic sequence with relatively simple spectrum, we suppose that the error of results shown in Table 1 is not much higher than 20%.

**Table 1.** Electron-, proton-, and doubly charged helium-impact broadening parameters for Ar VII spectral lines, for a perturber density of $10^{18}$ cm$^{-3}$ and temperatures from 20,000 to 500,000 K. The calculated wavelength of the transitions (in Å) and parameter $C$ are also given. When divided by the corresponding Stark width, this parameter gives an estimate for the maximal perturber density for which the line may be treated as isolated. $W_e$: electron-impact full width at half maximum of intensity; $d_e$: electron-impact shift; $W_p$: proton-impact full width at half maximum of intensity; $d_p$: proton-impact shift; $W_{He^{++}}$: doubly charged helium ion-impact full width at half maximum of intensity; $d_{He^{++}}$: doubly charged helium ion-impact shift.

| Transition | T(K) | $W_e$ (Å) | $d_e$ (Å) | $W_{H^+}$ (Å) | $d_{H^+}$ (Å) | $W_{He^{++}}$ (Å) | $d_{He^{++}}$ (Å) |
|---|---|---|---|---|---|---|---|
| $4s\ ^3S - 5p\ ^3P^o$ | 20,000 | 0.510E−01 | 0.174E−03 | 0.298E−03 | 0.197E−04 | 0.492E−03 | 0.272E−04 |
| 443.2 Å | 50,000 | 0.328E−01 | 0.217E−03 | 0.898E−03 | 0.636E−04 | 0.173E−02 | 0.117E−03 |
| $C = 0.47E+20$ | 100,000 | 0.243E−01 | 0.158E−03 | 0.148E−02 | 0.129E−03 | 0.291E−02 | 0.249E−03 |
| | 200,000 | 0.186E−01 | 0.215E−03 | 0.203E−02 | 0.223E−03 | 0.403E−02 | 0.444E−03 |
| | 300,000 | 0.162E−01 | 0.211E−03 | 0.220E−02 | 0.275E−03 | 0.439E−02 | 0.550E−03 |
| | 500,000 | 0.139E−01 | 0.227E−03 | 0.242E−02 | 0.352E−03 | 0.482E−02 | 0.709E−03 |
| $3p\ ^3P^o - 4s\ ^3S$ | 20,000 | 0.490E−02 | −0.779E−03 | 0.744E−06 | 0.135E−04 | 0.129E−05 | 0.186E−04 |
| 250.4 Å | 50,000 | 0.270E−02 | 0.615E−04 | 0.642E−05 | 0.432E−04 | 0.124E−04 | 0.792E−04 |
| $C = 0.32E+20$ | 100,000 | 0.191E−02 | 0.113E−03 | 0.295E−04 | 0.835E−04 | 0.579E−04 | 0.161E−03 |
| | 200,000 | 0.142E−02 | 0.131E−03 | 0.819E−04 | 0.133E−03 | 0.163E−03 | 0.265E−03 |
| | 300,000 | 0.121E−02 | 0.143E−03 | 0.113E−03 | 0.162E−03 | 0.227E−03 | 0.323E−03 |
| | 500,000 | 0.101E−02 | 0.136E−03 | 0.173E−03 | 0.196E−03 | 0.347E−03 | 0.396E−03 |
| $3p\ ^3P^o - 3d\ ^3D$ | 20,000 | 0.911E−02 | −0.564E−03 | 0.361E−05 | −0.107E−05 | 0.598E−05 | −0.148E−05 |
| 477.5 Å | 50,000 | 0.585E−02 | −0.436E−04 | 0.148E−04 | −0.348E−05 | 0.282E−04 | −0.639E−05 |
| $C = 0.48E+21$ | 100,000 | 0.417E−02 | −0.288E−04 | 0.396E−04 | −0.740E−05 | 0.767E−04 | −0.143E−04 |
| | 200,000 | 0.297E−02 | −0.165E−04 | 0.900E−04 | −0.150E−04 | 0.176E−03 | −0.298E−04 |
| | 300,000 | 0.245E−02 | −0.134E−04 | 0.129E−03 | −0.220E−04 | 0.253E−03 | −0.438E−04 |
| | 500,000 | 0.195E−02 | −0.237E−04 | 0.177E−03 | −0.333E−04 | 0.349E−03 | −0.669E−04 |
| $3p\ ^3P^o - 4d\ ^3D$ | 20,000 | 0.421E−02 | −0.445E−03 | 0.671E−05 | 0.621E−05 | 0.111E−04 | 0.858E−05 |
| 192.3 Å | 50,000 | 0.249E−02 | 0.896E−05 | 0.263E−04 | 0.200E−04 | 0.506E−04 | 0.366E−04 |
| $C = 0.94E+19$ | 100,000 | 0.181E−02 | 0.259E−04 | 0.589E−04 | 0.394E−04 | 0.115E−03 | 0.760E−04 |
| | 200,000 | 0.135E−02 | 0.428E−04 | 0.100E−03 | 0.646E−04 | 0.199E−03 | 0.128E−03 |
| | 300,000 | 0.115E−02 | 0.345E−04 | 0.127E−03 | 0.782E−04 | 0.252E−03 | 0.156E−03 |
| | 500,000 | 0.951E−03 | 0.379E−04 | 0.155E−03 | 0.971E−04 | 0.308E−03 | 0.196E−03 |
| $4p\ ^3P^o - 4d\ ^3D$ | 20,000 | 0.309 | −0.849E−02 | 0.811E−03 | 0.262E−03 | 0.134E−02 | 0.362E−03 |
| 1425.9 Å | 50,000 | 0.199 | 0.324E−03 | 0.292E−02 | 0.845E−03 | 0.562E−02 | 0.155E−02 |
| $C = 0.52E+21$ | 100,000 | 0.145 | 0.336E−03 | 0.577E−02 | 0.169E−02 | 0.113E−01 | 0.327E−02 |
| | 200,000 | 0.108 | 0.126E−02 | 0.877E−02 | 0.286E−02 | 0.173E−01 | 0.569E−02 |
| | 300,000 | 0.927E−01 | 0.595E−03 | 0.106E−01 | 0.346E−02 | 0.211E−01 | 0.691E−02 |
| | 500,000 | 0.777E−01 | 0.645E−03 | 0.120E−01 | 0.440E−02 | 0.237E−01 | 0.886E−02 |
| $3d\ ^3D - 4p\ ^3P^o$ | 20,000 | 0.184E−01 | −0.111E−02 | 0.352E−04 | 0.757E−05 | 0.584E−04 | 0.105E−04 |
| 416.0 Å | 50,000 | 0.116E−01 | 0.474E−04 | 0.133E−03 | 0.245E−04 | 0.256E−03 | 0.450E−04 |
| $C = 0.87E+20$ | 100,000 | 0.835E−02 | 0.115E−03 | 0.286E−03 | 0.513E−04 | 0.560E−03 | 0.991E−04 |
| | 200,000 | 0.612E−02 | 0.105E−03 | 0.469E−03 | 0.946E−04 | 0.926E−03 | 0.188E−03 |
| | 300,000 | 0.518E−02 | 0.121E−03 | 0.577E−03 | 0.126E−03 | 0.114E−02 | 0.251E−03 |
| | 500,000 | 0.427E−02 | 0.140E−03 | 0.679E−03 | 0.161E−03 | 0.135E−02 | 0.323E−03 |
| $3d\ ^3D - 5p\ ^3P^o$ | 20,000 | 0.139E−01 | −0.555E−03 | 0.900E−04 | 0.185E−04 | 0.148E−03 | 0.256E−04 |
| 240.6 Å | 50,000 | 0.868E−02 | 0.137E−03 | 0.271E−03 | 0.587E−04 | 0.523E−03 | 0.108E−03 |
| $C = 0.14E+20$ | 100,000 | 0.642E−02 | 0.162E−03 | 0.445E−03 | 0.109E−03 | 0.877E−03 | 0.211E−03 |
| | 200,000 | 0.490E−02 | 0.190E−03 | 0.616E−03 | 0.166E−03 | 0.122E−02 | 0.331E−03 |
| | 300,000 | 0.425E−02 | 0.199E−03 | 0.669E−03 | 0.204E−03 | 0.133E−02 | 0.406E−03 |
| | 500,000 | 0.362E−02 | 0.199E−03 | 0.739E−03 | 0.237E−03 | 0.147E−02 | 0.479E−03 |
| $4d\ ^3D - 5p\ ^3P^o$ | 20000 | 0.243 | −0.224E−03 | 0.163E−02 | 0.134E−03 | 0.269E−02 | 0.185E−03 |
| 952.0 Å | 50,000 | 0.158 | 0.166E−02 | 0.477E−02 | 0.431E−03 | 0.923E−02 | 0.790E−03 |
| $C = 0.22E+21$ | 100,000 | 0.118 | 0.174E−02 | 0.767E−02 | 0.856E−03 | 0.151E−01 | 0.165E−02 |
| | 200,000 | 0.902E−01 | 0.184E−02 | 0.104E−01 | 0.142E−02 | 0.207E−01 | 0.283E−02 |
| | 300,000 | 0.786E−01 | 0.221E−02 | 0.113E−01 | 0.172E−02 | 0.224E−01 | 0.343E−02 |
| | 500,000 | 0.672E−01 | 0.208E−02 | 0.124E−01 | 0.217E−02 | 0.246E−01 | 0.437E−02 |

Since the wavelengths in Table 1 are calculated ones, they are different from experimental wavelengths. However, we notice that they are correct in angular frequency units, because in such a case, relative and not absolute positions of energy levels are important for calculations. For the transformation of the Stark widths from Å-units to angular frequency units, the following formula can be used:

$$W(\mathring{A}) = \frac{\lambda^2}{2\pi c} W(s^{-1}) \tag{4}$$

where $c$ is the speed of light. For the correction of widths and/or shifts for the difference between calculated and experimental wavelength, one can use the expression:

$$W_{cor} = \left(\frac{\lambda_{\exp}}{\lambda}\right)^2 W. \tag{5}$$

The corresponding expressions for the shifts are analogous to Equations (4) and (5). Here, $W_{cor}$ denotes the corrected width, while $\lambda_{exp}$ is the experimental wavelength, $\lambda$ the calculated wavelength, and $W$ the width from Table 1.

Dividing the parameter $C$ [17] from Table 1 by the corresponding full width at half maximum, one obtains the maximal perturber density for which the line may be treated as isolated.

The obtained Ar VII Stark broadening parameters shown in Table 1 will be implemented in the STARK-B database [18,19]. This database contains Stark widths and shifts needed first of all for the investigations, modelling, and diagnostics of the plasma of stellar atmospheres, but also for diagnostics of laboratory plasmas and investigation of laser produced, inertial fusion plasma, and for plasma technologies.

The STARK-B database is one of the databases included in the virtual atomic and molecular data center—VAMDC [20,21], created in order to enable more effective search and mining of atomic and molecular data which are in different databases. Databases with atomic and molecular data which are in VAMDC—including STARK-B—can be accessed and searched through the VAMDC portal: http://portal.vamdc.org/.

## 4. Conclusions

The semiclassical perturbation calculation of Stark broadening parameters, widths, and shifts for spectral lines broadened by collisions of Ar VII ions with electrons, protons, and doubly charged helium ions have been performed for eight multiplets of Ar VII. The obtained values of Stark broadening parameters will be implemented in the STARK-B database—one of the databases included in the virtual atomic and molecular data center (VAMDC). Since Stark broadening data for Ar VII spectral lines considered here do not exist in the literature, we hope that they will be of interest for the relevant problems in astrophysical, laboratory, laser-produced, inertial fusion, and technological plasmas.

**Acknowledgments:** The support of the Ministry of Education, Science and Technological Development of the Republic of Serbia through project 176002 is gratefully acknowledged. This work has also been supported by the Paris Observatory, the PADC (Paris Data Center_OV), and the CNRS which are greatly acknowledged. We also acknowledge financial support from "Programme National de Physique Stellaire" (PNPS) of CNRS/INSU, CEA and CNES, France. This work has also been supported by the VAMDC (Virtual Atomic and Molecular Data Centre). VAMDC is funded under the Combination of Collaborative Projects and Coordination and Support Actions Funding Scheme of The Seventh Framework Program. Call topic: INFRA-2008-1.2.2 Scientific Data Infrastructure, Grant Agreement number: 239108. A part of this work has also been supported by the LABEX Plas@par project and received financial state aid managed by the Agence Nationale de la Recherche, as part of the programme "Investissements d'Avenir" under the reference, ANR11IDEX000402.

**Author Contributions:** These authors contributed equally to this work.

# References

1.  Taresch, G.; Kudritzki, R.P.; Hurwitz, M.; Bowyer, S.; Pauldrach, A.W.A.; Puls, J.; Butler, K.; Lennon, D.J.; Haser, S.M. Quantitative analysis of the FUV, UV and optical spectrum of the O3 star HD 93129A. *Astron. Astrophys.* **1997**, *321*, 531–548.
2.  Werner, K.; Rauch, T.; Kruk, J.W. Discovery of photospheric argon in very hot central stars of planetary nebulae and white dwarfs. *Astron. Astrophys.* **2007**, *466*, 317–322.
3.  Dimitrijević, M.S.; Sahal-Bréchot, S. On the Stark broadening of Ar VII spectral lines. *Univ. Thought* **2017**, *7*, 46–50.
4.  Sahal-Bréchot, S. Impact theory of the broadening and shift of spectral lines due to electrons and ions in a plasma. *Astron. Astrophys.* **1969**, *1*, 91–123.
5.  Sahal-Bréchot, S. Impact theory of the broadening and shift of spectral lines due to electrons and ions in a plasma (continued). *Astron. Astrophys.* **1969**, *2*, 322–354.
6.  Sahal-Bréchot, S. Stark broadening of isolated lines in the impact approximation. *Astron. Astrophys.* **1974**, *35*, 319–321.
7.  Sahal-Bréchot, S. Broadening of ionic isolated lines by interactions with positively charged perturbers in the quasistatic limit. *Astron. Astrophys.* **1991**, *245*, 322–330.
8.  Dimitrijević, M.S.; Sahal-Bréchot, S.; Bommier, V. Stark broadening of spectral lines of multicharged ions of astrophysical interest. I-C IV lines. *Astron. Astrophys.* **1991**, *89*, 581–590.
9.  Dimitrijević, M.S.; Sahal-Bréchot, S. Stark broadening of Li II spectral lines. *Phys. Scr.* **1996**, *54*, 50–55.
10. Sahal-Bréchot, S.; Dimitrijević, M.S.; Ben Nessib, N. Widths and Shifts of Isolated Lines of Neutral and Ionized Atoms Perturbed by Collisions with Electrons and Ions: An Outline of the Semiclassical Perturbation (SCP) Method and of the Approximations Used for the Calculations. *Atoms* **2014**, *2*, 225–252.
11. Fleurier, C.; Sahal-Bréchot, S.; Chapelle, J. Stark profiles of some ion lines of alkaline earth elements. *J. Quant. Spectrosc. Radiat. Transf.* **1977**, *17*, 595–603.
12. Saloman, E.B. Energy Levels and Observed Spectral Lines of Ionized Argon, Ar II through Ar XVIII. *J. Phys. Chem. Ref. Data* **2010**, *39*, 033101.
13. Bates, D.R.; Damgaard, A. The Calculation of the Absolute Strengths of Spectral Lines. *Philos. Trans. R. Soc. Lond. Ser. A Math. Phys. Sci.* **1949**, *242*, 101–122.
14. Oertel, G.K.; Shomo, L.P. Tables for the Calculation of Radial Multipole Matrix Elements by the Coulomb Approximation. *Astrophys. J. Suppl. Ser.* **1968**, *16*, 175–218.
15. Van Regemorter, H.; Hoang Binh, D.; Prud'homme, M. Radial transition integrals involving low or high effective quantum numbers in the Coulomb approximation. *J. Phys. B* **1979**, *12*, 1053–1061.
16. Griem, H.R. *Spectral Line Broadening by Plasmas*; Academic Press, Inc.: New York, NY, USA, 1974.
17. Dimitrijević, M.S.; Sahal-Bréchot, S. Stark broadening of neutral helium lines. *J. Quant. Spectrosc. Radiat. Transf.* **1984**, *31*, 301–313.
18. Sahal-Bréchot, S.; Dimitrijević, M.S.; Moreau, N. STARK-B Database, Observatory of Paris, LERMA and Astronomical Observatory of Belgrade, 2017. Available online: http://stark-b.obspm.fr (accessed on 24 August 2017).
19. Sahal-Bréchot, S.; Dimitrijević, M.S.; Moreau, N.; Ben Nessib, N. The STARK-B database VAMDC node: A repository for spectral line broadening and shifts due to collisions with charged particles. *Phys. Scr.* **2015**, *50*, 054008.
20. Dubernet, M.L.; Boudon, V.; Culhane, J.L.; Dimitrijevic, M.S.; Fazliev, A.Z.; Joblin, C.; Kupka, F.; Leto, G.; Le Sidaner, P.; Loboda, P.A.; et al. Virtual atomic and molecular data centre. *J. Quant. Spectrosc. Radiat. Transf.* **2010**, *111*, 2151–2159.
21. Dubernet, M.L.; Antony, B.K.; Ba, Y.A.; Babikov, Y.L.; Bartschat, K.; Boudon, V.; Braams, B.J.; Chung, H.K.; Daniel, F.; Delahaye, F.; et al. The virtual atomic and molecular data centre (VAMDC) consortium. *J. Phys. B* **2016**, *49*, 074003.

# 12

# Contribution of Lienard-Wiechert Potential to the Electron Broadening of Spectral Lines in Plasmas

**Mohammed Tayeb Meftah** [1,2,*], **Khadra Arif** [1], **Keltoum Chenini** [1,4], **Kamel Ahmed Touati** [1,3] **and Said Douis** [1,2]

[1] Laboratoire de Recherche de Physique des Plasmas et Surfaces, Ouargla 30000, Algérie; khadra.arif@gmail.com (K.A.); k1_chenini@yahoo.fr (K.C.); ktouati@yahoo.com (K.A.T.); fdouis@gmail.com (S.D.)

[2] Département de Physique, Faculté de Mathématiques et Sciences de la matière, Université Kasdi-Merbah, Ouargla 30000, Algérie

[3] Lycée professionnel les Alpilles, Rue des Lauriers, 13140 Miramas, France

[4] Département des Sciences et Technologies, Faculté des Sciences et Technologies, Université de Ghardaia, Ghardaia 47000, Algérie

* Correspondence: mewalid@yahoo.com

**Abstract:** Lienard-Wiechert or retarded electric and magnetic fields are produced by moving electric charges with respect to a rest frame. In hot plasmas, such fields may be created by high velocity free electrons. The resulting electric field has a relativistic expression that depends on the ratio of the free electron velocity to the speed of light in vacuum c. In this work, we consider the semi-classical dipole interaction between the emitter ions and the Lienard-Wiechert electric field of the free electrons and compute its contribution to the broadening of the spectral line shape in hot and dense plasmas.

**Keywords:** Stark effect; electron broadening; Lienard-Wiechert; retarded interaction; relativistic collision operator

## 1. Introduction

Line profiles and shifts are used to determine plasma parameters, especially in astrophysics where alternative methods (such as interferometry or Thomson scattering) are not possible. Doppler and pressure broadening (Stark broadening) are typically the two dominant mechanisms and we focus on the latter. In a number of hot astrophysical plasmas, electrons may be energetic enough that their thermal energy $K_B T$ ($K_B$ is the Boltzmann constant and $T$ the temperature) can be comparable to the rest mass. For the extreme densities encountered in some astrophysical objects, pressure broadening could dominate; however, for such objects the electrons may become relativistic due to the extreme temperatures and hence it makes sense to check the modifications to the pressure broadening by relativistic effects. Similarly, the laser-produced plasmas may be achieved in both very high densities and very high temperatures: the first leads to the dominance of Stark over Doppler broadening whereas the second leads to relativistic electron velocities. More specifically, plasma spectroscopy is used in a wide range of electron density from 10 particles per cm$^3$ (interstellar space) to $10^{25}$ particles per cm$^3$ (star interiors, inertial confinement fusion) and for temperatures between $10^7$ K and $10^{10}$ K. More precisely, as the Doppler effect is constant for a fixed temperature (about $10^8$ K in our case), we can neglect it for densities higher than $10^{20}$ cm$^{-3}$. In the present work, we investigate the region corresponding to the particular conditions of plasma: high density and high temperature. Under these conditions, (electron-ion) collisions will be, throughout this work, assumed binary and the dynamics of the electrons will be treated relativistically. Furthermore, in this work we shall use the statistical classical mechanics (not the quantum statistical mechanics as in the Fermi-Dirac distribution) because the condition $\lambda_{th} = h/\sqrt{2\pi m_e K_B T} < N_e^{-1/3}$ ($m_e$ is the electron mass, $\lambda_{th}$ is the De Broglie thermal length and $N_e$ is the electrons density) is fulfilled in the stated density and temperatures ranges. For example, if $N_e = 10^{24}$ cm$^{-3}$ and $T > 10^7$ K, it is easy to verify that this inequality is correct. This condition means that the wave function extent ($\lambda_{th}$) associated with the electron is smaller than the mean distance ($\sim N_e^{-1/3}$) between two free electrons. In the present work, we focus on electron broadening in the impact approximation [1,2]. We thus, reformulate the standard semiclassical collision operator by taking into account the relativistic effects of the Lienard-Wiechert retarded electric field [3] due to the free electron movement. In addition, we assume that the plasma is optically thin (the opacity phenomenon is neglected), for this raison the spectral line shape is not influenced by the absorption process. Furthermore, by neglecting the electrons recombinaison, we assume that the decoupling of the free electrons from the radiation field is satisfied. The units system used here (unless specified) is the CGS system. In many cases in line broadening, fast particles (typically electrons) are described by a collisional approach, while particles of which field have a weak variation during the inverse half width half maximum (HWHM) time scale are considered static and treated via a quasistatic microfield. For many applications, isolated lines have great importance. So, the calculations of the broadening of such a line in a plasma are normally made by using the impact approximation for electrons [1] in the semi-classical version [4], as the ionic contribution is typically negligible.

## 2. Theoretical Basis of the Electron Broadening

The Stark effect is important in plasmas of high degree of ionization and high temperatures. In all cases presented in the remainder of our work, the Stark effect is dominant. The two most popular approximations in the computations of the electronic collision operator $\Phi$ are the dipole approximation and the approximation of the classical path which considers the perturbing electrons in the impact approximation. Our departure point is the expression of the intensity of the spectral line shape [2,5]:

$$I(\omega) = -\frac{1}{\pi}\operatorname{Re}\left\{\boldsymbol{d}_{\alpha\beta}\langle\langle\alpha\beta\left|\left(i\omega - \frac{i(H_g - H_e)}{\hbar} + \Phi\right)^{-1}\right|\alpha'\beta'\rangle\rangle\boldsymbol{d}^*_{\alpha'\beta'}\right\} \tag{1}$$

where $\Phi$ is the relativistic collision operator (in Hz) which is independent from time and electric micro-field and has matrix elements given by

$$\langle\langle\alpha\beta\,|\Phi|\,\alpha'\beta'\rangle\rangle = \sum_{\alpha''}\boldsymbol{r}_{\alpha\alpha''}\boldsymbol{r}_{\alpha''\alpha'}\Phi_d\left(\omega_{\alpha\alpha''},\omega_{\alpha''\alpha'}\right)$$

$$+\sum_{\beta''}\boldsymbol{r}_{\beta\beta''}\boldsymbol{r}_{\beta''\beta'}\Phi_d\left(\omega_{\beta\beta''},\omega_{\beta''\beta'}\right) - \boldsymbol{r}_{\alpha\alpha'}\,\boldsymbol{r}_{\beta'\beta}\Phi_{int}\left(\omega_{\alpha\alpha'},\omega_{\beta'\beta}\right) \tag{2}$$

where $\alpha$, $\beta$ are the upper and lower levels respectively and $\boldsymbol{r}_{ab}$ ($\boldsymbol{d}_{ab}$ is dipole operator) is the matrix element of position operator of the bounded electron. We now aim to calculate the direct relativistic term $\Phi_d$ (in Hz/cm$^2$) and the relativistic term of interference $\Phi_{int} = 2\Phi_d$ [5]. Before starting this task, let us remind that the criterion of validity of the impact theory, according to Voslamber [6], is not applicable for any pair ($\omega_1, \omega_2$). However, for the isolated lines ($\omega_1 = -\omega_2$), this theoretical problem does not arise, and impact theory is valid. Specifically, the study of $\Phi_d$ is performed in the case of isolated lines for an ionic emitter and hyperbolic paths for free electrons. This treatment is based on the results obtained under the same conditions in the non-relativistic case. The relativistic collision operator is then given by

$$\Phi_d(\omega_1,\omega_2) = -\frac{2\pi N_e e^2}{3\hbar^2}\int_0^c v\,f(\beta)\,d\beta\int_{\rho_{min}}^{\rho_{max}}\rho_0 d\rho_0\int_{-\infty}^{+\infty}dt_1\int_{-\infty}^{t_1}dt_2\,e^{(i\omega_1 t_1 + i\omega_2 t_2)}\left[\boldsymbol{E}_{LW}(t_1)\cdot\boldsymbol{E}_{LW}(t_2)\right] \tag{3}$$

where c is the speed of light in vacuum, $v$ is the initial velocity of the colliding electron and $f(\beta)\,d\beta$ is the distribution of the velocities of Juttner-Maxwell (generalized Maxwell distribution of the velocities when the movement of particles is relativistic) given by

$$f(\beta)\, d\beta = \frac{\gamma^5 \beta^2 d\beta}{\theta K_2(1/\theta)} \exp(-\gamma/\theta) \tag{4}$$

where

$$\theta = \frac{K_B T}{m_e c^2}, \quad \gamma = 1/\sqrt{(1-\beta^2)}, \quad \beta = v/c \tag{5}$$

and $K_2(1/\theta)$ is a Bessel function, $T$ is the temperature, $m_e$ is the electron mass and $\rho_0$ is the impact parameter, whereas $\rho_{\min}$ and $\rho_{\max}$ are the limits of the last integral (Formula (3)) that will be chosen later. The electric field of Lienard-Wiechert is given by [3]

$$\mathbf{E}_{LW}(\mathbf{R}, \boldsymbol{\alpha}, t) = -e \frac{(\boldsymbol{\eta} - \boldsymbol{\alpha})(1 - \alpha^2)}{k^3 R^2(t')} - \frac{e}{c^2} \frac{\boldsymbol{\eta}}{k^3 R(t')} \times \left\{ (\boldsymbol{\eta} - \boldsymbol{\alpha}) \times \frac{dv(t')}{dt'} \right\} \tag{6}$$

then

$$\boldsymbol{\alpha} = \frac{v(t')}{c}, \boldsymbol{\eta} = \frac{\mathbf{R}(t')}{R(t')}, t' = t - R(t')/c \tag{7}$$

where the retarded time is given by: $t' = t - \frac{R(t')}{c}$, $e$ is the charge of the electron, $\mathbf{R}(t')$ is the electron position vector, and $\boldsymbol{\eta} = \frac{\mathbf{R}(t')}{R(t')}$ is a unit vector directed from the position of the moving charge (electron) towards the observation point (where the emitter is located), and $k$ is given by

$$k = \frac{dt}{dt'} = 1 + \frac{1}{c}\frac{dR(t')}{dt'} = 1 + \eta\alpha \tag{8}$$

The first term of the field (6), the velocity field, goes to the known Coulomb field when $v \to 0$ whereas the second term of the field, is the acceleration field or the radiation field. As the ratio of second term (the radiation field) of the field $E_{LW}$ on the first term is less than $v^2/c^2$, we can therefore neglect the second and use only the first part of the field in the subsequent development

$$\mathbf{E}_{LW}(\mathbf{R}, t) \simeq e \left[ \frac{(\boldsymbol{\eta} - \boldsymbol{\alpha})(1 - \alpha^2)}{k^3 R^2} \right] \tag{9}$$

By using the approximation $1 - \alpha^2 \simeq 1$, which is justified in our subsequent studies ($T = 8 \times 10^8$ K, the probable $\alpha$ is about 0.22), therefore the electric field becomes

$$\mathbf{E}_{LW}(\mathbf{R}, t) = e \left[ \frac{(\boldsymbol{\eta} - \boldsymbol{\alpha})}{k^3 R^2} \right] \tag{10}$$

If we neglect the fine structure ($\omega_1 = \omega_2 = 0$), we can write the collision operator as

$$\Phi_d(0,0) = -\frac{\pi N_e e^2}{3\hbar^2} \int_0^c v f(\beta) d\beta \int_{\rho_{\min}}^{\rho_{\max}} \rho_0 d\rho_0 \int_{-\infty}^{+\infty} \mathbf{E}_{LW}(t_1) dt_1 \int_{-\infty}^{+\infty} \mathbf{E}_{LW}(t_2)\, dt_2 \tag{11}$$

or equivalently

$$\Phi_d(0,0) = -\frac{\pi N_e e^2}{3\hbar^2} \int_0^c v f(\beta) d\beta \int_{\rho_{\min}}^{\rho_{\max}} \rho_0 d\rho_0 \mathbf{G}^2 \tag{12}$$

such as

$$\mathbf{G} = -e \int_{-\infty}^{+\infty} \frac{\left(\frac{\mathbf{R}(t')}{R(t')} - \frac{v(t')}{c}\right)}{\left(\frac{dt}{dt'}\right) k^2 R^2(t')} dt \tag{13}$$

or, by integrating over $t'$

$$\mathbf{G} = -e \int_{-\infty}^{+\infty} \frac{\left(\frac{\mathbf{R}(t')}{R(t')} - \frac{d\mathbf{R}(t')}{c\, dt'}\right)}{\left(1 + \frac{1}{c}\frac{dR(t')}{dt'}\right)^2 R^2(t')} dt' \tag{14}$$

$$G = -e \int_{-\infty}^{+\infty} \left( \frac{R(t')}{\left(1 + \frac{1}{c}\frac{dR(t')}{dt'}\right)^2 R^3(t')} - \frac{dR(t')}{\left(1 + \frac{1}{c}\frac{dR(t')}{dt'}\right)^2 cdt' R^2(t')} \right) dt' \tag{15}$$

In the following, we use the notations and the variable change

$$\epsilon = (1 + \frac{m^2 v^4 \rho_0^2}{Z_{em}^2 e^4})^{1/2}$$

$$t' = \frac{\rho_0}{v} (\epsilon \sinh(x) - x)$$

$$dt' = \frac{\rho_0}{v} (\epsilon \cosh(x) - 1) dx$$

$$R(t') = \rho_0 (\epsilon \cosh(x) - 1)$$

$$X = \rho_0 (\epsilon - \cosh(x))$$

$$\frac{dX}{dt'} = \frac{\frac{dX}{dx}}{\frac{dt'}{dx}} = -v\frac{\sinh(x)}{(\epsilon \cosh(x) - 1)}$$

$$Y = \rho_0 \sqrt{\epsilon^2 - 1} \sinh(x)$$

$$\frac{dY}{dt'} = \frac{\frac{dY}{dx}}{\frac{dt'}{dx}} = v\frac{\sqrt{\epsilon^2 - 1} \cosh(x)}{(\epsilon \cosh(x) - 1)}$$

$$R(t') = X\mathbf{i} + Y\mathbf{j}$$

$$\mathbf{v}(t') = \frac{dX}{dt'}\mathbf{i} + \frac{dY}{dt'}\mathbf{j}$$

where $Z_{em}$ is the charge of the ionic emitter and $(\mathbf{i}, \mathbf{j})$ are the basis of the cartesian coordinates. Then the function $G$ is given by

$$G = -\frac{e}{\rho_0^2} \int_{-\infty}^{+\infty} \frac{X(t')\mathbf{i} + Y(t')\mathbf{j}}{\left[(\epsilon \cosh(x) - 1 + \frac{v}{c}\epsilon \sinh(x))\right]^2 R(t')} dt'$$

$$-\frac{e}{c\rho_0^2} \int_{-\infty}^{+\infty} \frac{\frac{dX}{dt'}\mathbf{i} + \frac{dY}{dt'}\mathbf{j}}{\left[(\epsilon \cosh(x) - 1 + \frac{v}{c}\epsilon \sinh(x))\right]^2} dt' \tag{16}$$

Using (17), we jump to integrate over x:

$$G = -\frac{e}{v\rho_0^2} \int_{-\infty}^{+\infty} \frac{\rho_0 (\epsilon - \cosh(x))\mathbf{i} + \rho_0 \sqrt{\epsilon^2 - 1} \sinh(x)\mathbf{j}}{\left[\epsilon \cosh(x) - 1 + \frac{v}{c}\epsilon \sinh(x)\right]^2} dx$$

$$-\frac{e}{c\rho_0^2} \int_{-\infty}^{+\infty} \frac{-v\frac{\sinh(x)}{(\epsilon \cosh(x)-1)}\mathbf{i} + v\frac{\sqrt{\epsilon^2-1}\cosh(x)}{(\epsilon \cosh(x)-1)}\mathbf{j}}{\left[\epsilon \cosh(x) - 1 + \frac{v}{c}\epsilon \sinh(x)\right]^2} \frac{\rho_0}{v} (\epsilon \cosh(x) - 1) dx$$

$$= -\frac{e}{v\rho_0} \int_{-\infty}^{+\infty} \frac{(\epsilon - \cosh(x))\mathbf{i} + \sqrt{\epsilon^2 - 1} \sinh(x)\mathbf{j}}{\left[\epsilon \cosh(x) - 1 + \frac{v}{c}\epsilon \sinh(x)\right]^2} dx$$

$$-\frac{e}{c\rho_0} \int_{-\infty}^{+\infty} \frac{-\sinh(x)\mathbf{i} + \sqrt{\epsilon^2 - 1} \cosh(x)\mathbf{j}}{\left[\epsilon \cosh(x) - 1 + \frac{v}{c}\epsilon \sinh(x)\right]^2} dx \tag{17}$$

or in a more simplified form

$$\Phi_d(0,0) = -\frac{N_e e^4}{3\pi\hbar^2} \int_0^c \frac{\gamma^5 \beta d\beta}{\theta c K_2(1/\theta)} \exp(-\gamma/\theta) \int_{\rho_{min}}^{\rho_{max}} \frac{d\rho_0}{\rho_0} \left[\overline{A}^2 + (\epsilon^2 - 1)\overline{B}^2\right] \tag{18}$$

where

$$\overline{A} = \int_{-\infty}^{+\infty} \frac{-(\epsilon - \cosh(x)) + \beta \sinh(x)}{[\epsilon \cosh(x) - 1 + \beta\epsilon \sinh(x)]^2} dx$$

$$= 2\frac{(\epsilon^2 - 1)\sqrt{1 - \epsilon^2 + \beta^2\epsilon^2} - \beta^2\epsilon^2 \tanh^{-1}(\sqrt{1 - \epsilon^2 + \beta^2\epsilon^2})}{\epsilon(1 - \epsilon^2 + \beta^2\epsilon^2)^{3/2}} \qquad (19)$$

$$\overline{B} = -\int_{-\infty}^{+\infty} \frac{\sinh(x) + \beta \cosh(x)}{[\epsilon \cosh(x) - 1 + \beta\epsilon \sinh(x)]^2} dx = 0 \qquad (20)$$

Then the relativistic collision operator caused by the Lienard-Wiechert electric field $\Phi_{d,LW}$ is given by

$$\Phi_d(0,0)(Hz/cm^2) = \Phi_{d,LW} =$$

$$-\frac{4N_e e^4}{3\pi\hbar^2\theta c K_2(1/\theta)} \int_0^1 \frac{\exp(-\frac{1}{\theta\sqrt{1-\beta^2}})}{(1-\beta^2)^{5/2}} \beta d\beta \int_{\rho_{\min}}^{\rho_{\max}} \frac{[(\epsilon^2-1)\delta - \beta^2\epsilon^2 \tanh^{-1}(\delta)]^2}{\epsilon^2\rho_0\delta^6} d\rho_0 \qquad (21)$$

where we have put

$$\delta \sim \delta(\beta,\epsilon) = \sqrt{1 - \epsilon^2 + \beta^2\epsilon^2} \qquad (22)$$

By taking the maximum of the impact parameter $\rho_{max} = 0.68\lambda_D$ ($\lambda_D$ is the Debye length) [5] and the minimum of the impact parameter equal to Bohr radius (the Wiesskopf radius is much smaller than Bohr radius in our application), and after numerical integration of (31) we find the following result in Table 1.

We note that we have considered the lower limit of the integration over the impact parameter equal to the Bohr radius, because in our study we only intented to compare the two collision operators corresponding to Coulomb and Lienard-Wiechert interactions. In reality, as Formula (31) shows, by decreasing the lower limit of the impact parameter, the value of the collision operator increases and by increasing the lower limit, the value of the collision operator decreases. Another reason to mention is: by regarding Formula (3) of Ref. [5], we see that the minimum of the impact parameter (Wiesskopf radius $r_W$), in our conditions of the high temperature, high charge number $Z = 70$ and the upper and lower levels of the transition $n_a = 3$ (1s3d),$n_b = 2$ (1s2p)(in triplet case), is much smaller than the Bohr radius $a_0$ that is to say $r_W << a_0$. For example, for a density $10^{18}$ cm$^{-3}$, for the lower limit $r_W$, the value of the collision operator $\Phi_{d,LW}$ is $0.35 \times 10^{-2}$ eV (that is overestimated in our opinion) whereas it is $0.21 \times 10^{-2}$ eV for the lower limit equal to the Bohr radius $a_0$. In fact, in the region $r < a_0$, the quantum effects must be taken into account. For these reasons, we have only considered, in our calculation, the minimum of the impact parameter equal to the Bohr radius $a_0$. In addition, we remark from the Table 1 that the effect of the Lienard-Wiechert field is to reduce the amplitude of the collision operator. This reduction is more pronouced for the weak electron densities. We must also note that the criteria of the isolated lines becomes not valid for densities great than $10^{21}$ cm$^{-3}$ because the FWHM = 2 × HWHM of two neighboring spectral lines becomes of the same order of magnitude of the separation (about 3eV) between the line arising from 1s3d to 1s2p (triplet case that we study) transition and the neighboring line arising from 1s3p to 1s2s (singlet case) transition. Following the Table 1, and for densities less than $10^{21}$ cm$^{-3}$, the HWHM is small enough to be can considered that the studied line is isolated.

**Table 1.** Comparison between collision operators for Coulomb interaction $\Phi_{d,C}$ and for Lienard Wiechert interaction $\Phi_{d,LW}$ (multiplied by $\hbar$ in eV*s and by $a_0^2$ in cm$^2$ where $a_0$ is Bohr radius) at a temperature $T = 8 \times 10^8$ K and different densities and for Helium-like Ytterbium $Z_{em} = 68$, for the radiative transition 1s3d to 1s2p for triplet case. After this muliplication, the following results are in eV.

| Ne in cm$^{-3}$ | $\Phi_{d,C}$ in eV | $\Phi_{d,LW}$ in eV | HWHM$_C$ in eV | HWHM$_{LW}$ in eV | Percent |
|---|---|---|---|---|---|
| $10^{16}$ | $0.32 \times 10^{-4}$ | $0.24 \times 10^{-4}$ | $0.25 \times 10^{-4}$ | $0.17 \times 10^{-4}$ | 25 |
| $10^{18}$ | $0.27 \times 10^{-2}$ | $0.21 \times 10^{-2}$ | $0.20 \times 10^{-2}$ | $0.15 \times 10^{-2}$ | 22 |
| $10^{20}$ | 0.24 | 0.22 | 0.18 | 0.17 | 8 |
| $10^{22}$ | 26 | 25 | 20 | 19 | 4 |

Note: We have defined the percent to be equal $\left( \frac{\Phi_{d,C} - \Phi_{d,LW}}{\Phi_{d,C}} \cdot 100 \right)$.

## 3. Conclusions

In this work, we have investigated Lienard-Wiechert or retarded electric fields produced by moving electric charges with respect to a rest frame. Specifically, we have studied its contribution to the broadening of the spectral line shape of the Helium-like Ytterbium in hot and dense plasmas radiative (transition from 1s3d to 1s2p). The principal result, as the table shows is: the retarded Lienard-Wiechert interaction, narrows the line shape comparatively to the pure Coulomb interaction. The narrowing is more pronouced for the low densities.

**Acknowledgments:** We wish to acknowledge the support of LRPPS laboratory and Fethi Khelfaoui, by offering us the encouragement, and some technical materials for developing this work.

**Author Contributions:** All authors M.T.M., K.A., K.C., K.A.T. and S.D. were participated equivalently to this theoretical work.

## References

1.   Anderson, P.W. Pressure Broadening in the Microwave and Infra-Red Regions. *Phys. Rev.* **1949**, *76*, 647, doi:10.1103/PhysRev.76.647.

2.   Griem, H.R.; Baranger, M.; Kolb, A.C.; Oertel, G. Stark Broadening of Neutral Helium Lines in a Plasma. *Phys. Rev.* **1962**, *125*, 177, doi:10.1103/PhysRev.125.177.

3.   Jackson, J.D. Chapter 11: Special Theory of Relativity. In *Classical Electrodynamics*, 3rd ed.; John Wiley: New York, NY, USA, 1962; pp. 360–364.

4.   Griem, H.R.; Kolb, A.; Shen, Y. Stark Broadening of Hydrogen Lines in a Plasma. *Phys. Rev.* **1959**, *116*, 4, doi:10.1103/PhysRev.116.4.

5.   Alexiou, S. Collision operator for isolated ion lines in the standard Stark-broadening theory with applications to the Z scaling in the Li isoelectronic series 3P-3S transition. *Phys. Rev. A* **1994**, *49*, 106–119

6.   Voslamber, D.A. Non-Markovian impact theory comprehending partially overlapping lines. *J. Quant. Spectrosc. Radiat. Transf.* **1970**, *10*, 939–943

# 13

# Models of Emission-Line Profiles and Spectral Energy Distributions to Characterize the Multi-Frequency Properties of Active Galactic Nuclei

**Giovanni La Mura** [1,*,†], **Marco Berton** [1,2,‡], **Sina Chen** [1,‡], **Abhishek Chougule** [1,‡], **Stefano Ciroi** [1,‡], **Enrico Congiu** [1,2,‡], **Valentina Cracco** [1,‡], **Michele Frezzato** [1,‡], **Sabrina Mordini** [1,‡] and **Piero Rafanelli** [1,‡]

[1]    Department of Physics and Astronomy, University of Padua, Vicolo dell'Osservatorio 3, 35122 Padova, Italy; marco.berton@unipd.it (M.B.); sina.chen@phd.unipd.it (S.C.); abhishek.chougule@student.uibk.ac.at (A.C.); stefano.ciroi@unipd.it (S.C.); enrico.congiu@phd.unipd.it (E.C.); valentina.cracco@unipd.it (V.C.); michele.frezzato@phd.unipd.it (M.F.); sabrina.mordini@studenti.unipd.it (S.M.); piero.rafanelli@unipd.it (P.R.)

[2]    Astronomical Observatory of Brera, National Institute of Astrophysics (INAF), Via Bianchi 46, 23807 Merate, Italy

\*    Correspondence: giovanni.lamura@unipd.it

†    Current address: Vicolo dell'Osservatorio 3, 35122 Padova, Italy.

‡    These authors contributed equally to this work.

**Abstract:** The spectra of active galactic nuclei (AGNs) are often characterized by a wealth of emission lines with different profiles and intensity ratios that lead to a complicated classification. Their electromagnetic radiation spans more than 10 orders of magnitude in frequency. In spite of the differences between various classes, the origin of their activity is attributed to a combination of emitting components, surrounding an accreting supermassive black hole (SMBH), in the *unified model*. Currently, the execution of sky surveys, with instruments operating at various frequencies, provides the possibility to detect and to investigate the properties of AGNs on very large statistical samples. As a result of the spectroscopic surveys that allow the investigation of many objects, we have the opportunity to place new constraints on the nature and evolution of AGNs. In this contribution, we present the results obtained by working on multi-frequency data, and we discuss their relations with the available optical spectra. We compare our findings with the AGN unified model predictions, and we present a revised technique to select AGNs of different types from other line-emitting objects. We discuss the multi-frequency properties in terms of the innermost structures of the sources.

**Keywords:** galaxies: active; galaxies: nuclei; line: profiles; quasars: emission lines; catalogs; surveys

## 1. Introduction

When discussing active galactic nuclei (AGNs), we generally refer to the nuclei of galaxies for which a supermassive black hole (SMBH), with a mass ranging between $10^6$ and $10^{10}$ M$_\odot$, is fed by a continuous flow of matter from the surrounding environment. This process, denoted as *accretion*, leads to the conversion of the gravitational binding energy of the accreted material into heat and radiative energy, through the effects of the viscous interactions that arise in the accreted matter as it is accelerated up to several thousand kilometers per second by the strong gravitational pull of the black hole [1]. In spite of this common interpretation, AGNs present a wide range of striking observational differences in their spectra, in their total power and in the frequency range for which most of their energy is radiated away. While the first to be clearly identified were located in galaxies with exceptionally bright optical nuclei [2], nearly 10% of the total population were subsequently found to radiate large fractions of their power in the radio and the high-energy domains [3]. Their total luminosities can change over a wide range and are typically considered to lie between $10^{40}$ erg s$^{-1}$, in the case of low-luminosity sources, and some $10^{46}$ erg s$^{-1}$ for the most powerful sources. They are distributed from the local universe, where the low-luminosity objects are more common, all the way up to very high redshifts ($z \geq 7$), where powerful activity becomes more frequent.

In terms of the characteristics of the optical-UV spectra, AGNs are generally characterized by the presence of a non-thermal continuum, often well represented by a power-law shaped spectrum of the form $L_\nu \propto \nu^{-\alpha}$ and sometimes accompanied by prominent emission lines with different profiles. In some cases, we additionally observe a hot thermal excess, with a peak that likely falls in the far UV, or different contributions from the host-galaxy stellar populations. Following a scheme that was first outlined by [4], we generally classify Type 1 AGNs as those objects whose spectra show broad recombination lines, with profiles corresponding to velocity fields exceeding 1000 km s$^{-1}$ in full width at half maximum (FWHM), from H, He, or from other permitted lines of heavy ions such as Fe II, C IV and Si IV, together with narrow forbidden lines (FWHM $\approx$ 300–500 km s$^{-1}$) from, for example, [Ne V], [Ne III], [O III], [O II], [O I], [N II] and [S II]. We classify Type 2 AGNs as those for which both permitted and forbidden lines only have narrow profiles. In general, it is observed that Type 1 sources are brighter and commonly show a thermal UV excess, while Type 2 objects are dimmer and more severely contaminated by the host-galaxy spectral contributions.

The most widely accepted way to explain the observations is to assume that the central accreting SMBH is surrounded by a hot accretion disk, radiating in the optical, UV and X-ray frequencies, and a compact region (less than 0.1 pc in size) of dense ionized gas ($N_e \geq 10^9$ cm$^{-3}$), which produces Doppler-broadened recombination lines due to the large gravitational acceleration, and is therefore called the *broad-line region* (BLR). The gas that is located at larger distances (1 pc $\leq r \leq$ 1 kpc); although still being ionized and producing emission lines, it has a considerably smaller velocity field and electron densities closer to typical nebular environments ($10^3$ cm$^{-3}$ $\leq N_e \leq 10^6$ cm$^{-3}$). It, therefore, radiates both permitted and forbidden lines with narrow Doppler profiles, giving rise to what we call the *narrow-line region* (NLR). If the central structure is partially obscured by an optically thick distribution of matter, as is supported by some observational evidence [5], the difference between Type 1 and Type 2 objects is consistently explained by the fact that our line of sight, respectively, may or may not reach the central regions, without being intercepted by the surrounding material, in what is called the AGN *unified model* [6]. When the accretion flow becomes coupled with the magnetic fields that develop close to the SMBH in such a way that a relativistically beamed jet of plasma is accelerated away from the nucleus, AGNs become powerful sources of radio and high-energy emission, eventually developing extended radio morphologies [7].

In spite of the fairly comprehensive interpretation, AGNs still pose many fundamental questions, because most of the relevant physical processes involved in the accretion and in the acceleration of jets are confined close to the SMBH, in a region that lies still beyond the resolution capabilities of present-day instruments. For this reason, a large part of our current knowledge concerning AGNs is based on the analysis of their spectra and on monitoring the correlations existing among their numerous spectral components. However, not all sources are equally good for such investigations, as Type 1 objects tend to be dominated by the emission of the AGN, while Type 2 sources are strongly affected by obscuration and stellar light contributions from the host galaxy. The main details to understand the physics of AGNs, therefore, can only be constrained by careful inspection of spectral properties that should be possibly extended over large statistical samples, in order to compare the predictions of different models with the available observations. In recent times, several campaigns have been carried out to monitor the sky at different frequencies and to obtain spectroscopic observations. In this contribution, we describe a revised scheme to select AGNs on the basis of the properties of their emission lines and colors, we illustrate the potential of multi-frequency models to relate the observed spectral energy distributions (SEDs) with optical spectra, and we discuss the information that we are able to extract on their central structures from the combined analysis of line profiles and multi-frequency data.

## 2. Results

The most common property shared by different types of AGNs is the emission of an intense continuum of non-thermal radiation that can possibly ionize diffuse gas and, thus, give rise to emission lines in the spectra. Because lines excited by non-thermal ionizing continua differ in intensity and distribution from lines excited by the continuum of hot thermal sources [8,9], we can apply a set of diagnostic diagrams, based on the intensity ratios of specific lines, in order to recognize the footprint of ionization from thermal and non-thermal sources in external galaxies. When the amount of spectroscopic data was dramatically increased, as a result of the execution of large spectroscopic surveys such as the *Sloan Digital Sky Survey (SDSS)* [10,11], this method was further refined, demonstrating its ability to recognize the presence of obscured AGN activity [12]. This kind of information is fundamental to assess the statistical relations existing between obscured and unobscured sources, which constrain the structure of the central source and its possible dependence on luminosity or age. In order to perform such a study, however, we need an instrument that is in principle able to detect different types of AGN activity, with the smallest possible influence of selection effects. We obtained such a tool by combining spectroscopic and photometric parameters, on the basis of an investigation of how different spectral classes are related with multi-frequency emission.

### 2.1. AGN Selection from Spectroscopic Surveys

The investigation of AGN statistical properties requires the revising of the selection techniques, which have been classically designed to detect specific types of sources on the basis of their characteristic properties. While Type 1 AGNs are generally well identified by the presence of prominent broad emission lines in the spectra, which are commonly accompanied by a hot thermal continuum excess sometimes referred to as the *big blue bump*, Type 2 AGNs are only characterized by narrow emission lines, which can also be present in the spectra of galaxies with strong star formation activity. In the case of narrow line-emitting sources, the methods based on diagnostic diagrams are fairly well suited to distinguish between AGN and stellar activity, but if performed on a selection of spectra simply based on the presence of emission lines, to collect different types of sources, the emission-line diagnostics alone are not straightforwardly applicable. An example of this effect is shown in Figure 1, where we compare different methods to distinguish AGNs from thermally excited spectral line emitters. The reason that the classic diagnostic ratios cannot be used on a sample of objects including Type 1 sources is attributed to the use of the recombination lines, which are needed to normalize the strength of the forbidden lines that probe the temperature and the ionization structure of the gas. The presence of a strong contribution in the broad component of the recombination lines in Type 1 objects, which is not balanced in the forbidden lines, forces Type 1 sources to populate the non-AGN region of the plots. This is apparent, because this type of diagnostic diagram is designed to work on the emission of the NLR alone and cannot account for the BLR component. It has been proposed that a different choice of the emission-line diagnostic ratios, involving only forbidden lines, might in principle solve the problem [13], but the available choices either involve the use of weak lines, or they are strongly subject to the effects of interstellar extinction.

If the direct sight of the central regions in Type 1 AGNs can give rise to problems in recognizing their spectral signature, on the other hand, we have the opportunity to take advantage of the strong blue and UV continuum, which is produced by the central source and is not obscured along the line of sight. It has been shown that Type 1 AGNs can effectively be selected by means of photometric criteria that compare their colors with those of non-active objects, to the extent that the SDSS uses a photometric pipeline to select QUASAR candidates for follow-up spectroscopy [14,15]. This method has some limitations when comparing objects at very different redshifts; however, in the low-redshift domain in which the diagnostic lines are still available in the optical frequency window ($z \leq 0.5$), it defines a well-established parameter space in which Type 1 AGNs can be effectively distinguished from other line-emitting sources. A projection of this parameter space on the $u - r$ versus $g - z$ color–color diagram is also illustrated in Figure 1. The choice of these extended color bands, which was based on the extinction-corrected magnitude measurements of all SDSS objects with emission lines of Hα, Hβ, [O III]λ5007 and [N II]λ6583 detected at more than the 5σ level, maximizes the effect of the blue continuum of Type 1 sources over the stellar continuum of other sources. As a consequence, we recover the possibility of detecting different types of nuclear activity by combining classic spectroscopic diagnostics and photometric colors into a multi-dimensional parameter space, in which AGNs populate a separate sequence with respect to other non-AGN powered sources.

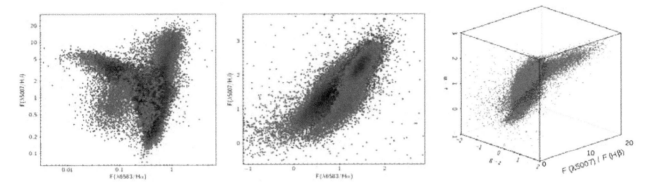

**Figure 1.** Distribution of active galactic nuclei (AGNs; blue points) and star formation (red points) powered objects with emission-line spectra on the [O III]λ5007/Hβ vs. [N II]λ6583/Hα diagnostic diagram (**left panel**), on the $u - r$ vs $g - z$ color–color diagram (**middle panel**) and on the three-dimensional parameter space combining the two-color photometry with the [O III]λ5007/Hβ diagnostic ratio. The distribution in the three-dimensional parameter space illustrates how Type 1 AGNs are best identified among line-emitting sources by means of photometric criteria, while Type 2 sources can be distinguished on the basis of their spectral properties.

## 2.2. Emission Lines and Models of AGN Spectral Energy Distributions

The selection of general samples of AGNs belonging to different spectral classes allows us to search for observations of the corresponding sources in multiple frequencies. By combining the available data, it is possible to reconstruct the AGN SEDs and to compare these with the associated optical spectra, as is illustrated for example in Figure 2. The plots show the different SEDs of two prototypical AGNs (3C 273 for Type 1 and NGC 1068 for Type 2) modeled through a combination of thermal and non-thermal radiation components, together with their characteristic spectra. We can immediately appreciate how the occurrence of a Type 1 spectrum, with broad lines and a blue continuum, is well associated with a strong dominance of the non-thermal contribution and a direct sight towards the hottest central regions, resulting in an excess of ionizing radiation, in agreement with the unified model predictions. Conversely, the Type 2 SED is totally consistent with an obscured central source, whose low-energy ionizing radiation is severely absorbed and reprocessed by a distribution of material that subsequently re-emits photons in the infrared (IR) domain. Only the more penetrating high-energy photons and the long-wavelength radio emission, which is practically unaffected by obscuration, can propagate directly from the central source, therefore resulting in an optical spectrum that is dominated by the host galaxy and the emission lines originating from the unobscured NLR.

In addition to providing observational evidence in support of the unified model, and possibly to associate different degrees of absorption and light reprocessing to various AGN classes, the combination of SED models and spectroscopic observations is a promising instrument to improve our knowledge of AGN distribution. As a result of the numerous efforts that have been devoted to surveying the whole sky at various frequencies, it is now possible to recognize AGN candidates from their SED, even in sky areas that have not yet been covered by detailed and publicly available spectroscopic programs. The further possibility that models of the observed SED may lead to a prediction of the expected AGN class, in addition to optimizing the execution of follow-up campaigns, provides statistical constraints to infer structural and evolutionary details in large samples of AGNs.

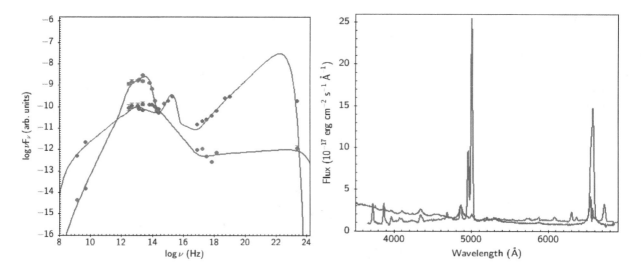

**Figure 2.** Examples of spectral energy distribution (SED) models (**left panel**) and optical spectra (**right panel**) for the prototypical Type 1 active galactic nucleus (AGN) 3C 273 (blue line and points) and the prototypical Type 2 AGN NGC 1068 (red line and points). The SEDs have been normalized to the same optical V-band magnitude and the models are based on the combination of two cut-off power laws and one black-body contribution. The spectrum of 3C 273 was observed at the Asiago Astrophysical Observatory, while the spectrum of NGC 1068 is taken from literature [16]. Both spectra were taken to rest-frame wavelength scale.

### 2.3. Multi-Frequency Analysis of the Central Engine

Because of the extremely compact size of the regions in which the continuum and the bulk of the emission lines are produced in AGNs, we do not yet have a fully developed interpretation of their innermost structures. Most of our current understanding is derived from the analysis of spectra and from models that carry out the inferred physical conditions. Therefore, the extension of spectroscopic analysis to different frequencies improves our ability to explore unresolved structures. Such an example is illustrated in Figure 3 for the case of the QSO PG 1114 + 445, for which we compare the *XMM Newton* X-ray spectrum with measurements of the normalized intensities of the broad components of the Balmer lines of hydrogen. This quantity is defined as

$$I_n = \frac{\lambda_{ul} I_{ul}}{g_u A_{ul}} \tag{1}$$

where $\lambda_{ul}$ is the wavelength corresponding to the transition from an upper level $u$ to a lower level $l$ (with $l = 2$ for H Balmer lines), $I_{ul}$ is the measured intensity, $g_u$ is the statistical weight of the upper level, and $A_{ul}$ is the spontaneous transition probability. In the case of an optically thin line, we can use the general expression:

$$I_{ul} = \frac{1}{4\pi} \frac{hc}{\lambda_{ul}} A_{ul} \int_0^{s^*} N_u \mathrm{d}s \tag{2}$$

where $h$ and $c$ are the Planck constant and the vacuum speed of light, while $N_u$ is the concentration of atoms in the upper level, and the integration is executed throughout the extension of the source to obtain

$$I_n = \frac{1}{4\pi} \frac{hc}{g_u} \int_0^{s^*} N_u \mathrm{d}s \tag{3}$$

If we now assume that the distribution of emitting atoms is spatially constant within the source and that it can be represented by a Boltzmann formula at least in the high-excitation stages (a condition known as *partial local thermodynamic equilibrium*—PLTE), we obtain

$$I_n = \frac{1}{4\pi} \frac{hc}{g_1} N_1 s^* \exp\left(-\frac{\Delta E_u}{k_B T_e}\right) \qquad (4)$$

where $\Delta E_u$ is the upper-level excitation energy. It is therefore clear that, under the assumed conditions, we expect the following:

$$\log I_n = \log\left(\frac{1}{4\pi} \frac{hc}{g_1} N_1 s^*\right) - \frac{\log e}{k_B T_e} \Delta E_u \qquad (5)$$

which is a linear function of the upper-level excitation energy [17].

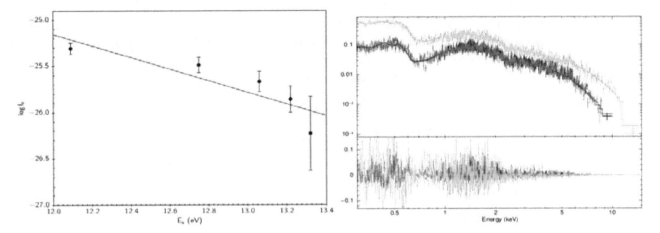

**Figure 3.** Suppression of high-order Balmer lines, observed through the normalized intensities of the broad components in the optical spectrum of PG 1114 + 445 (**left panel**), compared with the soft X-ray spectrum (**right panel**) obtained from the *XMM Newton* EPIC instrument (green points) and from the twin MOS cameras (red and black points). The X-ray spectrum is modeled with a power-law spectrum, a soft X-ray thermal excess and an ionized absorber. Absorption from the neutral medium within the Milky Way is also taken into account.

The deviation from the expected linear behavior observed in Figure 3 is an indication that the flux of high-order Balmer-line photons is lower than the prediction, which can happen in the presence of a dense layer of recombining plasma. In this case, indeed, Equation (2) would be modified in as

$$I_{ul} = \frac{1}{4\pi} \frac{hc}{\lambda_{ul}} A_{ul} \int_0^{s^*} N_u e^{-k_{lu}s} ds \qquad (6)$$

where

$$k_{lu} = N_l \sigma_{lu} \qquad (7)$$

is the line absorption coefficient, controlled by the density of ions in the lower level $N_l$ and the line photon absorption cross-section $\sigma_{lu}$. Under typical nebular conditions, this would be $k_{lu} << 1$ for any $l > 1$; however, in the presence of a thick layer of ionized plasma, such as that responsible for the observed absorption in the soft X-ray spectrum, which has an estimated column density of $N_C = 4.896 \times 10^{21}$ cm$^{-2}$, this could no longer be negligible. Under such circumstances, indeed, we expect the lower level of the Balmer series to be overpopulated by recombination processes and by Ly$\alpha$ photon trapping. If, therefore, this X-ray absorber is located outside the BLR, it could very likely also be responsible for the absorption of Balmer photons.

To further explore the possibility that the presence of ionized gas layers could affect the broad emission lines, we began an investigation into the Balmer-line profiles in the broad component of the spectra of a sample of narrow-line Seyfert 1 galaxies (NLS1s), selected from the SDSS Data Release 7 [18]. Because of the relatively small width of their broad-line components (FWHM$_{H\beta}$ < 2000 km s$^{-1}$ [19]), the profiles of these lines are largely unaffected by blends with other broad lines, although the narrow components are still to be accounted for. These, however, are much simpler to model, taking the narrow forbidden lines of [O III] as templates to constrain their widths and to be subtracted from the global profiles. By taking the broad-line components in a velocity scale, we are able to compare the resulting profiles, as illustrated in Figure 4. Using a narrow emission-line width fixed at 0.75 times the width of [O III] $\lambda$5007, in order to account for the larger velocity dispersion of the high-ionization gas [20], we modeled the narrow components of H$\beta$, H$\alpha$ and the [N II] $\lambda\lambda$6548, 6583 doublet, attempting to isolate the broad emission-line profiles. Regarding the resulting FWHM of H$\alpha$ and H$\beta$, we find that most of the line profiles are very similar, favoring the interpretation of an optically thin gas, although we still observe deviations from this behaviour. It has already been noted that the profiles of these lines can be different in some objects and that they may even exhibit different reverberation lags [21,22]. This could point towards a displacement of their emission sites or to a relevant role of dust absorption in the central regions of the source [23,24], as there is convincing evidence that a substantial amount of dust can exist in the central regions of AGNs, at smaller scales than the NLR [25].

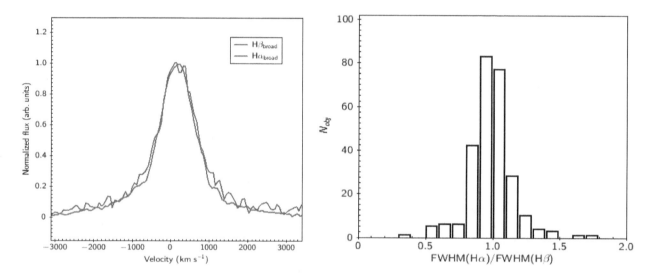

**Figure 4. (Left panel)** comparison of the H$\beta$ (blue) and H$\alpha$ (red) broad line components, after subtraction of the narrow emission lines, plotted in velocity scale and normalized to the same peak intensity in the spectrum of 1RXS J113247.0 + 062626. **(Right panel)** distribution of the ratio FWHM(H$\alpha$)/FWHM(H$\beta$) in the broad components of the emission lines from the spectra NLS1 galaxies investigated in [18]. While the distribution is strongly peaked around 1, favoring the interpretation of an optically thin gas, some differences in the line profiles can arise as a consequence of dust absorption or emission-line radiation transfer effects.

## 3. Discussion

The processes that occur in the unresolved central regions of AGNs leave characteristic signatures in the emission and absorption components of the observed radiation, and they also control which parts of the source are visible. Although a preliminary analysis of the relationships existing between multi-frequency data and optical spectra argue in favor of the unified model, a systematic study that collects the huge amount of available observational material still has to be carried out. This type of investigation is highly desirable because of the invaluable constraints that it could place on the nature and evolution of the innermost structures of AGNs, but it presents obvious difficulties related to the amount of data that should be considered. With our work, we present a technique to select different types of AGNs, with available spectro-photometric measurements, and we provide some examples of how the comparison of their spectra with the overall SED lead to supporting the unified picture, as well as to the intriguing possibility that SEDs built on the basis of public data could be used to select targets for follow-up spectroscopy or even to attempt preliminary classifications.

A particularly interesting result is that based on the combination of optical and X-ray spectroscopic observations. In the presence of a significant layer of ionized plasma, such as that observed in the X-ray spectrum of PG 1114 + 445, optical depth effects become important for determining the relative strengths of the recombination lines. In particular, these could explain the various emission-line intensity ratios that are known to deviate significantly from the standard recombination predictions in Type 1 sources. A substantial increase in optical depth gives rise to two important effects. On the one hand, the Hα photon has a larger absorption cross-section with respect to higher-order Balmer photons. On the other hand, it corresponds to a transition between two adjacent levels and, therefore, has a smaller probability to decay through other spontaneous transition channels, with respect to higher-order Balmer photons. Therefore, an increase in the line optical depth might lead to a reduction of the high-order Balmer line intensities with respect to Hα, but Hα could averagely emerge from an outer layer of the line-emitting region. In a dynamical configuration dominated by Keplerian motions, this would imply a narrower line profile. On the other hand, the role played by dust in the emission-line regions would possibly lead to the opposite result, suppressing the short-wavelength, high-order emission lines more severely than Hα, therefore favoring an enhancement of this line from deeper regions. Comparing the measurements of the width of the broad emission-line profiles, as illustrated in Figure 4, shows that neither effect is likely to severely affect the emission in the vast majority of sources, although some exceptions may still exist. This result favors the possibility that the optically thin interpretation can be reasonably applied to the broad Balmer lines, suggesting that only a small amount of material should lie along the path of the photons. We can therefore conclude that a thorough investigation of the line intensity ratios along the profiles, which requires data with very high signal-to-noise ratios, will certainly provide further insight into the as yet not completely understood problem of the BLR structure.

## 4. Materials and Methods

The data and results presented in this paper are based on publicly available services and databases, such as the *VizieR* catalogue service[1] and the SDSS *SkyServer*[2]. The plots have been produced with *TOPCAT*[3] [26] and *XSPEC*[4] software. Most of the illustrated measurements, involving line intensities and photometric colors, were directly extracted from public databases using their standard Web interfaces. Exceptions to the illustrated picture are the following:

1. Measurements of emission-line normalized intensities were obtained from public SDSS spectra after extracting the BLR contribution by means of multi-Gaussian fits of the observed line profiles performed with *IRAF*[5].
2. Measurements of the ionized-plasma column density, obtained through XSPEC models combining a power-law emission, a soft X-ray thermal excess, an absorption contribution from neutral gas constrained by the H I column density within the Milky Way and an ionized absorber, applied to a **40** ks observation carried out by *XMM Newton* and reduced with the Science Analysis Software (SAS), version 10.
3. The models of SED in multi-frequency AGN data, obtained by applying Levemberg–Marquardt non-linear $\chi^2$ minimization to combinations of cut-off power laws and black-body contributions.

Table 1 contains a summary of the data sources from which we obtained the SED points, while Table 2 reports a summary of the parameters used in the spectral energy distribution (SED) fitting models.

---

**Table 1.** Data sources for the selection of multiple-wavelength spectral energy distribution (SED) points.

| Instr. / Catalogue | $\log \nu$ (Hz) | Band | Reference |
|---|---|---|---|
| NVSS | 9.15 | Radio | [27] |
| IRAS | 12.48–13.40 | FIR | [28] |
| WISE | 13.13–13.94 | FIR | [29] |
| 2MASS | 14.14–14.38 | NIR | [30] |
| GALEX | 15.00–15.30 | UV | [31] |
| XMM | 16.86–17.68 | X-ray | [32] |
| INTEGRAL | 18.70–19.00 | Soft $\gamma$ ray | [33] |
| Fermi/LAT | 23.00–26.00 | $\gamma$ ray | [34] |

**Table 2.** Spectral energy distribution (SED) fitting models.

| Object | Function | $\log \nu_{min}^{(a)}$ | $\log \nu_{max}^{(a)}$ | Norm. | Index[b] | Temp. | $\chi_{red}^2$ |
|---|---|---|---|---|---|---|---|
| 3C 273 | Power law | 8.6 | 13.1 | $2.30 \times 10^{-18}$ | 0.60 | – | |
| | Power law | 13.1 | 17.6 | $1.31 \times 10^{-6}$ | −0.31 | – | |
| | Power law | 17.1 | 22.6 | $1.26 \times 10^{-23}$ | 0.70 | – | |
| | Black body | – | – | $1.03 \times 10^{-12}$ | – | 25, 800 K | 1.11 |
| NGC 1068 | Power law | 8.5 | 13.0 | $2.49 \times 10^{-28}$ | 1.52 | – | |
| | Power law | 12.8 | 17.5 | $7.068 \times 10^2$ | −0.85 | – | |
| | Power law | 17.0 | 24.0 | $4.82 \times 10^{-13}$ | 0.05 | – | |
| | Black body | – | – | $1.04 \times 10^{-4}$ | – | 540 K | 1.17 |

[a] Logarithm of exponential cut-off frequency given in Hz; [b] The power-law index is given according to the notation $\nu F_\nu \propto \nu^\alpha$.

**Acknowledgments:** Funding for this research program has been provided by the Department of Physics and Astronomy of the University of Padua and by the European Space Agency (ESA), under Express Procurement Contract No. 4000111138/14/NL/CB/gp. Additional support from ASTROMUNDUS mobility grants iss also gratefully acknowledged. Funding for the Sloan Digital Sky Survey IV has been provided by the Alfred P. Sloan Foundation, the U.S. Department of Energy Office of Science, and the Participating Institutions. SDSS-IV acknowledges support and resources from the Center for High-Performance Computing at the University of Utah. The SDSS Web site is www.sdss.org; SDSS-IV is managed by the Astrophysical Research Consortium for the Participating Institutions of the SDSS Collaboration including the Brazilian Participation Group, the Carnegie Institution for Science, Carnegie Mellon University, the Chilean Participation Group, the French Participation Group, Harvard-Smithsonian Center for Astrophysics, Instituto de Astrofísica de Canarias, the Johns Hopkins University, the Kavli Institute for the Physics and Mathematics of the Universe (IPMU)/University of Tokyo, Lawrence Berkeley National Laboratory, Leibniz Institut für Astrophysik Potsdam (AIP), Max-Planck-Institut für Astronomie (MPIA Heidelberg), Max-Planck-Institut für Astrophysik (MPA Garching), Max-Planck-Institut für Extraterrestrische Physik (MPE), National Astronomical Observatories of China, New Mexico State University, New York University, the University of Notre Dame, Observatário Nacional / MCTI, the Ohio State University, Pennsylvania State University, Shanghai Astronomical Observatory, United Kingdom Participation Group, Universidad Nacional Autónoma de México, the University of Arizona, the University of Colorado Boulder, the University of Oxford, the University of Portsmouth, the University of Utah, the University of Virginia, the University of Washington, the University of Wisconsin, Vanderbilt University, and Yale University. This research has made use of the VizieR catalogue access tool, CDS, Strasbourg, France. The original description of the VizieR service was published in A&AS 143, 23; This work is based on observations collected with the 1.22 m Galileo telescope of the Asiago Astrophysical Observatory, operated by the Department of Physics and Astronomy "G. Galilei" of the University of Padova. We thank the reviewers for useful suggestions leading to the improvement of this work.

**Author Contributions:** G.L.M. is the contribution speaker, the main author of the text and the investigator of AGN multi-frequency properties; M.B. adapted the multi-frequency archive with timing capabilities; S.C. works on samples of NLS1 galaxies; A.C. contributed to the selection of all emission-line spectra with diagnostic lines from SDSS Data Release 14; S.C. supervised spectroscopic data reduction procedures; E.C. investigated advanced models of NLR emission lines; V.C. selected and analyzed the SDSS DR7 sample of NLS1; M.F. developed models of NLR structure; S.M. worked on multiple Gaussian fits and intensity ratios of optical lines; P.R. conceived the multi-frequency AGN archive and supervised the physical interpretation of optical spectra.

## Abbreviations

The following abbreviations are used in this manuscript:

AGN     Active galactic nucleus
BLR     Broad-line region
FWHM    Full width at half maximum
NGC     New General Catalogue
NLR     Narrow-line region
PLTE    Partial local thermodynamic equilibrium
SED     Spectral energy distribution

## References

1.  Blandford, R.D. Black hole models of quasars. In *Quasars*; Swarup, G., Kapahi, V.K., Eds.; D. Reidel Publishing Co.: Dordrecht, The Netherlands, 1986; pp. 359–369.
2.  Seyfert, C.K. Nuclear Emission in Spiral Nebulae. *Astrophys. J.* **1943**, *97*, 28–40.
3.  Elvis, M.; Wilkes, B.J.; McDowell, J.C.; Green, R.F.; Bechtold, J.; Willner, S.P.; Oey, M.S.; Polomski, E.; Cutri, R. Atlas of quasar energy distributions. *Astrophys. J. Suppl.* **1994**, *95*, 1–68.
4.  Khachikian, E.Y.; Weedman, D.W. An atlas of Seyfert galaxies. *Astrophys. J.* **1974**, *192*, 581–589.
5.  Antonucci, R.R.J.; Miller, J.S. Spectropolarimetry and the nature of NGC 1068. *Astrophys. J.* **1985**, *297*, 621–632.
6.  Antonucci, R. Unified models for active galactic nuclei and quasars. *Ann. Rev. Astron. Astrophys.* **1993**, *31*, 473–521.
7.  Urry, C.M.; Padovani, P. Unified Schemes for Radio-Loud Active Galactic Nuclei. *Publ. Astron. Soc. Pac.* **1995**, *107*, 803–845.
8.  Baldwin, J.A.; Phillips, M.M.; Terlevich, R. Classification parameters for the emission-line spectra of extragalactic objects. *Publ. Astron. Soc. Pac.* **1981**, *93*, 15–19.
9.  Veilleux, S.; Osterbrock, D.E. Spectral classification of emission-line galaxies. *Astrophys. J. Suppl.* **1987**, *63*, 295–310.
10. York, D.G.; Adelman, J.; Anderson, J.E., Jr.; Anderson, S.F.; Annis, J.; Bahcall, N.A.; Bakken, J.A.; Barkhouser, R.; Bastian, S.; Berman, E.; et al. The Sloan Digital Sky Survey: Technical Summary. *Anim. Jam* **2000**, *120*, 1579–1587.
11. Blanton, M.R.; Bershady, M.A.; Abolfathi, B.; Albareti, F.D.; Allende Prieto, C.; Almeida, A.; Alonso-García, J.; Anders, F.; Anderson, S.F.; Andrews, B.; et al. Sloan Digital Sky Survey IV: Mapping the Milky Way, Nearby Galaxies, and the Distant Universe. *Anim. Jam* **2017**, *154*, 28–62.
12. Kewley, L.J.; Groves, B.; Kauffmann, G.; Heckman, T. The host galaxies and classification of active galactic nuclei. *Mon. Not. R. Astron. Soc.* **2006**, *372*, 961–976.
13. Vaona, L.; Ciroi, S.; Di Mille, F.; Cracco, V.; La Mura, G.; Rafanelli, P. Spectral properties of the narrow-line region in Seyfert galaxies selected from the SDSS-DR7. *Mon. Not. R. Astron. Soc.* **2012**, *427*, 1266–1283.
14. Richards, G.T.; Fan, X.; Newberg, H.J.; Strauss, M.A.; Vanden Berk, D.E.; Schneider, D.P.; Yanny, B.; Boucher, A.; Burles, S.; Frieman, J.A.; et al. Spectroscopic Target Selection in the Sloan Digital Sky Survey: The Quasar Sample. *Anim. Jam* **2002**, *123*, 2945–2975.
15. Schneider, D.P.; Richards, G.T.; Hall, P.B.; Strauss, M.A.; Anderson, S.F.; Boroson, T.A.; Ross, N.P.; Shen, Y.; Brandt, W.N.; Fan, X.; et al. The Sloan Digital Sky Survey Quasar Catalogue. V. Seventh Data Release. *Anim. Jam* **2010**, *139*, 2360–2373.
16. Moustakas, J.; Kennicutt, R.C., Jr. An Integrated Spectrophotometric Survey of Nearby Star-forming Galaxies. *Astrophys. J. Suppl.* **2006**, *164*, 81–98.
17. Popović, L.Č. Balmer Lines as Diagnostics of Physical Conditions in Active Galactic Nuclei Broad Emission Line Regions. *Astrophys. J.* **2003**, *599*, 140–146.
18. Cracco, V.; Ciroi, S.; Berton, M.; Di Mille, F.; Foschini, L.; La Mura, G.; Rafanelli, P. A spectroscopic analysis of a sample of narrow-line Seyfert 1 galaxies selected from the Sloan Digital Sky Survey. *Mon. Not. R. Astron. Soc.* **2016**, *462*, 1256–1280.
19. Osterbrock, D.E.; Pogge, R.W. The spectra of narrow-line Seyfert 1 galaxies. *Astrophys. J.* **1985**, *297*, 166–176.

20. Berton, M.; Foschini, L.; Ciroi, S.; Cracco, V.; La Mura, G.; Lister, M.L.; Mathur, S.; Peterson, B.M.; Richards, J.L.; Rafanelli, P. Parent population of flat-spectrum radio-loud narrow-line Seyfert 1 galaxies. *Automob. Assoc.* **2015**, *578*, A28.

21. Shapovalova, A.I.; Popović, L.Č.; Burenkov, A.N.; Chavushyan, V.H.; Ilić, D.; Kollatschny, W.; Kovačević, A.; Bochkarev, N.G.; Carrasco, L.; León-Tavares, J.; et al. Spectral optical monitoring of 3C 390.3 in 1995–2007. I. Light curves and flux variation in the continuum and broad lines. *Automob. Assoc.* **2010**, *517*, A42.

22. Shapovalova, A.I.; Popović, L.Č.; Burenkov, A.N.; Chavushyan, V.H.; Ilić, D.; Kovačević, A.; Bochkarev, N.G.; León-Tavares, J. Long-term variability of the optical spectra of NGC 4151. II. Evolution of the broad Hα and Hβ emission-line profiles. *Automob. Assoc.* **2010**, *509*, A106.

23. Gaskell, C.M. The case for cases B and C: Intrinsic hydrogen line ratios of the broad-line region of active galactic nuclei, reddenings, and accretion disc sizes. *Mon. Not. R. Astron. Soc.* **2017**, *467*, 226–238.

24. Dong, X.; Wang, T.; Wang, J.; Yuan, W.; Zhou, H.; Dai, H.; Zhang, K. Broad-line Balmer decrements in blue active galactic nuclei. *Mon. Not. R. Astron. Soc.* **2008**, *383*, 581–592.

25. Heard, C.Z.P.; Gaskell, C.M. The location of the dust causing internal reddening of active galactic nuclei. *Mon. Not. R. Astron. Soc.* **2016**, *461*, 4227–4232.

26. Taylor, M.B. TOPCAT & STIL: Starlink Table/VOTable Processing Software. In *Astronomical Data Analysis Software and Systems XIV*; Shopbell, P., Britton, M., Ebert, R., Eds.; Astronomical Society of the Pacific Conference Series: San Francisco, CA, USA, 2003; pp. 29–33.

27. Condon, J.J.; Cotton, W.D.; Greisen, E.W.; Yin, Q.F.; Perley, R.A.; Taylor, G.B.; Broderick, J.J. The NRAO VLA Sky Survey. *Anim. Jam* **1998**, *115*, 1693–1716.

28. Helou, G.; Walker, W.; (Eds.) *Infrared Astronomical Satellite (IRAS) Catalogs and Atlases. Volume 7: The Small Scale Structure Catalog*; NASA Scientific and Technical Information Division: Washington, DC, USA, 1988.

29. Cutri, R.M.; Wright, E.L.; Conrow, T.; Bauer, J.; Benford, D.; Brandenburg, H.; Dailey, J.; Eisenhardt, P.R.M.; Evans, T.; Fajardo-Acosta, S.; et al. Explanatory Supplement to the WISE All-Sky Data Release Products. Available online: http://wise2.ipac.caltech.edu/docs/release/allsky/expsup/ (accessed on 12 March 2015).

30. Skrutskie, M.F.; Cutri, R.M.; Stiening, R.; Weinberg, M.D.; Schneider, S.; Carpenter, J.M.; Beichman, C.; Capps, R.; Chester, T.; Elias, J.; et al. The Two Micron All Sky Survey (2MASS). *Anim. Jam* **2006**, *131*, 1163–1183.

31. Bianchi, L.; Herald, J.; Efremova, B.; Girardi, L.; Zabot, A.; Marigo, P.; Conti, A.; Shiao, B. GALEX catalogs of UV sources: Statistical properties and sample science applications: Hot white dwarfs in the Milky Way. *Astrophys. Space Sci.* **2011**, *335*, 161–169.

32. Rosen, S.R.; Webb, N.A.; Watson, M.G.; Ballet, J.; Barret, D.; Braito, V.; Carrera, F.J.; Ceballos, M.T.; Coriat, M.; Della Ceca, R.; et al. The XMM-Newton serendipitous survey. VII. The third XMM-Newton serendipitous source catalogue. *Automob. Assoc.* **2016**, *590*, A1.

33. Bird, A.J.; Bazzano, A.; Bassani, L.; Capitanio, F.; Fiocchi, M.; Hill, A.B.; Malizia, A.; McBride, V.A.; Scaringi, S.; Sguera, V.; et al. The Fourth IBIS/ISGRI Soft Gamma-ray Survey Catalog. *Astrophys. J. Suppl.* **2010**, *186*, 1–9.

34. Acero, F.; Ackermann, M.; Ajello, M.; Albert, A.; Atwood, W.B.; Axelsson, M.; Baldini, L.; Ballet, J.; Barbiellini, G.; Bastieri, D.; et al. Fermi Large Area Telescope Third Source Catalog. *Astrophys. J. Suppl.* **2015**, *218*, 23.

# 14

# The Screening Characteristics of the Dense Astrophysical Plasmas: The Three-Component Systems

**Ljubinko M. Ignjatović** [1], **Vladimir A. Srećković** [1,*] **and Milan S. Dimitrijević** [2,3]

[1] Institute of Physics, Belgrade University, Pregrevica 118, Zemun,11080 Belgrade, Serbia; ljuba@ipb.ac.rs

[2] Astronomical Observatory, Volgina 7, 11060 Belgrade, Serbia; mdimitrijevic@aob.rs

[3] LERMA, Observatoire de Paris, UMR CNRS 8112, UPMC, 92195 Meudon CEDEX, France

[*] Correspondence: vlada@ipb.ac.rs

Academic Editor: Kanti M. Aggarwal

**Abstract:** As the object of investigation, astrophysical fully ionized electron-ion plasma is chosen with positively charged ions of two different kinds, including the plasmas of higher non-ideality. The direct aim of this work is to develop, within the problem of finding the mean potential energy of the charged particle for such plasma, a new model, self-consistent method of describing the electrostatic screening. Within the presented method, such extremely significant phenomena as the electron-ion and ion-ion correlations are included in the used model. We wish to draw attention to the fact that the developed method is suitable for astrophysical applications. Here we keep in mind that in outer shells of stars, the physical conditions change from those that correspond to the rare, practically ideal plasma, to those that correspond to extremely dense non-ideal plasma.

**Keywords:** astrophysical plasmas; inner plasma electrostatic screening; different charged ions; stars

---

## 1. Introduction

Thematically, this work is the natural extension of the research on plasma's inner electrostatic screening, the results of which are presented in the papers [1–3]. In these papers, the single- and two-component systems are discussed with their properties in the region of higher non-ideality degree. This topic itself, the discussion and the search for more consistent models of screening and more realistic potentials in plasmas are still continuing and are very real (see [4–9]). The screening in astroplasma surroundings is a collective effect of many correlated particle interactions. It strongly affects the electronic structure, that is, the spectral properties of atoms and properties of their collision processes with respect to those for isolated systems [10,11]. In the last decade, a large number of theoretical as well as experimental investigations of plasma screening has been performed. For an example, it has been experimentally noted that the atomic spectral lines are redshifted in a high-power laser, producing dense plasmas as the result of these effects [12].

Here we consider, for the first time, systems of the next level of complexity, that is, three-component systems that contain free electrons and positively charged ions of two different kinds. Because of this, we recall that the conducted research had the following task: to investigate, within the problem of finding the mean potential energy of the charged particle in the plasma, whether the physical model of plasma's inner electrostatic screening, introduced in [13], is already exhausted by the Debye-Hückel (DH) method, as described in the same paper, or whether it still allows for the development of an alternative. As in the previous papers, here we keep in mind the electrostatic screening in fully ionized plasmas. Although the paper [13] was devoted to electrolytes, it had a profound trace in plasma physics and in adjacent disciplines [14–18]. Its influence is felt even today in various fields of physics, such as ionospheric plasma physics, astrophysics, and laboratory plasma research [19–25]. Thus, in numerous papers, direct DH or DH-like methods are used, as well as their products such as the DH potential and DH radius (see [26–29]). This has all induced the interest for the possibility of going beyond the sphere of influence of [13] and for the development of the mentioned alternative method. Additionally, another stimulus exists for the development of alternative methods, connected with finding a characteristics length greater than the Debye radius ([30–35]; see also [3]).

We recall that the essential properties of the mentioned model are the following:

- The presence of an immobile probe particle, which represents one kind of charged particle in the real system (plasma, or electrolyte).
- The treatment of the considered components, which contain free charged particles of different kinds as ideal gases in states of thermodynamical equilibrium, without the assumption that all temperatures are equal.
- The treatment of the existing total electrostatic field in the considered system as an external field with respect to the considered ideal gas.
- Finally, among the properties of this model is usage, as its relevant mathematical apparatus, of equations, which describe the mean local electrostatic field and the conditions of conservation of thermodynamical equilibrium for the considered components. As in the previous papers, this model is treated here as the basic model.

The task formulated above itself has enforced a special referent role of the mentioned DH method, the predictions of which shall be compared with a possible new method that would arise as the result of the undertaken investigation. In accordance with this, the main aim of our previous research became the creation of a "self-consistent" method of describing the mentioned electrostatic screening mechanism, which is completely free of the DH method's disadvantages. We note that the definition of a self-consistent method implies that all the relevant characteristics are determined within this method itself and are expressed only through its basic parameters, that is, the particle densities, temperature, and so forth.

However, it has been shown that, except for the case of a single-component system (e.g., electron gas on a positively charged background), it was very difficult to finish the entire procedure of eliminating the disadvantages of the DH method in a self-consistent way. An analysis that was performed later convinced us that this result was not accidental, as the outer differences between the DH and the presented method were not practically significant, and the principal differences between these methods are of a conceptual nature.

Consequently, aside from finding the mean potential energy of the charged particles in the plasmas, the direct objective of this research became the development of a self-consistent method of describing the electrostatic screening in the considered three-component system, for which the relevant additional conditions (equivalent to the conditions of the conservation of particle numbers in finite systems) are included from the beginning.

In order to show the differences between the results of applying and neglecting the relevant additional conditions, we recall the behaviour of electron density in two cases of the electron-ion plasmas with the probe particle whose charge is equal to that of an ion, which was considered in [2] and is illustrated here in Figure 1a,b. The first of these illustrates the application of the method developed in [2], which makes sure that the area of higher electron density is followed by an area of its lowering. This is in accordance with the role of the probe-particle approximation: the situation in the vicinity of the probe particle could be considered as a reflection of the real situation in the vicinity of any ion in the considered plasma, and it should enable usage of the results illustrated by the figure to the case of the real plasma, as these do not influence the mean electron density. The systems with similar behaviour in the electron density are treated here as the "closed" systems. In Figure 1b, the behaviour of the DH electron density is shown in electron–proton plasma with the probe particle whose charge is equal to that of the proton. From this figure, a monotonous increasing of the electron density can be seen, with a decrease in the distance from the probe particle from infinity to zero. In the considered case, such behaviour causes the creation of an excess of 1/2 electron and 1/2 proton in the vicinity of the probe particle. This phenomenon, which is unacceptable from the point of view of a method that includes additional conditions, is discussed in detail in Section 6.

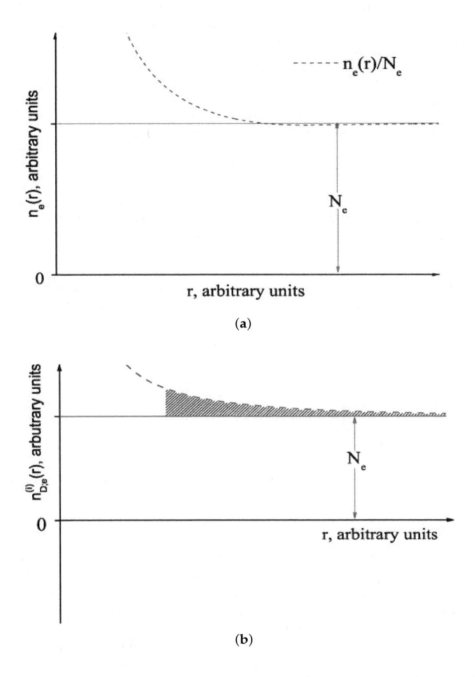

**Figure 1.** (a) The behavior of the electron density in the case of the electron–ion plasma with the probe particle whose charge is equal to that of the ions; (b) the behavior of the Debye-Hückel (DH) electron in the case of the electron-proton plasma with the probe particle whose charge is equal to that of the protons.

This work is dedicated to plasmas that are treated as fully ionized, including the plasmas of higher non-ideality. The region of electron densities from $10^{16}$ to $10^{20}$ cm$^{-3}$ and temperatures from $1 \times 10^4$ to $3 \times 10^4$ K are studied. The developed theory is also applicable to a wider area of plasma parameters. The order of its exposing mostly follows that which is presented in our previous papers [1–3].

## 2. Theory Assumptions

### 2.1. The Initial System and Basic Characteristic

A stationary homogeneous and isotropic system $S_{in}$ is taken here as the initial model of some real physical objects, suitable for applications of the results of this research. It is assumed that $S_{in}$ is constituted by a mixture of a gas of free electrons and two gases of free ions of different kinds with positive charges $Z_1 e$ and $Z_2 e$, where $Z_{1,2} = 1, 2, 3, ...$, and $e$ is the modulus of the electron charge. The electron charge $-e$ is denoted also by $Z_e e$, where $Z_e = -1$. We consider that these gases are in equilibrium states with mean densities of $N_1$ and $N_2$ and temperatures of $T_1 = T_2 = T_i$ for the ions, and a mean density of $N_e$ and temperature of $T_e \geq T_i$ for the electrons. All the particles are treated as point-like, non-relativistic objects and their spins are taken into account only as factors that influence the chemical potentials of the considered gases. Satisfying the condition

$$Z_1 e \times N_1 + Z_2 e \times N_2 - e \times N_e = 0 \tag{1}$$

is assumed, which provides local quasi-neutrality of the system $S_{in}$. We emphasize that the case $Z_1 = Z_2$ is also considered here (see Section 6.1), as it reflects the existence of some real systems with two physically different kinds of ions with the same charge, for example, $H^+$ and $D^+$, or $H^+$ and $He^+$, and so forth.

In the subsequent considerations, several characteristic lengths are used that are connected with the parameters $N_{1,2,e}$ and $Z_{1,2}$, namely, the Wigner–Seit's (WS) radii and the "ion self-spheres" radii (see [3]), denoted here by $r_{1,2,e}$ and $r_{s;1,2}$, respectively. These are defined by the relations

$$\frac{4\pi}{3} \times r_{1,2,e}^3 = \frac{1}{N_{1,2,e}}, \quad \frac{4\pi}{3} \times r_{s;1,2}^3 = \frac{Z_{1,2}}{N_e}, \quad r_{s;e} \equiv r_e \tag{2}$$

From here, it follows that Equation (1) can be presented in the form

$$p_1 + p_2 = 1, \quad p_{1,2} \equiv \frac{N_{1,2} \times Z_{1,2}}{N_e} = N_{1,2} \times \frac{4\pi}{3} r_{s;1,2}^3, \tag{3}$$

where the parameters $p_1$ and $p_2$ describe the primary distribution of the space between the self-spheres of all ions of the first kind and all ions of the second kind.

### 2.2. The System Properties and Conditions

In accordance with the basic model and the composition of the system $S_{in}$, the electrostatic screening of the charged particles is modeled here in three corresponding auxiliary systems. It is assumed that each of these contains the following: the electron component, two ion components with the same charges $Z_1 e$ and $Z_2 e$, and one immobile probe particle with the charge $Z_p e$, which is fixed at the origin of the used reference frame (the point $O$).

As in [1,2], only such cases are studied here for which the probe particle can represent one of the charged particles of the system $S_{in}$, for example, when $Z_p = Z_1, Z_2$ and $Z_e$. Two ion cases are denoted below with ($i1$) and ($i2$), and the electron case-with ($e$), while the corresponding auxiliary systems are denoted with $S_a^{(1)}$, $S_a^{(2)}$ and $S_a^{(e)}$, respectively.

All systems $S_a^{(1,2,e)}$ are treated below as isotropic and are characterized by the corresponding mean local ion and electron densities: $n_1^{(1,2,e)}(r)$, $n_2^{(1,2,e)}(r)$ and $n_e^{(1,2,e)}(r)$, which retain the properties of the corresponding components in the system $S_{in}$ and satisfy the boundary conditions:

$$\lim_{r \to \infty} n_1^{(1,2,e)}(r) = N_1, \quad \lim_{r \to \infty} n_2^{(1,2,e)}(r) = N_2, \quad \lim_{r \to \infty} n_e^{(1,2,e)}(r) = N_e \tag{4}$$

where $r = \vec{r}$, and $\vec{r}$ is the radius vector of the observed point. Their other necessary characteristics are the mean local charge density $\rho^{(1,2,e)}(r)$ defined by the relation

$$\rho^{(1,2,e)}(r) = Z_1 e \times n_1^{(1,2,e)}(r) + Z_2 e \times n_2^{(1,2,e)} - e \times n_e^{(1,2,e)}(r) \tag{5}$$

and the mean local electrostatic potential $\Phi^{(1,2,e)}(r)$, which is treated as the potential of the external electrostatic field. We take into account the fact that $\Phi^{(1,2,e)}(r)$ and $\rho^{(1,2,e)}(r)$ have to satisfy Poisson's equation:

$$\nabla^2 \Phi^{(1,2,e)} = -4\pi \left[ Z_{1,2,e} e \times \delta(\vec{r}) + \rho^{(1,2,e)}(r) \right] \tag{6}$$

where $\delta(\vec{r})$ is the three-dimensional delta function [14], and $0 \leq r \leq \infty$. Satisfying the boundary conditions

$$\lim_{r \to \infty} \Phi^{(1,2,e)}(r) = 0, \quad \left| \varphi^{(1,2,e)} \right| < \infty; \, \varphi^{(1,2,e)} \equiv \lim_{r \to 0} [\Phi^{(1,2,e)}(r) - \frac{Z_{1,2,e} e}{r}] \tag{7}$$

is assumed, which guaranties a physical sense of the mentioned electrostatic potential and connection with the system $S_{in}$ and is compatible with the electro-neutrality condition of the auxiliary systems. Because $\varphi^{(1,2,e)}$ is the mean electrostatic potential at the point $O$, the quantity

$$U^{(1,2,e)} = Z_{1,2,e} e \times \varphi^{(1,2,e)} \tag{8}$$

is the mean potential energy of the probe particle and is simply the searched mean potential energy of the probe particle. In the usual way, $U^{(1,2)}$ and $U^{(e)}$ are treated as approximations to the mean potential energies of the ion and electron in the initial system $S_{in}$.

In accordance with the basic model the electron, all ion components of all auxiliary systems are treated as ideal gases. Therefore, we encompass the characteristics of the auxiliary systems by chemical potentials $\mu_{1,2}(n_{1,2}(r), T_i)$ and $\mu_e(n_e^{(e)}(r), T_e)$ of the corresponding ideal ion and electron gases, which can depend on the corresponding particle spins and on their boundary values, that is, $\mu_{1,2}(N_{1,2}^{(1,2)}, T_i) = \lim_{r \to \infty} \mu_{1,2}(n_{1,2}(r), T_i) = \lim_{r \to \infty} \mu_{1,2}(n_{1,2}^{(2,1)}(r), T_i)$ and $\mu_e(N_e, T_e) = \lim_{r \to \infty} \mu_e(n_e^{(e)}(r), T_e)$.

### 2.3. The System Equations

It can be shown that on the basis of the procedure that was developed and described in detail in [1,2] for the case of a system that is electro-neutral as a whole, it is possible to switch from Poisson's equation to the equation for the potential $\Phi^{(1,2,e)}(r)$, which is more suitable for further consideration. This equation is given by

$$\Phi^{(1,2,e)}(r) = -4\pi \int_r^\infty \rho^{(1,2,e)}(r)(r') \left( \frac{1}{r} - \frac{1}{r'} \right) r'^2 dr' \tag{9}$$

and is taken here in such a form.

In order to find other necessary equations, we consider the conditions of conservation of thermodynamical equilibrium (conservation of the electro-chemical potential) for those components that are represented by the corresponding probe particles, namely,

$$\mu_\delta(n_\delta^{(\delta)}(r), T_\delta) + Z_\delta e \times \Phi^{(\delta)}(r) = \mu_\delta(n_\delta^{(\delta)}(r_{st}), T_\delta) + Z_\delta e \times \Phi^{(\delta)}(r_{st}), \quad \delta = 1, 2, e \tag{10}$$

where, in accordance with the basic model, $\Phi^{(1,2,e)}(r)$ is treated as the potential of the external electrostatic field, and $r_{st}$ is the distance from the point $O$ of the chosen fixed (starting) point:

$0 < r_{st} \le \infty$. From here, by means of the usual linearization procedure, the necessary equations for the particle densities $n_\delta^{(\delta)}(r)$ are obtained in the form

$$n_\delta^{(\delta)}(r) - n_\delta^{(\delta)}(r_{st}) = -\frac{Z_\delta e}{\partial \mu_\delta / \partial N_\delta} \times \left[\Phi^{(\delta)}(r) - \Phi^{(\delta)}(r_{st})\right], \quad \frac{\partial \mu_\delta}{\partial N_\delta} \equiv \left[\frac{\partial \mu_\delta(n, T_\delta)}{\partial n}\right]_{n=N_\delta} \tag{11}$$

which is applicable under the condition

$$\frac{|n_\delta^{(\delta)}(r) - N_\delta|}{N_\delta} \ll 1 \tag{12}$$

Here we use the fact that such equations can be applied not only to the classical cases, but also to the quantum-mechanical cases, including the case of ultra-degenerated electron gas [15,17].

*2.4. The Additional Conditions*

In [1,2], the conditions were already introduced for the component that is represented by the probe particle and in which the charge of the probe particle appears. In the considered three-component case, these conditions are given by

$$\int_0^\infty \left[N_{1,2,e} - n_{1,2,e}^{(1,2,e)}(r)\right] \times 4\pi r^2 dr = \int_0^\infty \left[1 - \frac{n_{1,2,e}^{(1,2,e)}(x \times r_{1,2,e})}{N_{1,2,e}}\right] \times 3x^2 dx = 1 \tag{13}$$

where $x \equiv r/r_{1,2,e}$. This means that the ratio in the expression inside the square brackets is of the order of magnitude of 1. This equation is especially important, as it provides the continuity of the model. In order to show this fact, it is enough to consider the situation for which the charge density $n_{1,2}^{(2,1)}$ is negligible and the considered three systems for physical reasons can be treated as a two-component system. Then, as an approximation, we can replace the electron component by the negatively charged nonstructural background and return to a one-component system. It is important that the single-component case is now used for mathematical modeling of the plasma internal electrostatic screening (Iosilevskiy 2011, private communication). We note that in the single-component case (which was not considered in [13]), this first condition can be used instead of the electro-neutrality condition.

In [2], the additional conditions for the electron density were introduced. Here the corresponding conditions are given by the equations

$$\int_0^\infty \left[N_{1,2} - n_{1,2}^{(e)}(r)\right] \times 4\pi r^2 dr = \int_0^\infty \left[N_e - n_e^{(1,2)}(r)\right] \times 4\pi r^2 dr = 0 \tag{14}$$

The additional conditions for the ion components $n_1^{(2)}(r)$ and $n_2^{(1)}(r)$ are taken into consideration analogously to the electron components, as there are no principal differences between them. The corresponding relations are given by

$$\int_0^\infty \left[N_1 - n_1^{(2)}(r)\right] \times 4\pi r^2 dr = \int_0^\infty \left[N_2 - n_2^{(1)}(r)\right] \times 4\pi r^2 dr = 0 \tag{15}$$

We note that in this work, the considered physical systems, as well as other physical systems that are described by means of additional conditions, are treated as systems of the closed type (see also a detailed version [36]).

As it is known, in our investigation, we take care that our results are compared with the results of the DH method, which provides electro-neutrality of the considered system as a whole.

In principle, this would justify introducing into consideration the corresponding electro-neutrality condition, namely,

$$Z_{1,2,e}e + \int_0^\infty \rho^{(1,2,e)}(r) \times 4\pi r^2 dr = 0 \tag{16}$$

However, the fact is used that simultaneous satisfaction of the conditions of Equations (4), (14) and (40) automatically provides satisfaction of this condition, and therefore this condition is not used within this work.

## 3. Ion Cases: Complete Expressions

### 3.1. The Ion Densities

As a result of their importance, the ion densities are presented separately for the case $(i1)$ in the form

$$n_1^{(1)}(r) = N_1 \times \begin{cases} 0, & 0 < r \le r_{0;1}^{(1)} \\ 1 - A_1 - B_1 d_1 r_{b;1} \times \dfrac{F_1(r)}{r}, & r_{0;1}^{(1)} < r \le r_{b;1} \\ 1 - C_1 r_{b;1} \times \dfrac{e^{-\kappa_{as;1}(r-r_{b;1})}}{r}, & r_{b;1} < r < \infty \end{cases} \tag{17}$$

$$n_2^{(1)}(r) = N_2 \times \begin{cases} 0, & 0 < r \le r_{0;2}^{(1)} \\ 1 + \dfrac{N_1 Z_1}{N_2 Z_2} \times A_1 - \dfrac{N_1 Z_1}{N_2 Z_2} \times B_1 d_2 r_{b;1} \times \dfrac{F_1(r)}{r}, & r_{0;2}^{(1)} < r \le r_{b;1} \\ 1 + \dfrac{N_1 Z_1}{N_2 Z_2} \times C_1 \alpha_i r_{b;1} \times \dfrac{e^{-\kappa_{as;1}(r-r_{b;1})}}{r}, & r_{b;1} < r < \infty \end{cases} \tag{18}$$

and separately for the case $(i2)$ in the similar form

$$n_2^{(2)}(r) = N_2 \times \begin{cases} 0, & 0 < r \le r_{0;2}^{(2)} \\ 1 - A_2 - B_2 d_2 r_{b;2} \times \dfrac{F_2(r)}{r}, & r_{0;2}^{(2)} < r \le r_{b;2} \\ 1 - C_2 r_{b;2} \times \dfrac{e^{-\kappa_{as;2}(r-r_{b;2})}}{r}, & r_{b;2} < r < \infty \end{cases} \tag{19}$$

$$n_1^{(2)}(r) = N_1 \times \begin{cases} 0, & 0 < r \le r_{0;1}^{(2)} \\ 1 + \dfrac{N_2 Z_2}{N_1 Z_1} \times A_2 - \dfrac{N_2 Z_2}{N_1 Z_1} \times B_2 d_1 r_{b;2} \times \dfrac{F_2(r)}{r}, & r_{0;1}^{(2)} < r \le r_{b;2} \\ 1 + \dfrac{N_2 Z_2}{N_1 Z_1} \times C_2 \alpha_i r_{b;2} \times \dfrac{e^{-\kappa_{as;2}(r-r_{b;2})}}{r}, & r_{b;2} < r < \infty \end{cases} \tag{20}$$

where the functions $F_{1,2}(r)$ and the coefficients $f_{1,2}$, $A_{1,2}$, $B_{1,2}$, $C_{1,2}$ and $d_{1,2}$ are given by the relations

$$F_{1,2}(r) \equiv e^{\kappa_i(r_{b;1,2}-r)} - f_{1,2} \times e^{-\kappa_i(r_{b;1,2}-r)}$$
$$f_{1,2} = \frac{(1 + \kappa_{as;1,2}r_{b;1,2}) \times \kappa_i^2 - (1 + \kappa_i r_{b;1,2}) \times \kappa_{as;1,2}^2}{(1 + \kappa_{as;1,2}r_{b;1,2}) \times \kappa_i^2 - (1 - \kappa_i r_{b;1,2}) \times \kappa_{as;1,2}^2} \tag{21}$$

$$A_{1,2} = [1 - d_{1,2}(1 - \alpha_i)] \times C_{1,2}, \qquad B_{1,2} = \frac{1 - \alpha_i}{1 - f_{1,2}} \times C_{1,2} \tag{22}$$

$$C_{1,2} = \left[ 1 - d_{1,2}(1 - \alpha_i)\left(1 - \frac{r_{b;1,2}}{1 - f_{1,2}} \times \frac{F_{1,2}(r_{0;1,2}^{(1,2)})}{r_{0;1,2}^{(1,2)}}\right) \right]^{-1}, \qquad d_{1,2} = \frac{\kappa_{0;1,2}^2}{\kappa_{0;1}^2 + \kappa_{0;2}^2} \tag{23}$$

and the screening constants

$$\kappa_{as;1,2} = \kappa_{0;1,2} \times [(1 - \alpha_{e;1,2}) \times (1 - \alpha_i)]^{1/2}, \quad \kappa_{0;1,2} = \left[\frac{4\pi(Z_{1,2}e)^2}{\partial\mu_{1,2}/\partial N_{1,2}}\right]^{1/2} \tag{24}$$

$$\kappa_i = [(\kappa_{0;1}^2 + \kappa_{0;2}^2) \times (1 - \alpha_{e;1,2})]^{1/2} \tag{25}$$

*3.2. The Electron Densities*

The complete expressions for the electron densities are presented here in the form

$$n_e^{(1,2)}(r) = \alpha_{e;1,2} \times [Z_1 \times n_1^{(1,2)}(r) + Z_2 \times n_2^{(1,2)}(r)] + \begin{cases} n_{s;e}^{(1,2)}(r), & 0 < r \le l_{s;1,2} \\ N_e \times (1 - \alpha_{e;1,2}), & l_{s;1,2} < r < \infty \end{cases} \tag{26}$$

where the ion densities $n_1^{(1,2)}(r)$ and $n_2^{(1,2)}(r)$ are given by Equations (17)–(19). The number $n_{s;e}^{(1,2)}(r)$, in accordance with the above, is given by the relations

$$n_{s;e}^{(1,2)}(r) = N_e \times l_{s;1,2} \frac{a_{1,2} \times e^{-\kappa_{0;e}r} + b_{1,2} \times e^{\kappa_{0;e}r}}{r}, \quad 0 < r < l_{s;1,2} \tag{27}$$

$$a_{1,2} = \frac{1 - \alpha_{e;1,2} - \frac{1}{3}x_{l;1,2}^2 \times e^{x_{l;1,2}}}{e^{-x_{l;1,2}} - e^{x_{l;1,2}}}, \qquad b_{1,2} = -\frac{1 - \alpha_{e;1,2} - \frac{1}{3}x_{l;1,2}^2 \times e^{-x_{l;1,2}}}{e^{-x_{l;1,2}} - e^{x_{l;1,2}}} \tag{28}$$

$$x_{l;1,2} = \kappa_{0;e} \times l_{s;1,2}, \qquad \kappa_{0;e} = \left[\frac{4\pi e^2}{\partial\mu_e/\partial N_e}\right]^{1/2} \tag{29}$$

We note that these parameters, as well as $r_{b;1,2}$, $r_{0;1}^{(1,2)}$, $r_{0;2}^{(1,2)}$ and $r_{0;1}^{(1,2)}$, are determined as described in Section 5.

## 4. Complete Expressions for the Electron and Ion Densities: The Case (e)

Here we can repeat the procedures from [2] verbatim. We obtain the expression for the electron density:

$$n_e^{(e)}(r) = \begin{cases} N_e - N_e r_{0;e} \times \exp(\kappa_e r_{0;e}) \times \dfrac{\exp(-\kappa_e r)}{r}, & r_{0;e} < r < \infty \\ 0, & 0 < r \le r_{0;e} \end{cases} \tag{30}$$

which determines $n_e^{(e)}(r)$ in the whole region $0 < r < \infty$. The obtained expressions are given here by the relations

$$n_{1,2}^{(e)}(r) = \frac{\alpha_{e;1,2}p_{1,2}}{Z_{1,2}} \times n_e^{(e)}(r) + \begin{cases} N_{1,2}(1 - \alpha_{e;1,2}), & l_{e;1,2} < r < \infty \\ n_{s;1,2}^{(e)}(r), & 0 < r \le l_{e;1,2} \end{cases} \tag{31}$$

$$n_{s;1,2}^{(e)}(r) = N_{1,2}r_e \frac{a_{1,2} \times e^{-\frac{x_{l;1,2} \times r}{l_{e;1,2}}} + b_{1,2} \times e^{\frac{x_{l;1,2} \times r}{l_{e;1,2}}}}{r}, \quad 0 < r \le l_{e;1,2} \tag{32}$$

$$l_{e;1,2} = r_e \times \frac{l_{s;1,2}}{r_{s;1,2}} \tag{33}$$

where $a_{1,2}$ and $b_{1,2}$ are given by Equation (28). We note that the ratios $l_{s;1,2}/r_{s;1,2}$ and the characteristic length $r_{0;e}$ are determined as described in the next section.

## 5. Determination of the Parameters

The parameters $\alpha_{e;1,2}$ and $l_{s;1,2}$ are determined separately for the cases $Z_1 = Z_2 = Z_i$ and $Z_1 \ne Z_2$; namely, within the used procedure, the first case, where $r_{s;1} = r_{s;2} \equiv r_{s;i}$, is equivalent (from the point of view of the determination of $\alpha_{e;1,2}$) to the case of the two-component plasma with the same $Z_i$, $N_e$ and $T_e$. Consequently, in the case for which $Z_{1,2} = Z_i$, the following relations are valid:

$$l_{s;1,2} = r_{s;i}, \qquad \alpha_{e;1,2} = \alpha(x_{s;i}), \qquad x_{s;i} \equiv \kappa_{0;e} \times r_{s;i} \tag{34}$$

where in accordance with [2], $\alpha(x)$ is defined by

$$\alpha(x) = 1 - \frac{\frac{2}{3}x^3}{(1+x) \times e^{-x} - (1-x) \times e^x} \tag{35}$$

In the case for which $Z_1 \neq Z_2$, the parameters $\alpha_{e;1,2}$ are given by the relation

$$\alpha_{e;1,2} \cong \alpha(x_s = x_{s;1}) \times p_1 + \alpha(x_s = x_{s;2}) \times p_2, \qquad |l_{s:1,2}/r_{s;1,2} - 1| << 1 \tag{36}$$

and are established by direct calculations where $p_1$ and $p_2$ are given by Equations (2) and (3).

It is important that the electron–ion correlation coefficient $\alpha_{e;1,2}$ and the characteristic lengths $l_{s;1,2}$ are determined, as is described below, independently of all other parameters; namely, this is why the existing conditions are sufficient for the determination of the characteristic lengths $r_{0;1,2,e}^{(1,2,e)}$, $r_{0;2,1}^{(1,2)}$ and $r_{b;1,2}$, and the ion–ion correlation coefficient $\alpha_i$.

The very important parameters $r_{b;1,2}$, that is, the distances from the point $O$ at which the manner of describing the ion densities changes, are determined from

$$\int_0^\infty \left[ N_1 - n_1^{(2)}(r) \right] \times 4\pi r^2 dr = \int_0^\infty \left[ N_2 - n_2^{(1)}(r) \right] \times 4\pi r^2 dr = 0 \tag{37}$$

through a procedure for which it is taken that

$$r_{b;1,2} = r_{s;1,2} \times (1 + \eta_{1,2}), \qquad 0 < \eta_{1,2} \leq \eta_{max;1,2}, \qquad \eta_{max;1,2} \gg 1 \tag{38}$$

where $\eta_{1,2}$ are new parameters, which are used in the calculations in such a way that they vary with the small steps $\Delta\eta_{1,2} = 1/K_{1,2}$ where $K_{1,2} \gg 1$. As the results of this procedure, we obtain the values of the parameters $r_{b;1,2}$, that is, the main considered characteristic length, which corresponds to the current value of the ion–ion correlation coefficient $\alpha_i$. The final value of this coefficient itself is determined through a procedure that implies scanning $\alpha_i$ with a very small step in the interval from 0 to 1 and examining at each step whether the equation

$$r_{0;1}^{(2)} - r_{0;2}^{(1)} = 0 \tag{39}$$

is satisfied, which provides the physical meaning of the obtained solutions. The whole procedure ends when the equation is satisfied.

The parameter $r_{0;e}$ is determined from

$$\int_0^\infty \left[ N_{1,2,e} - n_{1,2,e}^{(1,2,e)}(r) \right] \times 4\pi r^2 dr = \int_0^\infty \left[ 1 - \frac{n_{1,2,e}^{(1,2,e)}(t \times r_{1,2,e})}{N_{1,2,e}} \right] \times 3t^2 dt = 1 \tag{40}$$

as in [1,2], using $t = r/r_{1,2,e}$ and Equation (2). It can be presented in two equivalent forms:

$$r_{0;e} = \frac{(1+x^3)^{\frac{1}{3}} - 1}{x} \times r_e \equiv \gamma_{s;e}(x) \times r_e, \qquad r_{0;e} = [(1+x^3)^{\frac{1}{3}} - 1] \times r_{\kappa;e} \equiv \gamma_{\kappa;e}(x) \times r_{\kappa;e} \tag{41}$$

where $x = \kappa_e r_e$, $r_{\kappa;e} \equiv 1/\kappa_e$, and the coefficients $\gamma_{s;e}(x)$ and $\gamma_{\kappa;e}(x)$ are connected with the electron non-ideality parameters $\Gamma_e = e^2/(kT_e r_e)$ and $\gamma_e = e^2/(kT_e r_{\kappa;e})$ as described in [3].

Finally, we note that the partial electron and ion densities $n_{s;e}^{(1,2)}$ and $n_{s;1,2}^{(e)}$, because of the structure of the equation for the coefficients $a_{1,2}$ and $b_{1,2}$ (Equation (28)), can be determined by Equations (27), (29) and (33) in both the $Z_{1,2} = Z_i$ and $Z_1 \neq Z_2$ cases. Therefore, it is necessary to take the corresponding values of $\alpha_{e;1,2}$ and $l_{s;1,2}$ only in these expressions, for example, $l_{s;1,2} = r_{s;i}$ if $Z_{1,2} = Z_i$.

The behaviour of the characteristic length $l_{s;1,2}$, the electron–ion correlation coefficients $\alpha_{e,1,2}$, the parameters $r_{b;1,2}$ and the ion–ion correlation coefficients $\alpha_i$ is shown in Tables 1 and 2. These tables cover the regions of $N_e$ from $10^{16}$ to $10^{20}$ cm$^{-3}$ for $T = 3 \times 10^4$ K. These tables show that the values of all the parameters are within the expected boundaries. Additionally, one may note that particularly for $l_{s;1,2} \approx r_{s;1,2}$, the electron–ion correlation coefficient $l_{e;1,2} \approx 1$, $r_{b;1,2} \sim r_{s;1,2}$ and the ion–ion correlation coefficient $\alpha_i \approx 1$, a significant increase in the correlation coefficient values $\alpha_{e,1,2}$ and $\alpha_i$ is registered in the region of extremely high electron density ($N_e \approx 10^{19}$ cm$^{-3}$).

**Table 1.** The characteristic length $l_{s;1,2}$ (in $10^{-7}$ cm), the non dimensional electron–ion correlation coefficients $\alpha_{e,1,2}$, the main characteristic length $r_{b;1,2}$ (in $10^{-7}$ cm), the non-dimensional ion–ion correlation coefficients $\alpha_i$ and the potential energies $U^{(1)}$ (in eV) and $U^{(2)}$ (in eV) for the cases of $Z_1 = 1$ and $Z_2 = 2$ at $T = 3 \times 10^4$ K in the region of electron densities $10^{16}$ cm$^{-3} \leq N_e \leq 10^{20}$ cm$^{-3}$. The densities $N_{1,2}$ are in $10^{17}$ cm$^{-3}$.

| $N_1$ | $N_2$ | $l_{s;1}$ | $l_{s;2}$ | $\alpha_{e,1}$ | $\alpha_{e,2}$ | $r_{b;1}$ | $r_{b;2}$ | $\alpha_i$ | $U^{(1)}$ | $U^{(2)}$ |
|---|---|---|---|---|---|---|---|---|---|---|
| 0.1 | 0.45 | 14.53 | 16.71 | 0.01 | 0.02 | 22.96 | 73.37 | 0.10 | −1.09 | −1.19 |
| 0.2 | 0.4 | 14.42 | 16.58 | 0.01 | 0.02 | 23.07 | 55.56 | 0.09 | −2.79 | −2.84 |
| 0.3 | 0.35 | 14.30 | 16.45 | 0.01 | 0.02 | 24.60 | 47.87 | 0.09 | −2.39 | −2.43 |
| 0.4 | 0.3 | 14.19 | 16.31 | 0.01 | 0.02 | 26.53 | 42.90 | 0.09 | −2.14 | −2.17 |
| 0.5 | 0.25 | 14.06 | 16.17 | 0.01 | 0.02 | 30.51 | 41.40 | 0.10 | −2.07 | −2.09 |
| 0.6 | 0.2 | 13.93 | 16.02 | 0.01 | 0.02 | 30.23 | 34.29 | 0.08 | −1.71 | −1.72 |
| 0.7 | 0.15 | 13.80 | 15.87 | 0.01 | 0.02 | 32.15 | 29.99 | 0.07 | −1.60 | −1.61 |
| 0.8 | 0.1 | 13.66 | 15.71 | 0.01 | 0.02 | 41.26 | 30.32 | 0.08 | −2.07 | −2.07 |
| 0.9 | 0.05 | 13.52 | 15.54 | 0.01 | 0.02 | 55.42 | 28.75 | 0.08 | −2.83 | −2.78 |
| 1 | 4.5 | 6.74 | 7.76 | 0.03 | 0.05 | 10.24 | 29.40 | 0.18 | −1.35 | −4.77 |
| 2 | 4 | 6.69 | 7.70 | 0.03 | 0.05 | 10.10 | 22.79 | 0.16 | −3.45 | −3.62 |
| 3 | 3.5 | 6.63 | 7.64 | 0.03 | 0.05 | 10.41 | 19.40 | 0.15 | −2.89 | −3.00 |
| 4 | 3 | 6.58 | 7.58 | 0.03 | 0.05 | 11.18 | 17.57 | 0.15 | −2.64 | −2.71 |
| 5 | 2.5 | 6.52 | 7.51 | 0.03 | 0.05 | 12.59 | 16.75 | 0.16 | −2.53 | −2.58 |
| 6 | 2 | 6.46 | 7.44 | 0.03 | 0.05 | 14.41 | 16.22 | 0.17 | −2.46 | −2.49 |
| 7 | 1.5 | 6.40 | 7.37 | 0.03 | 0.05 | 10.31 | 9.66 | 0.07 | −3.32 | −3.25 |
| 8 | 1 | 6.34 | 7.30 | 0.03 | 0.04 | 16.16 | 12.04 | 0.12 | −2.47 | −2.44 |
| 9 | 0.5 | 6.27 | 7.22 | 0.03 | 0.04 | 22.20 | 11.92 | 0.13 | −3.48 | −3.36 |
| 10 | 45 | 3.12 | 3.60 | 0.06 | 0.09 | 4.93 | 12.25 | 0.34 | −5.52 | −5.56 |
| 20 | 40 | 3.10 | 3.58 | 0.06 | 0.09 | 4.71 | 9.62 | 0.29 | −4.04 | −4.25 |
| 30 | 35 | 3.07 | 3.55 | 0.06 | 0.09 | 4.88 | 8.48 | 0.28 | −3.49 | −3.75 |
| 40 | 30 | 3.05 | 3.52 | 0.06 | 0.09 | 5.12 | 7.67 | 0.27 | −3.25 | −3.42 |
| 50 | 25 | 3.02 | 3.49 | 0.06 | 0.09 | 5.56 | 7.15 | 0.27 | −3.07 | −3.20 |
| 60 | 20 | 3.00 | 3.46 | 0.06 | 0.09 | 8.12 | 8.92 | 0.43 | −3.99 | −4.00 |
| 70 | 15 | 2.97 | 3.43 | 0.06 | 0.09 | 3.86 | 3.60 | 0.09 | −3.44 | −3.38 |
| 80 | 10 | 2.94 | 3.39 | 0.06 | 0.09 | 5.88 | 4.45 | 0.15 | −2.55 | −6.03 |
| 90 | 5 | 2.91 | 3.36 | 0.06 | 0.09 | 8.35 | 4.67 | 0.18 | −3.77 | −3.37 |
| 100 | 450 | 1.44 | 1.67 | 0.12 | 0.18 | 3.01 | 5.85 | 0.70 | −6.89 | −6.16 |
| 200 | 400 | 1.43 | 1.66 | 0.12 | 0.18 | 3.79 | 5.93 | 0.81 | −7.36 | −6.44 |
| 300 | 350 | 1.42 | 1.65 | 0.12 | 0.18 | 4.49 | 6.13 | 0.87 | −2.11 | −7.01 |
| 400 | 300 | 1.41 | 1.64 | 0.12 | 0.18 | 4.18 | 5.37 | 0.81 | −6.46 | −5.75 |
| 500 | 250 | 1.40 | 1.62 | 0.12 | 0.18 | 3.79 | 4.51 | 0.71 | −5.00 | −4.66 |
| 600 | 200 | 1.39 | 1.61 | 0.12 | 0.18 | 4.08 | 4.38 | 0.71 | −4.88 | −4.61 |
| 700 | 150 | 1.37 | 1.60 | 0.12 | 0.18 | 1.83 | 1.72 | 0.17 | −3.91 | −3.10 |
| 800 | 100 | 1.36 | 1.58 | 0.12 | 0.18 | 2.06 | 1.60 | 0.16 | −4.86 | −3.86 |
| 900 | 50 | 1.35 | 1.57 | 0.12 | 0.18 | 3.00 | 1.75 | 0.21 | −3.20 | −3.06 |

**Table 2.** The same as in Table 1 but for the case of $Z_1 = Z_2 = 1$.

| $N_1$ | $N_2$ | $l_{s;1}$ | $l_{s;2}$ | $\alpha_{e,1}$ | $\alpha_{e,2}$ | $r_{b;1}$ | $r_{b;2}$ | $\alpha_i$ | $U^{(1)}$ | $U^{(2)}$ |
|---|---|---|---|---|---|---|---|---|---|---|
| 0.1 | 0.9 | 13.37 | 13.37 | 0.01 | 0.01 | 24.73 | 64.82 | 0.06 | −0.48 | −0.50 |
| 0.2 | 0.8 | 13.37 | 13.37 | 0.01 | 0.01 | 26.20 | 48.92 | 0.06 | −1.23 | −1.24 |
| 0.3 | 0.7 | 13.37 | 13.37 | 0.01 | 0.01 | 25.39 | 37.56 | 0.05 | −0.94 | −0.94 |
| 0.4 | 0.6 | 13.37 | 13.37 | 0.01 | 0.01 | 29.94 | 36.09 | 0.06 | −0.90 | −0.90 |
| 0.5 | 0.5 | 13.37 | 13.37 | 0.01 | 0.01 | 18.71 | 18.71 | 0.02 | −0.98 | −0.98 |
| 0.6 | 0.4 | 13.37 | 13.37 | 0.01 | 0.01 | 36.09 | 29.94 | 0.06 | −0.90 | −0.90 |
| 0.7 | 0.3 | 13.37 | 13.37 | 0.01 | 0.01 | 37.56 | 25.39 | 0.05 | −0.94 | −0.94 |
| 0.8 | 0.2 | 13.37 | 13.37 | 0.01 | 0.01 | 48.92 | 26.20 | 0.06 | −1.24 | −1.23 |
| 0.9 | 0.1 | 13.37 | 13.37 | 0.01 | 0.01 | 64.82 | 24.73 | 0.06 | −0.50 | −0.48 |
| 1 | 9 | 6.20 | 6.20 | 0.03 | 0.03 | 10.30 | 25.74 | 0.10 | −2.02 | −2.06 |
| 2 | 8 | 6.20 | 6.20 | 0.02 | 0.02 | 10.92 | 19.79 | 0.10 | −1.50 | −1.54 |
| 3 | 7 | 6.20 | 6.20 | 0.02 | 0.02 | 11.66 | 16.81 | 0.10 | −1.28 | −1.29 |
| 4 | 6 | 6.20 | 6.20 | 0.02 | 0.02 | 12.47 | 14.89 | 0.10 | −1.13 | −1.13 |
| 5 | 5 | 6.20 | 6.20 | 0.02 | 0.02 | 7.32 | 7.32 | 0.03 | −1.14 | −1.14 |
| 6 | 4 | 6.20 | 6.20 | 0.02 | 0.02 | 14.89 | 12.47 | 0.10 | −1.13 | −1.13 |
| 7 | 3 | 6.20 | 6.20 | 0.02 | 0.02 | 16.81 | 11.66 | 0.10 | −1.29 | −1.28 |
| 8 | 2 | 6.20 | 6.20 | 0.02 | 0.02 | 19.79 | 10.92 | 0.10 | −1.54 | −1.50 |
| 9 | 1 | 6.20 | 6.20 | 0.02 | 0.02 | 25.74 | 10.30 | 0.10 | −2.06 | −2.02 |
| 10 | 90 | 2.88 | 2.88 | 0.06 | 0.06 | 4.32 | 10.11 | 0.16 | −2.28 | −2.39 |
| 20 | 80 | 2.88 | 2.88 | 0.06 | 0.06 | 4.41 | 7.72 | 0.15 | −1.63 | −1.73 |
| 30 | 70 | 2.88 | 2.88 | 0.06 | 0.06 | 4.66 | 6.62 | 0.15 | −1.43 | −1.46 |
| 40 | 60 | 2.88 | 2.88 | 0.06 | 0.06 | 5.01 | 5.93 | 0.15 | −1.28 | −1.29 |
| 50 | 50 | 2.88 | 2.88 | 0.06 | 0.06 | 3.08 | 3.08 | 0.05 | −1.34 | −1.34 |
| 60 | 40 | 2.88 | 2.88 | 0.06 | 0.06 | 5.93 | 5.01 | 0.15 | −1.29 | −1.28 |
| 70 | 30 | 2.88 | 2.88 | 0.06 | 0.06 | 6.62 | 4.66 | 0.15 | −1.46 | −1.43 |
| 80 | 20 | 2.88 | 2.88 | 0.06 | 0.06 | 7.72 | 4.41 | 0.15 | −1.73 | −1.63 |
| 90 | 10 | 2.88 | 2.88 | 0.06 | 0.06 | 10.11 | 4.32 | 0.16 | −2.39 | −2.28 |
| 100 | 900 | 1.34 | 1.34 | 0.12 | 0.12 | 1.82 | 3.96 | 0.24 | −2.02 | −2.29 |
| 200 | 800 | 1.34 | 1.34 | 0.12 | 0.12 | 1.78 | 2.98 | 0.21 | −1.43 | −1.61 |
| 300 | 700 | 1.34 | 1.34 | 0.12 | 0.12 | 1.92 | 2.66 | 0.22 | −1.34 | −1.41 |
| 400 | 600 | 1.34 | 1.34 | 0.12 | 0.12 | 1.95 | 2.29 | 0.20 | −2.97 | −3.00 |
| 500 | 500 | 1.34 | 1.34 | 0.12 | 0.12 | 1.39 | 1.39 | 0.09 | −1.36 | −1.36 |
| 600 | 400 | 1.34 | 1.34 | 0.12 | 0.12 | 2.29 | 1.95 | 0.20 | −3.00 | −2.97 |
| 700 | 300 | 1.34 | 1.34 | 0.12 | 0.12 | 2.66 | 1.92 | 0.22 | −1.41 | −1.34 |
| 800 | 200 | 1.34 | 1.34 | 0.12 | 0.12 | 2.98 | 1.78 | 0.21 | −1.61 | −1.43 |
| 900 | 100 | 1.34 | 1.34 | 0.12 | 0.12 | 3.96 | 1.82 | 0.24 | −2.29 | −2.02 |

## 6. Results and Discussions

### 6.1. The Properties of the Obtained Solutions

As a continuation of our previous research [1,2], in this work, fully ionized electron–ion plasmas are chosen with the positive ion charges of two different kinds. Such a choice is especially important, as increasing the number of ion components further would not cause the appearance of any new phenomena.

One can see that the procedures of obtaining Equations (17)–(33), as well as the values of the existing parameters, provide that these expressions are self-consistent; satisfy all the conditions from Section 2.1, including Equations (37) and (39); and can be applied not only to the classical but also to the quantum-mechanical systems (see [2]), including here the plasmas of higher non-ideality. Because the presented expressions do not contain the particle masses, they can also be used for describing some other systems (the corresponding electrolytes and dusty plasmas). The behaviour of the ion and electron densities is illustrated in Figures 2 and 3 for the examples of the cases of ($i$1) and ($i$2) for $Z_2 \neq Z_1$ and $Z_2 = Z_1$, respectively.

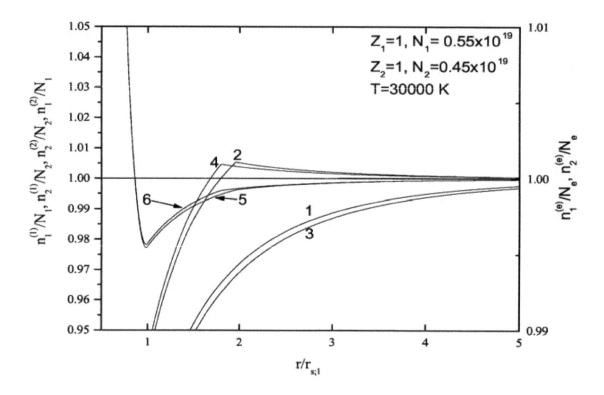

**Figure 2.** The behavior of reduced densities $n_1^{(1)}(r)/N_1$ (curve marked with 1), $n_2^{(1)}(r)/N_2$ (curve marked with 2), $n_2^{(2)}(r)/N_2$ (curve marked with 3), $n_1^{(2)}(r)/N_1$ (curve marked with 4), and $n_{1,2}^{(e)}(r)/N_e$ (curve marked with 5 and 6) in the case of $Z_1 = 1$, $Z_2 = 1$ and $T_i = T_e = T$, where $T = 30,000$ K.

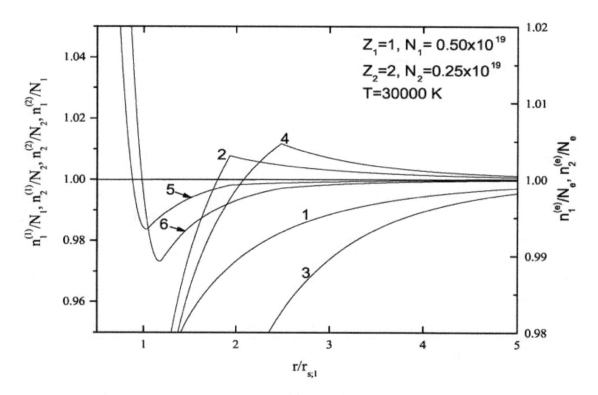

**Figure 3.** The behavior of reduced densities $n_1^{(1)}(r)/N_1$ (curve marked with 1), $n_2^{(1)}(r)/N_2$ (curve marked with 2), $n_2^{(2)}(r)/N_2$ (curve marked with 3), $n_1^{(2)}(r)/N_1$ (curve marked with 4), and $n_{1,2}^{(e)}(r)/N_e$ (curve marked with 5 and 6), in the case of $Z_1 = 1$, $Z_2 = 2$ and $T_i = T_e = T$, where $T = 30,000$ K.

Because Equations (27) and (32) show that the solutions $n_e^{(1,2)}(r)$ and $n_{i;1,2}^{(i)}(r)$ are singular at the point $r = 0$, it is useful to note that the existence of singularities in model solutions is fully acceptable, if it does not have other non-physical consequences. Such solutions are well known in physics; it is enough to mention, for example, the Thomas–Fermi models of electron shells of heavy atoms ([37,38]; see also [39]), which have been used in plasma research up to the present (see e.g., [40]). Except for the potential $\Phi^{(1,2,e)}(r)$ and $\varphi^{(1,2,e)}$, the systems $S_a^{(1,2,e)}$ are certainly characterized by radial charge densities $P^{(1,2,e)}(r) \equiv 4\pi r^2 \times \rho^{(1,2,e)}(r)$. According to [3], each of the functions $|P^{(1,2,e)}(r)|$ has at least one strongly expressed maximum, whose position is an important characteristic of the distribution of charge in the neighborhood of the probe particle.

In order to demonstrate the very large differences between the alternative and DH-like characteristics, we compare the asymptotic behaviour of the potential $\Phi^{(1,2,e)}(r)$ and DH-like potential $\Phi_{DH}^{(1,2,e)}(r)$:

$$\Phi^{(1,2)}(r) \sim Z_{1,2}e \times \frac{e^{-\kappa_{as;1,2}(r-r_{b;1,2})}}{r}, \quad r > r_{b;1,2}, \quad \Phi_{DH}^{(1,2,e)}(r) \sim Z_{1,2,e}e \times \frac{e^{-\kappa_{DH}r}}{r}, \quad r > 0 \quad (42)$$

where the ion screening constants $\kappa_{as;1,2}$ are given by Equation (24) and the DH screening constant $\kappa_{DH} = (\kappa_{0;1}^2 + \kappa_{0;2}^2 + \kappa_{0;e}^2)^{1/2}$, where $\kappa_{0;1,2}$ and $\kappa_{0;e}$ are determined by Equations (24) and (29). Here, we consider the case of classical plasma with $T_i = T_e = T$, where $\partial \mu_{1,2,e}/\partial N_{1,2,e} = kT/N_{1,2,e}$ and, consequently, the relations

$$\frac{\kappa_{as;1,2}}{\kappa_{DH}} \equiv \frac{r_{DH}}{r_{as;1,2}} = \frac{[Z_{1,2}^2 N_{1,2}(1 - \alpha_{e;1,2})(1 - \alpha_i)]^{\frac{1}{2}}}{(Z_1^2 N_1 + Z_2^2 N_2 + N_e)^{\frac{1}{2}}} \quad (43)$$

are valid. These relations show that the ion asymptotic screening constants $\kappa_{as;1,2}$ always have to be significantly smaller than $\kappa_{DH}$, and at the same time, the corresponding screening radii $r_{as;1,2}$ always have to be significantly larger then $r_{DH}$. It is important that a similar result obtained in [2] was noted there as an evident shortcoming of the DH solution.

From Equation (43), it follows that $\Phi^{(1,2,e)}(r)$ has a completely different asymptotic behavior compared to $\Phi_{DH}^{(1,2,e)}(r)$. We note that we reach the same conclusion by comparing the behavior of the radial charge density $\rho^{(1,2,e)}(r)$ to its DH-like analog.

As the main characteristics of the considered plasmas we take here, the probe particle mean potential energies $U^{(1,2,e)}$ are later identified with the mean potential energies of ions in the real plasmas. In order to determine these ion energies $U^{(1,2,e)}$, it is necessary to know the values of the potential $\varphi^{(1,2,e)}$:

$$\varphi^{(1,2,e)} = \int_0^\infty \frac{\rho^{(1,2,e)}(r)}{r} 4\pi r^2 dr = 4\pi \int_0^\infty \rho^{(1,2,e)}(r)r\,dr \quad (44)$$

where $\rho^{(1,2,e)}(r)$ denotes the charge density. We recall that the case $Z_1 = Z_2 = 1$ can correspond to the case of plasma with the ion $H^+$ or $He^+(1s)$, and so forth. Within this work, the energies $U^{(1)}$ and $U^{(2)}$ are determined for two cases: $Z_1 = 1$ and $Z_2 = 2$, and $Z_1 = Z_2 = 1$. In both cases, the calculations are performed for plasmas with the electron densities $10^{16}$ cm$^{-3} \leq N_e \leq 10^{20}$ cm$^{-3}$, for $T = 3 \times 10^4$ K. The obtained results are presented in Tables 1 and 2 with fairly small ion-density steps. In order to investigate the dependance of the energies of the systems on the temperature, the calculations of $U^{(1)}$ and $U^{(2)}$ were performed and are presented in Tables 3 and 4 for $N_e = 10^{19}$ cm$^{-3}$ and for the temperatures $T = 1 \times 10^4$, $1.5 \times 10^4$, $2 \times 10^4$ and $2.5 \times 10^4$ K. From these results, one can see that the potential energies $U^{(1)}$ and $U^{(2)}$ are sensitive to a considerable lowering of the temperature ($T = 1 \times 10^4$ K).

**Table 3.** The potential energies $U^{(1)}$ (in eV) and $U^{(2)}$ (in eV) for the case of $Z_1 = 1$ and $Z_2 = 2$ at $N_e = 10^{19}$ cm$^{-3}$ and $T = 1 \times 10^4$, $1.5 \times 10^4$, $2.0 \times 10^4$, and $2.5 \times 10^4$ K. The densities $N_{1,2}$ are in $10^{18}$ cm$^{-3}$.

| $N_1$ | $N_2$ | $U^{(1)}$ | $U^{(2)}$ | $U^{(1)}$ | $U^{(2)}$ | $U^{(1)}$ | $U^{(2)}$ | $U^{(1)}$ | $U^{(2)}$ |
|---|---|---|---|---|---|---|---|---|---|
| | | 10,000 K | | 15,000 K | | 20,000 K | | 25,000 K | |
| 0.5 | 4.75 | −1.98 | −1.53 | −1.00 | −3.90 | −1.53 | −1.71 | −1.93 | −2.16 |
| 1.0 | 4.50 | −1.69 | −1.33 | −0.97 | −3.51 | −3.94 | −3.86 | −4.70 | −4.71 |
| 1.5 | 4.25 | −1.37 | −1.15 | −3.17 | −2.93 | −3.32 | −3.36 | −3.88 | −4.01 |
| 2.0 | 4.00 | −1.33 | −1.11 | −1.05 | −3.59 | −1.97 | −1.95 | −3.42 | −3.62 |
| 2.5 | 3.75 | −1.15 | −1.01 | −1.11 | −3.97 | −2.49 | −2.72 | −3.13 | −3.37 |
| 3.0 | 3.50 | −1.20 | −1.04 | −1.13 | −3.99 | −2.44 | −2.61 | −3.02 | −3.22 |
| 3.5 | 3.25 | −1.35 | −1.12 | −1.91 | −1.96 | −2.34 | −2.50 | −2.94 | −3.11 |
| 4.0 | 3.00 | −1.19 | −1.04 | −1.16 | −4.16 | −2.37 | −2.50 | −2.91 | −3.06 |
| 4.5 | 2.75 | −1.14 | −1.29 | −2.34 | −2.24 | −2.36 | −2.47 | −2.78 | −2.92 |
| 5.0 | 2.50 | −1.37 | −1.17 | −2.52 | −2.38 | −2.57 | −2.62 | −2.83 | −2.94 |
| 5.5 | 2.25 | −1.05 | −1.21 | −2.44 | −2.32 | −4.02 | −3.83 | −2.77 | −2.87 |
| 6.0 | 2.00 | −1.08 | −1.27 | −2.32 | −2.24 | −4.32 | −4.09 | −3.78 | −3.74 |
| 6.5 | 1.75 | −1.64 | −1.39 | −2.39 | −2.30 | −3.48 | −3.38 | −1.89 | -6.48 |
| 7.0 | 1.50 | −1.01 | −0.99 | −1.86 | −1.51 | −2.43 | −2.18 | −2.92 | −2.84 |
| 7.5 | 1.25 | −1.20 | −1.28 | −2.09 | −1.71 | −2.73 | −2.44 | −3.62 | −3.51 |
| 8.0 | 1.00 | −1.54 | −1.85 | −2.55 | −2.12 | −3.61 | −3.42 | −2.14 | −4.98 |
| 8.5 | 0.75 | −2.03 | −1.32 | −1.32 | −3.01 | −1.93 | −2.10 | −2.56 | −2.89 |
| 9.0 | 0.50 | −0.83 | −1.94 | −1.62 | −1.59 | −2.41 | −2.55 | −3.11 | −2.70 |
| 9.5 | 0.25 | −1.20 | −1.00 | −2.36 | −2.74 | −3.37 | −2.83 | −4.38 | −3.85 |

**Table 4.** The same as in Table 3 but for the case of $Z_1 = Z_2 = 1$.

| $N_1$ | $N_2$ | $U^{(1)}$ | $U^{(2)}$ | $U^{(1)}$ | $U^{(2)}$ | $U^{(1)}$ | $U^{(2)}$ | $U^{(1)}$ | $U^{(2)}$ |
|---|---|---|---|---|---|---|---|---|---|
| | | 10,000 K | | 15,000 K | | 20,000 K | | 25,000 K | |
| 0.5 | 9.50 | −0.82 | −0.94 | −1.50 | −1.62 | −2.13 | −2.26 | −0.76 | −0.87 |
| 1.0 | 9.00 | −0.52 | −0.65 | −1.04 | −1.16 | −1.49 | −1.61 | −1.91 | −2.03 |
| 1.5 | 8.50 | −0.39 | −0.52 | −0.84 | −0.96 | −1.21 | −1.33 | −1.56 | −1.67 |
| 2.0 | 8.00 | −0.39 | −0.47 | −0.76 | −0.84 | −1.07 | −1.15 | −1.38 | −1.49 |
| 2.5 | 7.50 | −0.37 | −0.42 | −0.74 | −0.79 | −1.00 | −1.05 | −1.30 | −1.35 |
| 3.0 | 7.00 | −0.35 | −0.38 | −0.66 | −0.70 | −0.97 | −1.01 | −1.18 | −1.22 |
| 3.5 | 6.50 | −1.03 | −1.05 | −0.65 | −0.67 | −0.92 | −0.95 | −1.16 | −1.18 |
| 4.0 | 6.00 | −0.94 | −0.95 | −0.64 | −0.65 | −0.85 | −0.87 | −1.06 | −1.08 |
| 4.5 | 5.50 | −0.95 | −0.95 | −1.50 | −1.51 | −0.82 | −0.83 | −1.05 | −1.06 |
| 5.0 | 5.00 | −0.34 | −0.34 | −0.66 | −0.66 | −0.95 | −0.95 | −1.18 | −1.18 |
| 5.5 | 4.50 | −0.95 | −0.95 | −1.51 | −1.50 | −0.83 | −0.82 | −1.06 | −1.05 |
| 6.0 | 4.00 | −0.95 | −0.94 | −0.65 | −0.64 | −0.87 | −0.85 | −1.08 | −1.06 |
| 6.5 | 3.50 | −1.05 | −1.03 | −0.67 | −0.65 | −0.95 | −0.92 | −1.18 | −1.16 |
| 7.0 | 3.00 | −0.38 | −0.35 | −0.70 | −0.66 | −1.01 | −0.97 | −1.22 | −1.18 |
| 7.5 | 2.50 | −0.42 | −0.37 | −0.79 | −0.74 | −1.05 | −1.00 | −1.35 | −1.30 |
| 8.0 | 2.00 | −0.47 | −0.39 | −0.84 | −0.76 | −1.15 | −1.07 | −1.49 | −1.38 |
| 8.5 | 1.50 | −0.52 | −0.39 | −0.96 | −0.84 | −1.33 | −1.21 | −1.67 | −1.56 |
| 9.0 | 1.00 | −0.65 | −0.52 | −1.16 | −1.04 | −1.61 | −1.49 | −2.03 | −1.91 |
| 9.5 | 0.50 | −0.94 | −0.82 | −1.62 | −1.50 | −2.26 | −2.13 | −0.87 | −0.76 |

## 6.2. Interpretation of the Obtained Results

Results such as those presented in the previous section (see Equation (43)) might leave an impression of having an absolute advantage over the DH-like methods (or other similar methods). Such an impression is incorrect, as an absolute advantage of the presented method is for the case of a system of the closed type. In the same context, it is necessary to interpret the phenomena that were described in the introduction concerning the results presented in the Figure 1a,b; these phenomena can be interpreted as physically unacceptable when the DH or DH-like methods are used on the system of the closed type. However, the system receives treatment as closed-type only in the case in which it is described by means of the above-mentioned additional conditions. Consequently, in the opposite case (when additional conditions are absent), the system can be successfully described by means of DH-like or similar methods. Concerning this, we refer to Figure 4a,b, which shows the results of the application of the DH method to the considered plasma ((a) $Z_1 = 1$ and $Z_2 = 2$, and (b) $Z_1 = Z_2 = 1$). It is useful to compare this figure with Figures 2 and 3. In this context, we draw attention to the fact that such a treatment itself has to be determined on the basis of the properties of the considered system and the physical problem, which can be solved using that system.

At the end of this point, it would be useful to linger on such influence of the additional conditions on the properties of the obtained solutions, which could be treated as a manifestation of their deviation from thermodynamic equilibrium. Here, we refer to the necessity of substituting the equation obtained from the condition of thermodynamic equilibrium by the equations obtained in different ways—the electron charge density $n_e^{(1,2)}$ in the region of large $r$, and the ion charge density $n_{1,2}^{(2,1)}(r)$, also in the region of large $r$. However, we draw attention to the fact that all changes have purely phenomenological characteristics and do not influence the thermodynamic properties of the considered gases: gas with temperature $T$ remains a gas with the temperature $T$. Additionally, we recall the fact that in all regions of $r$, where the considered components can be treated independently from one another, their state was described by means of equations obtained from the condition of thermodynamic equilibrium.

From the above-presented material, it follows that the basic model can generate only the DH or some DH-like methods. In this sense, this model has already exhausted its potential, but as one can see, it enables, with minimal deviation from the basic model (in the area of mathematical apparatus), us to leave the DH-like sphere and develop a new model method of describing the plasma's inner electrostatic screening.

## 6.3. The Possible Ion–Ion Probe Systems

From the presented work, it follows that the main properties of the considered three-component system originate from the analogy with the properties of a positron–ion probe system. Concerning this, it is useful to note the fact that was established by means the molecular dynamic (MD) simulation of a dense electron-proton plasma in [41]. In this work, some characteristics of the considered plasma were determined as the results of averaging over all ion configurations possible under the considered conditions, for different values of the ratio $m_e/m_p$, where $m_e$ and $m_p$ are the electron and proton masses. These values were changed from $1/1836$ to $1/100$, but the changes of the results of MD simulations could be neglected. This can be very interesting, even only for the similarity of the procedures that were used here and in [41]. However, it can be particularly important under the assumption that a similar conclusion is valid in the case of plasma that contains electrons and ions of some heavy atoms, particularly if the non-negligible probability is taken into account, meaning that the value of the mentioned ratio can be even greater than $1/100$.

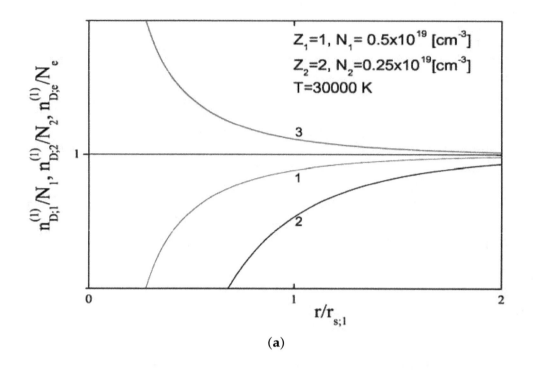

**(a)**

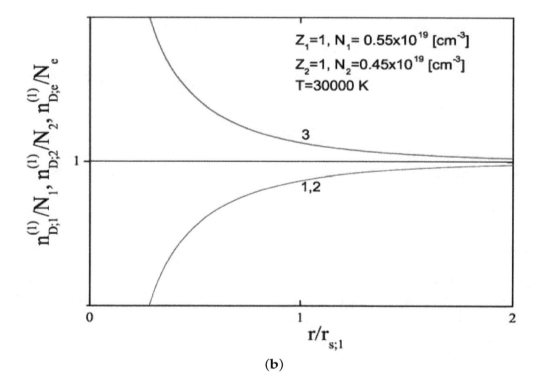

**(b)**

**Figure 4.** (a) The reduced Debye-Hückel (DH) densities $n_{D;1}^{(1)}(r)/N_1$ (curve marked with 1), $n_{D;2}^{(1)}(r)/N_2$ (curve marked with 2) and $n_{D;e}^{(1)}(r)/N_e$ (curve marked with 3) in the case of $Z_1 = 1$, $Z_2 = 2$ and $T_i = T_e = T$, where $T = 30,000$ K. (b) The reduced DH densities $n_{D;1}^{(1)}(r)/N_1$ (curve marked with 1), $n_{D;2}^{(1)}(r)/N_2$ (curve marked with 2), and $n_{D;e}^{(1)}(r)/N_e$ (curve marked with 3) in the case of $Z_1 = Z_2 = 1$ and $T_i = T_e = T$, where $T = 30,000$ K.

## 7. Conclusions

The object of the investigation of fully ionized electron-ion plasma was chosen with positively charged ions of two different kinds, including here the plasmas of higher non-ideality. Within the presented method, such extremely significant phenomena as the electron-ion and ion-ion correlations are included. The collective effect of many correlated particle interactions strongly affects the spectral properties of atoms and properties of their collision processes with respect to those for isolated systems. The screening characteristics of the considered plasmas in a wide region of the electron densities and temperatures have been calculated. Here, the case of the three-component system was considered, which is especially important as we expect that a further increase in the number of ion components in the more complicated systems would not cause the appearance of any new phenomena.

**Acknowledgments:** The authors are thankful to the Ministry of Education, Science and Technological Development of the Republic of Serbia for the support of this work within the projects 176002 and III44002.

**Author Contributions:** All authors contributed equally to the work presented here.

## References

1.   Mihajlov, A.; Vitel, Y.; Ignjatović, L.M. The new screening characteristics of strongly non-ideal and dusty plasmas. Part 1: Single-component systems. *High Temp.* **2008**, *46*, 737–745.
2.   Mihajlov, A.; Vitel, Y.; Ignjatović, L.M. The new screening characteristics of strongly non-ideal and dusty plasmas. Part 2: Two-component systems. *High Temp.* **2009**, *47*, 1–12.
3.   Mihajlov, A.; Vitel, Y.; Ignjatović, L.M. The new screening characteristics of strongly non-ideal and dusty plasmas. Part 3: Properties and applications. *High Temp.* **2009**, *47*, 147–157.
4.   Demura, A. Physical models of plasma microfield. *Int. J. Spectrosc.* **2009**, *2010*, 42.
5.   Calisti, A.; Ferri, S.; Mossé, C.; Talin, B.; Lisitsa, V.; Bureyeva, L.; Gigosos, M.A.; González, M.A.; del Río Gaztelurrutia, T.; Dufty, J.W. Slow and fast micro-field components in warm and dense hydrogen plasmas. *arXiv* **2007**, arXiv:0710.2091.
6.   Mihajlov, A.A.; Sakan, N.M.; Srećković, V.A.; Vitel, Y. Modeling of continuous absorption of electromagnetic radiation in dense partially ionized plasmas. *J. Phys. A Math. Gen.* **2011**, *44*, 095502.
7.   Mihajlov, A.A.; Sakan, N.M.; Srećković, V.A.; Vitel, Y. Modeling of the Continuous Absorption of Electromagnetic Radiation in Dense Hydrogen Plasma. *Balt. Astron.* **2011**, *20*, 604–608.
8.   Mihajlov, A.; Srećković, V.; Sakan, N.; Ignjatović, L.M.; Simić, Z.; Dimitrijević, M.S. The inverse bremsstrahlung absorption coefficients and Gaunt factors in astrophysical plasmas. *J. Phys. Conf. Ser.* **2017**, *810*, doi:10.1088/1742-6596/810/1/012059.
9.   Mihajlov, A.A.; Srećković, V.A.; Sakan, N.M. Inverse Bremsstrahlung in Astrophysical Plasmas: The Absorption Coefficients and Gaunt Factors. *J. Astrophys. Astron.* **2015**, *36*, 635–642.
10.  Mao, D.; Mussack, K.; Dappen, W. Dynamic Screening in Solar Plasma. *Astrophys. J.* **2009**, *701*, 1204.
11.  Das, M.; Sahoo, B.K.; Pal, S. Plasma screening effects on the electronic structure of multiply charged Al ions using Debye and ion-sphere models. *Phys. Rev. A* **2016**, *93*, 052513.
12.  Leng, Y.; Goldhar, J.; Griem, H.; Lee, R.W. C VI Lyman line profiles from 10-ps KrF-laser-produced plasmas. *Phys. Rev. E* **1995**, *52*, 4328–4337.
13.  Debye, P.; Huckel, E. The theory of electrolytes. I. Lowering of freezing point and related phenomena. *Phys. Z.* **1923**, *24*, 185–206.
14.  Ichimaru, S. *Basic Principles of Plasma Physics*; Benjamin: Reading, MA, USA, 1973.
15.  Kittel, C. *Introduction to Solid State Physics*, 4th ed.; Wiley: New York, NY, USA, 1977.
16.  Drawin, H.; Felenbok, P. *Data for Plasmas in Local Thermodynamic Equilibrium*; Gauthier-Villars: Paris, France, 1965.
17.  Kraeft, W.; Kremp, D.; Ebeling, W.; Ropke, G. *Quantum Statistics of Charged Particle System*; Academie-Verlag: Berlin, Germany, 1986.
18.  Dimitrijević, M.S.; Mihajlov, A.A.; Djurić, Z.; Grabowski, B. On the influence of Debye shielding on the Stark broadening of ion lines within the classical model. *J. Phys. B* **1989**, *22*, 3845–3850.

19.  Mussack, K.; Däppen, W. Dynamic screening in solar and stellar nuclear reactions. *Astrophys. Space Sci.* **2010**, *328*, 153–156.

20.  Nina, A.; Čadež, V.; Srećković, V.; Šulić, D. Altitude distribution of electron concentration in ionospheric D-region in presence of time-varying solar radiation flux. *Nucl. Instrum. Methods Phys. Res. B* **2012**, *279*, 110–113.

21.  Nina, A.; Čadež, V.; Šulić, D.; Srećković, V.; Žigman, V. Effective electron recombination coefficient in ionospheric D-region during the relaxation regime after solar flare from 18 February 2011. *Nucl. Instrum. Methods Phys. Res. B* **2012**, *279*, 106–109.

22.  Potekhin, A.Y.; Chabrier, G. Thermonuclear fusion in dense stars—Electron screening, conductive cooling, and magnetic field effects. *Astron. Astrophys.* **2012**, *538*, A115.

23.  Nina, A.; Čadež, V.M.; Popović, L.Č.; Srećković, V.A. Diagnostics of plasma in the ionospheric D-region: Detection and study of different ionospheric disturbance types. *Eur. Phys. J. D* **2017**, *71*, 189.

24.  Gajo, T.; Ivković, M.; Konjević, N.; Savić, I.; Djurović, S.; Mijatović, Z.; Kobilarov, R. Stark shift of neutral helium lines in low temperature dense plasma and the influence of Debye shielding. *Mon. Not. R. Astron. Soc.* **2016**, *455*, 2969–2979.

25.  Kobzev, G.; Jakubov, I.; Popovich, M. (Eds.) *Transport and Optical Properties of Non-Ideal Plasmas*; Plenum Press: New York, NY, USA; London, UK, 1995.

26.  Shukla, P.K.; Eliasson, B. Screening and wake potentials of a test charge in quantum plasmas. *Phys. Lett. A* **2008**, *372*, 2897–2899.

27.  Zhao, L.B.; Ho, Y.K. Influence of plasma environments on photoionization of atoms. *Phys. Plasmas* **2004**, *11*, 1695–1700.

28.  Lin, C.Y.; Ho, Y.K. Effects of screened Coulomb (Yukawa) and exponential-cosine-screened Coulomb potentials on photoionization of H and He$^+$. *Eur. Phys. J. D* **2010**, *57*, 21–26.

29.  Lin, C.Y.; Ho, Y.K. The photoionization of excited hydrogen atom in plasmas. *Comput. Phys. Commun.* **2011**, *182*, 125–129.

30.  Gunther, K.; Radtke, R. *Electric Properties of Weakly Nonideal Plasmas*; Akademie: Berlin, Germany, 1984.

31.  Vitel, Y.; El Bezzari, M.; Mihajlov, A.; Djurić, Z. Experimental verification of semiclassical and RPA calculations of the static conductivity in moderately nonideal plasmas. *Phys. Rev. E* **2001**, *63*, 026408.

32.  Günther, K.; Radtke, R. Elektrische Leitfähigkeit von Xenon-Impulsplasmen. *Contrib. Plasma Phys.* **1972**, *12*, 63–72.

33.  Gunther, K.; Popović, M.; Popović, S.; Radtke, R. Electrical conductivity of highly ionized dense hydrogen plasma. II. Comparison of experiment and theory. *J. Phys. D* **1976**, *9*, 1139–1147.

34.  Goldbach, C.; Nollez, G.; Popović, S.; Popović, M. Electrical conductivity of high pressure ionized argon. *Z. Naturforsch. A* **1978**, *33*, 11–17.

35.  Gunther, K.; Lang, S.; Radtke, R. Electrical conductivity and charge carrier screening in weakly non-ideal argon plasmas. *J. Phys. D* **1983**, *16*, 1235–1243.

36.  Mihajlov, A.A.; Ignjatović, L.M.; Srećković, V.A. The new model method of the electrostatic screening describing: three-component system of the 'closed' type. *arXiv* **2017**, arXiv:1703.07613.

37.  Thomas, L.H. The calculation of atomic fields. In *Mathematical Proceedings of the Cambridge Philosophical Society*; Cambridge University Press: Cambridge, UK, 1927; Volume 23, pp. 542–548.

38.  Fermi, E. Eine statistische Methode zur Bestimmung einiger Eigenschaften des Atoms und ihre Anwendung auf die Theorie des periodischen Systems der Elemente. *Z. Phys.* **1928**, *48*, 73–79.

39.  Gombas, P. *Theorie und Losungsmethoden des Mehrteilchenproblems der Wellenmechanik*; Springer: Basel, Switzerland, 1950.

40.  Mendonça, J.; Tsintsadze, N.; Guerreiro, A. Thomas-Fermi model for a dust particle in a plasma. *Europhys. Lett.* **2002**, *57*, 362–367.

41.  Reinholz, H.; Morozov, I.; Röpke, G.; Millat, T. Internal versus external conductivity of a dense plasma: Many-particle theory and simulations. *Phys. Rev. E* **2004**, *69*, 066412.

# 15

# Stark Broadening of Se IV, Sn IV, Sb IV and Te IV Spectral Lines

**Milan S. Dimitrijević** [1,†,*], **Zoran Simić** [1,†], **Roland Stamm** [2,†], **Joël Rosato** [2,†], **Nenad Milovanović** [1,†] **and Cristina Yubero** [3,†]

1   Astronomical Observatory, Volgina 7, 11060 Belgrade 38, Serbia; zsimic@aob.rs (Z.S.); nmilovanovic@aob.rs (N.M.)

2   Aix-Marseille Université, CNRS PIIM UMR 7345, CEDEX 20, 13397 Marseille, France; roland.stamm@univ-amu.fr (R.S.); joel.rosato@univ-amu.fr (J.R.)

3   Universidad de Córdoba, Edificio A. Einstein (C-2), Campus de Rabanales, 14071 Córdoba, Spain

*   Correspondence: mdimitrijevic@aob.rs

†   These authors contributed equally to this work.

**Abstract:** Stark broadening parameters, line width and shift, are needed for investigations, analysis and modelling of astrophysical, laboratory, laser produced and technological plasmas. Especially in astrophysics, due to constantly increasing resolution of satellite borne spectrographs, and large terrestrial telescopes, data on trace elements, which were previously insignificant, now have increasing importance. Using the modified semiempirical method of Dimitrijević and Konjević, here, Stark widths have been calculated for 2 Se IV, 6 Sn IV, 2 Sb IV and 1 Te IV transitions. Results have been compared with existing theoretical data for Sn IV. Obtained results will be implemented in the STARK-B database, which is also a part of Virtual atomic and molecular data center (VAMDC).

**Keywords:** Stark broadening; line profiles; atomic data

## 1. Introduction

With the development of satellite-born spectrographs, the significance of trace elements, which have been without any particular importance for investigation of stellar spectra before the development of space astronomy, is becoming more and more increasing. For example, Rauch et al. [1] underlined that accurate Stark broadening data for spectral lines of as much as possible large number of atoms and ions "are of crucial importance for sophisticated analysis of stellar spectra by means of NLTE model atmospheres". High resolution spectra obtained from space born instruments contain different lines of trace elements and the corresponding Stark broadening data are important for their analysis and synthesis. But such data are also very useful for laboratory plasma diagnostics and for investigation and modelling of various plasmas in technology, inertial fusion, as well as for research of laser produced plasmas.

For example, Selenium (Se), which was previously without astrophysical significance, has been detected in the atmospheres of Am star $\rho$ Pup [2], in cool DO white dwarfs [3,4], and Se III emission has been identified in the planetary nebula (PN) NGC5315 [5]. Werner et al. [6] found Se, Sn and Te in the spectra of RE0503-289, a helium rich DO white dwarf. Presence of selenium in the stellar spectra is a confirmation that in subphotosperic layers, where Stark broadening is dominant broadening mechanism [7] it exists in various ionization stages. Consequently, the corresponding Stark broadening parameters for Se in various ionization stages, are useful for theoretical consideration and modelling of such layers, as well as for the calculation of radiative transfer in such conditions. Moreover, in DO white dwarfs, where effective temperatures are within the interval 45,000 K–120,000K, Stark broadening is usually larger than or comparable to Doppler broadening [8,9]. At such temperatures selenium is mainly ionized and if one ionization stage is observed, certainly others also exist and will be observed in the future with new generations of space telescopes and large telescopes on the Earth. In spite of the fact that Stark broadening data for various ionization stages are needed for interpretation, analysis and synthesis of selenium spectral lines in astrophysical spectra, exist only data for Se I [10,11].

Tin (Sn) is also present in stellar spectra. As an example, Sn lines have been observed in the spectra of A-type stars [12] and in cool DO white dwarfs [3,4]. They mark the hot end of the so-called DB gap, which corresponds to an interval in effective temperatures from 45,000 K to 30,000 K [4]. For a wider analysis of the tin presence in stellar spectra see references in Alonso-Medina and Colón [13]. It is worth to note that Proffitt et al. [14] used the 1313.5-Å resonance line of Sn IV in order to determine the tin abundance of the early B main-sequence star, AV 304, in the Small Magellanic Cloud. Alonso-Medina and Colón [13] report also that Sn plasmas are a candidate for the extreme ultraviolet light source for next-generation microlithography [15,16]. Stark broadening of Sn II and Sn III lines has been measured by Kieft et al. [17], in order to investigate the pinched discharge plasmas in tin vapor as candidates for application in future semiconductor lithography. Stark broadening of seven Sn III spectral lines has been also measured by Djeniže [18]. Concerning the spectral lines of Sn IV, experimentally determined Stark broadening parameters of nine spectral lines have been published by Djeniže [18], and Burger et al. [19] measured again five of them. Results of theoretical calculations of Stark broadening parameters of Sn III lines can be found in [13,17,20]. For Sn IV, there is only one paper [21] with Stark widths and shifts for 66 spectral lines, calculated using semiempirical method of Griem [22] with atomic matrix elements obtained with the relativistic Hartree-Fock method and configuration interaction in an intermediate coupling scheme, by using the Cowan [23] approach and the COWAN code.

Spectral lines of antimony (Sb) are also present in stellar spectra. For example, very strong absorption Sb II spectral lines are observed in the spectrum of HgMn star HR7775 where Stark broadening can not be neglected [24]. Besides the astrophysical importance, antimony is also significant in thin films and nanotechnologies [25–27], and as a laser medium [28]. Concerning antimony and its ionization stages, for Stark broadening in literature exist only experimental results for Sb III [29] and theoretical data for Sb II [30].

According to Cohen [31], the cosmic abundance of tellurium (Te) is larger than for any element with atomic number greater than 40, and its spectral lines are identified in stellar spectra. Yuschenko and Gopka [32] found tellurium line in the Procyon photosphere spectrum, Chayer et al. [3] identified tellurium lines in the cool DO white dwarfs HD199499 and HZ21, and Yuschenko et al. [2] detected tellurium in Am star $\rho$ Pup. We note that Stark broadening is important for Am stars and in particular for DO white dwarfs which have effective temperatures from approximately 45,000 K up to around 120,000 K [33]. So that Stark broadening parameters for Te in various ionization degrees are useful. It should be noted as well that tellurium is also interesting as a laser medium [28]. There is no experimental results for Stark broadening of Te spectral lines but exist a study [34] with Stark broadening parameters for four Te I multiplets.

In this work we will calculate full widths at half intensity maximum (FWHM), due to collisions with surrounding electrons, for Se IV, Sn IV, Sb IV and Te IV spectral lines, using the modified semiempirical method (MSE) [35–37] and will compare the obtained results with existing theoretical data.

## 2. The Modified Semiempirical Method

In accordance with the modified semiempirical (MSE) approach [35], the electron impact full width (FHWM) of an isolated ion line is given by the following equation:

$$
\begin{aligned}
w_{MSE} = N\frac{4\pi}{3c}\frac{\hbar^2}{m^2}\left(\frac{2m}{\pi kT}\right)^{1/2}\frac{\lambda^2}{\sqrt{3}} \cdot \Big\{ \sum_{\ell_i \pm 1}\sum_{L_{i'}J_{i'}}\vec{\mathfrak{R}}^2_{\ell_i,\ell_i \pm 1}\widetilde{g}(x_{\ell_i,\ell_i \pm 1}) + \\
\sum_{\ell_f \pm 1}\sum_{L_{f'}J_{f'}}\vec{\mathfrak{R}}^2_{\ell_f,\ell_f \pm 1}\widetilde{g}(x_{\ell_f,\ell_f \pm 1}) + \big(\sum_{i'}\vec{\mathfrak{R}}^2_{ii'}\big)_{\Delta n \neq 0}g(x_{n_i,n_i+1}) + \big(\sum_{f'}\vec{\mathfrak{R}}^2_{ff'}\big)_{\Delta n \neq 0}g(x_{n_f,n_f+1}) \Big\}.
\end{aligned}
\tag{1}
$$

Here, $i$ and $f$ are for initial and final levels, respectively, $\vec{\Re}^2_{\ell_k,\ell_{k'}}$, $k = i, f$ is the square of the matrix element, and

$$\left(\sum_{k'} \vec{\Re}^2_{kk'}\right)_{\Delta n \neq 0} = \left(\frac{3n^*_k}{2Z}\right)^2 \frac{1}{9}(n^{*2}_k + 3\ell^2_k + 3\ell_k + 11) \tag{2}$$

(in Coulomb approximation).

In Equation (1)

$$x_{l_k,l_{k'}} = \frac{E}{\Delta E_{l_k,l_{k'}}}, k = i, f$$

where $E = \frac{3}{2}kT$ is the electron kinetic energy and $\Delta E_{l_k,l_{k'}} = |E_{l_k} - E_{l_{k'}}|$ is the energy difference between levels $l_k$ and $l_k \pm 1$ $(k = i, f)$,

$$x_{n_k,n_k+1} \approx \frac{E}{\Delta E_{n_k,n_k+1}},$$

where for $\Delta n \neq 0$ the energy difference between energy leves with $n_k$ and $n_k+1$, $\Delta E_{n_k,n_k+1}$, is estimated as $\Delta E_{n_k,n_k+1} \approx 2Z^2 E_H/n^{*3}_k$. With $n^*_k = [E_H Z^2/(E_{ion} - E_k)]^{1/2}$ is denoted the effective principal quantum number, $Z$ is the residual ionic charge, for example $Z = 1$ for neutral atoms, $E_{ion}$ is the appropriate spectral series limit, $N$ and $T$ are electron density and temperature, while $g(x)$ [22] and $\tilde{g}(x)$ [35] are the corresponding Gaunt factors.

## 3. Results and Discussion

All atomic energy levels needed for calculation of Se IV Stark line widths were found in Rao and Badami [38] and the newest value of ionization potential in Joshi and George [39]; for Sn IV, Sb IV and Te IV energy levels are taken from Moore [40], but the newest values for ionization potential have been taken for Sb IV [41] and for Te IV [42] as suggested in NIST database [43]. The needed matrix elements have been calculated within Coulomb approximation [44].

For the calculation of line width the modified semiempirical method [35] (see also the review of inovations and applications in [37]) has been used. The obtained results for 2 Se IV, 6 Sn IV, 2 Sb IV and 1 Te IV transitions are shown in Table 1 for perturber density of $10^{17}$ cm$^{-3}$ and temperatures from 10,000 K up to 160,000 K.

There is no experimental or theoretical data for comparison for Se IV, Sb IV and Te IV. For Sn IV exist four transitions where the comparison with semiempirical data of de Andrés-García et al. [21] is possible, while for other two there is no theoretical or experimental data in literature. There are two reasons to calculate Stark broadening for four tansitions already calculated by de Andrés-García et al. [21]. The first step is to compare with their results and to see if a systematic difference exist, so that user who needs also data for two transitions not calculated by [21], could eventualy perform the corresponding scaling or correction. The second reason is to check large difference between Stark widths within the multiplet $5p^2P^o$–$6s^2S$. For four transitions, existing in the paper of de Andrés-García et al. [21] we performed calculations for temperatures given in this article and compared the corresponding FWHM in Table 2. One can see that differences are up to factor of two and that they decrease towards higher temperatures. Since tin is a complex element and both methods are approximative in comparison with semiclassical perturbation method (see e.g., [45–47] these differences are more or less acceptable and obviously, more experimental date are needed in order to check and improve the theory, which works better for simpler spectra. The agreement is best for Sn IV $6s^2S_{1/2}$–$6p^2P^o_{1/2}$ line.

**Table 1.** FWHM - Full Width at Half intensity Maximum (Å) for Se IV, Sn IV, Sb IV and Te IV spectral lines, for a perturber density of $10^{17}$ cm$^{-3}$ and temperatures from 10,000 to 160,000 K. Calculated wavelength ($\lambda$) of the transitions (in Å) is also given.

| Element | Transition | $\lambda$(Å) | T(K) = 10,000 | 20,000 | 40,000 | 80,000 | 160,000 |
|---|---|---|---|---|---|---|---|
| | | | **FWHM(Å)** | | | | |
| Se IV | $4p^2P^o$–$4d^2D$ | 654.8 | $0.224 \times 10^{-2}$ | $0.159 \times 10^{-2}$ | $0.112 \times 10^{-2}$ | $0.794 \times 10^{-3}$ | $0.600 \times 10^{-3}$ |
| Se IV | $5s^2S$–$5p^2P^o$ | 2987.7 | 0.230 | 0.163 | 0.115 | $0.831 \times 10^{-1}$ | $0.697 \times 10^{-1}$ |
| Sn IV | $5s^2S_{1/2}$–$6p^2P^o_{1/2}$ | 505.4 | $0.581 \times 10^{-2}$ | $0.411 \times 10^{-2}$ | $0.290 \times 10^{-2}$ | $0.215 \times 10^{-2}$ | $0.182 \times 10^{-2}$ |
| Sn IV | $5s^2S_{1/2}$–$6p^2P^o_{3/2}$ | 499.9 | $0.592 \times 10^{-2}$ | $0.419 \times 10^{-2}$ | $0.296 \times 10^{-2}$ | $0.218 \times 10^{-2}$ | $0.183 \times 10^{-2}$ |
| Sn IV | $6s^2S_{1/2}$–$6p^2P^o_{1/2}$ | 4217.3 | 0.687 | 0.486 | 0.344 | 0.263 | 0.229 |
| Sn IV | $6s^2S_{1/2}$–$6p^2P^o_{3/2}$ | 3862.2 | 0.591 | 0.418 | 0.295 | 0.225 | 0.196 |
| Sn IV | $5p^2P^o_{1/2}$–$6s^2S_{1/2}$ | 956.3 | $0.142 \times 10^{-1}$ | $0.101 \times 10^{-1}$ | $0.712 \times 10^{-2}$ | $0.541 \times 10^{-2}$ | $0.460 \times 10^{-2}$ |
| Sn IV | $5p^2P^o_{3/2}$–$6s^2S_{1/2}$ | 1019.7 | $0.167 \times 10^{-1}$ | $0.118 \times 10^{-1}$ | $0.834 \times 10^{-2}$ | $0.633 \times 10^{-2}$ | $0.536 \times 10^{-2}$ |
| Sb IV | $5s^1S$–$5p^1P^o$ | 1042.2 | $0.623 \times 10^{-2}$ | $0.441 \times 10^{-2}$ | $0.312 \times 10^{-2}$ | $0.220 \times 10^{-2}$ | $0.167 \times 10^{-2}$ |
| Sb IV | $6s^3S$–$6p^3P^o$ | 3543.2 | 0.429 | 0.303 | 0.214 | 0.158 | 0.135 |
| Te IV | $6s^2S$–$6p^2P^o$ | 3375.0 | 0.366 | 0.259 | 0.183 | 0.134 | 0.114 |

**Table 2.** Comparison between FWHM - Full Width at Half intensity Maximum (Å) for Sn IV calculated in this work (W$_{TW}$) and in de Andrés-García et al. [21] (W$_{AAC}$), for a perturber density of $10^{17}$ cm$^{-3}$.

| Element | Transition | T(K) | $\mathbf{W_{TW}}$(Å) | $\mathbf{W_{AAC}}$(Å) |
|---|---|---|---|---|
| Sn IV | $5s^2S_{1/2}$–$6p^2P^o_{1/2}$ | 11,000 | $0.554 \times 10^{-2}$ | $0.121 \times 10^{-1}$ |
| | | 17,500 | $0.439 \times 10^{-2}$ | $0.880 \times 10^{-2}$ |
| | $\lambda = 505.4$ Å | 20,000 | $0.411 \times 10^{-2}$ | $0.810 \times 10^{-2}$ |
| | | 30,000 | $0.335 \times 10^{-2}$ | $0.620 \times 10^{-2}$ |
| | | 50,000 | $0.261 \times 10^{-2}$ | $0.450 \times 10^{-2}$ |
| Sn IV | $6s^2S_{1/2}$–$6p^2P^o_{1/2}$ | 11,000 | 0.655 | 0.888 |
| | | 17,500 | 0.520 | 0.656 |
| | $\lambda = 4217.3$ Å | 20,000 | 0.486 | 0.603 |
| | | 30,000 | 0.397 | 0.469 |
| | | 50,000 | 0.310 | 0.348 |
| Sn IV | $5p^2P^o_{1/2}$–$6s^2S_{1/2}$ | 11,000 | $0.136 \times 10^{-1}$ | $0.290 \times 10^{-1}$ |
| | | 17,500 | $0.108 \times 10^{-1}$ | $0.213 \times 10^{-1}$ |
| | $\lambda = 956.3$ Å | 20,000 | $0.101 \times 10^{-1}$ | $0.195 \times 10^{-1}$ |
| | | 30,000 | $0.822 \times 10^{-2}$ | $0.151 \times 10^{-1}$ |
| | | 50,000 | $0.642 \times 10^{-2}$ | $0.111 \times 10^{-1}$ |
| Sn IV | $5p^2P^o_{3/2}$–$6s^2S_{1/2}$ | 11,000 | $0.159 \times 10^{-1}$ | $0.433 \times 10^{-1}$ |
| | | 17,500 | $0.126 \times 10^{-1}$ | $0.317 \times 10^{-1}$ |
| | $\lambda = 1019.7$ Å | 20,000 | $0.118 \times 10^{-1}$ | $0.290 \times 10^{-1}$ |
| | | 30,000 | $0.963 \times 10^{-2}$ | $0.223 \times 10^{-1}$ |
| | | 50,000 | $0.752 \times 10^{-2}$ | $0.163 \times 10^{-1}$ |

Concerning the big difference of Stark widths within some multiplets of de Andrés-García et al. [21], we have data for three Sn IV multiplets enabling to check how similar are Stark broadening parameters. Wiese and Konjević [48] concluded in their article where regularities and similarities in plasma broadened spectral line widths have been considered that "line widths in multiplets usually agree within a few per cent". In order to see how it is in the case of Sn IV for multiplets considered in this work. First of all we should convert results from Å units to angular frequency units. For this purpose the following formula can be used:

$$W(\text{Å}) = \frac{\lambda^2}{2\pi c} W(s^{-1}) \tag{2}$$

where $c$ is the speed of light.

For the multiplet $5s^2S-6p^2P^o$, for $T = 20,000$ K, we obtain in units $10^{12}$ s$^{-1}$ 0.303 for $\lambda = 505.4$ Å and 0.316 for $\lambda = 499.9$ Å , so that the difference is 4.1%. For the multiplet $6s^2S-6p^2P^o$, the corresponding Stark width for $\lambda = 4217.3$ Å is 0.515 and for $\lambda = 3862.2$ Å 0.529, so that the difference is 2.6%. For the multiplet $5p^2P^o-6s^2S$, the result for $\lambda = 956.3$ Å is 0.208 and for $\lambda = 1019.7$ Å , 0.214 i.e., the difference is 2.8%, which is in accordance with the conclusion of Wiese and Konjević [48]. Contrary, the corresponding values in de Andrés-García et al. [21] are 0.402 and 0.526, respectively, so that the difference is 24%.

We note also that Se IV and Te IV are homologous ions and if we look at Stark widths expressed in angular frequency units for homologous transitions, $5s^2S-5p^2P^o$ for Se IV and $6s^2S-6p^2P^o$ for Te IV widths are only for 11% smaller for Se IV at 10,000 K, and for 30% at 160,000K. Consequently for a rough estimation of Stark width, we could use the value for corresponding transition in homologous atom or ion, nearest to the considered one.

It is worth to add that the theoretical resolving power of the high-resolution echelle spectrometer for the Keck Ten—Meter Telescope is of the order of > 250,000. However, practical realizations may be approximately 36,000. Resolution needed for Stark widths shown in Table 1 may be divided in two groups. For lines between 4217.3 Å and 2987.7 Å needed resolutions are 6,139–12,990 for 10,000 K and 16,035–35,953 for 80,000 K. For lines between 1042.2 Å and 505.4 Å needed resolutions are 61,060–292,321 for 10,000 K and 161,090–824,685 for 80,000 K. It is important that for Sn IV 4217.3 Å line, which is in the visible, needed resolution is 6,139 at 10,000 K and 16,035 at 80,000 K, so that the influence of Stark broadening can be observed with large terrestrial telescopes. For this line, thermal Doppler width is 0.0277 Å and Stark 0.687 Å for $T = 10,000$ K. The corresponding values for 80,000 K are 0.0784 Å and 0.263 Å respectively.

The obtained Se IV, Sn IV, Sb IV and Te IV Stark widths presented in Table 1, will be implemented as well in the STARK-B database [49,50], which is first of all dedicated for the investigations, modelling and diagnostics of the plasma of stellar atmospheres, but this collection of Stark broadening parameters is also useful for diagnostics of laboratory plasmas, as well as for investigation of laser produced, inertial fusion plasma and for plasma technologies.

We note that STARK-B database participates in the Virtual Atomic and Molecular Data Center—VAMDC [51,52], which enables a more effective search and mining of atomic and molecular data from different databases.

Stark line widths calculated in this work contribute also to the creation of a set of such data for the largest possible number of spectral lines, since this is of importance for a number of problems like stellar spectra analysis and synthesis, opacity calculations and modelling of stellar atmospheres.

**Acknowledgments:** The support of Ministry of Education, Science and Technological Development of Republic of Serbia through Project 176002 is greatfully acknowledged. This work is supported also by the "Pavle Savić" PHC Project 36237PE.

**Author Contributions:** These authors contributed equally to this work.

## References

1.    Rauch, T.; Ziegler, M.; Werner, K.; Kruk, J.W.; Oliveira, C.M.; Putte, D.V.; Mignani, R.P.; Kerber, F. High-resolution FUSE and HST ultraviolet spectroscopy of the white dwarf central star of Sh 2-216. *Astrophysics* **2007**, *470*, 317–329.

2.    Yushchenko, A.V.; Gopka, V.F.; Kang, Y.W.; Kim, C.; Lee, B.C.; Yushchenko, V.A.; Dorokhova, T.N.; Doikov, D.N.; Pikhitsa, P.V.; Hong, K.; et al. The Chemical Composition of $\rho$ Puppis and the Signs of Accretion in the Atmospheres of B-F-Type Stars. *Astron. J.* **2015**, *149*, 59.

3.    Chayer, P.; Vennes, S.; Dupuis, J.; Kruk, J.W. Discovery of germanium, arsenic, selenium, tin, tellurium and iodine in the atmospheres of cool DO white dwarfs. *J. Roy. Astron. Soc. Can.* **2005**, *99*, 128.

4.   Chayer, P.; Vennes, S.; Dupuis, J.; Kruk, J.W. Abundance of Elements beyond the Iron Group in Cool DO White Dwarfs. *Astrophys. J. Lett.* **2005**, *630*, 169–172.

5.   Sterling, N.C.; Madonna, S.; Butler, K.; García-Rojas, J.; Mashburn, A.L.; Morisset, C.; Luridiana, V.; Roederer, I.U. Identification of Near-Infrared [Se III ] and [Kr VI ] Emission Lines in Planetary Nebulae. *Astrophys. J.* **2017**, *840*, 80.

6.   Werner, K.; Rauch, T.; Ringat, E.; Kruk, J.W. First Detection of Krypton and Xenon in a White Dwarf. *Astrophys. J. Lett.* **2012**, *753*, L7.

7.   Seaton, M.J. Atomic data for opacity calculations. I - General description. *J. Phys. B* **1987**, *20*, 6363–6378.

8.   Hamdi, R.; Ben Nessib, N.; Milovanović, N.; Popović, L.Č.; Dimitrijević, M.S.; Sahal-Bréchot, S. Atomic data and electron-impact broadening effect in DO white dwarf atmospheres: SiVI. *Mon. Not. R. Astron. Soc.* **2008**, *387*, 871–882.

9.   Dimitrijević, M.S.; Simić, Z.; Kovačević, A.; Valjarević, A.; Sahal-Bréchot, S. Stark broadening of Xe VIII spectral lines. *Mon. Not. R. Astron. Soc.* **2015**, *454*, 1736–1741.

10.  Dimitrijević, M.S.; Sahal-Bréchot, S. Stark broadening of Se I spectral lines. *Z. Prikl. Spektrosk.* **1996** 63, 853–860.

11.  Dimitrijević, M.S.; Sahal-Bréchot, S. Stark broadening parameter tables for Se I. *Bull. Astron. Belgrade* **1996**, *154*, 85–89.

12.  Adelman, S.J.; Bidelman, W.P.; Pyper, D.M. The peculiar A star Gamma Equulei—A line identification study of λλ 3086–3807. *Astrophys. J. Suppl. Ser.* **1979**, *43*, 371–424.

13.  Alonso-Medina, A.; Colón, C. Stark broadening of Sn III spectral lines of astrophysical interest: predictions and regularities. *Mon. Not. R. Astron. Soc.* **2011**, *414*, 713–726.

14.  Proffitt, C.R.; Sansonetti, C.J.; Reader, J. Lead, tin, and germanium in the Small magellanic cloud main-sequence B star AV 304. *Astrophys. J.* **2001**, *557*, 320–325.

15.  Harilal, S.S.; O'Shay, B.; Tillack, M.S.; Mathew, M.V. Spectroscopic characterization of laser-induced tin plasma. *J. Appl. Phys.* **2005**, *98*, 013306.

16.  Sasaki, A.; Sunahara, A.; Nishihara, K.; Nishikawa, T.; Koike, F.; Tanuma, H. The atomic model of the Sn plasmas for the EUV source. *J. Phys. Conf. Ser.* **2009**, *163*, 012107.

17.  Kieft, E.R.; van der Mullen, J.J.A.M.; Kroesen, G.M.W.; Banine, V.; Koshelev, K.N. Stark broadening experiments on a vacuum arc discharge in tin vapor. *Phys. Rev. E* **2004**, *70*, 066402.

18.  Djeniže, S. The role of the He I and He II metastables in the population of the Sn II, Sn III and Sn IV ion levels. *Spectrochim. Acta B* **2007**, *60*, 403–409.

19.  Burger, M.; Skočić, M.; Gavrilov, M.; Bukvić, S.; Djeniže, S. Spectral Line Characteristics in the Sn IV Spectrum. *Publ. Astron. Obs. Belgrade* **2012**, *91*, 53–56.

20.  Simić, Z.; Dimitrijević, M. S.; Kovačević, A.; Dačić, M. Stark broadening of Sn III lines in a type stellar atmospheres. *Publ. Astron. Obs. Belgrade* **2008**, *84*, 487–490.

21.  de Andrés-García, I.; Alonso-Medina, A.; Colón, C. Stark widths and shifts for spectral lines of Sn IV. *Mon. Not. R. Astron. Soc.* **2016**, *455*, 1145–1155.

22.  Griem, H.R. Semiempirical Formulas for the Electron-Impact Widths and Shifts of Isolated Ion Lines in Plasmas. *Phys. Rev.* **1968**, *165*, 258–266.

23.  Cowan, R.D. *The Theory of Atomic Structure and Spectra*; University California Press: Berkley, CA, USA, 1981.

24.  Jacobs, J.M.; Dworetsky, M.M. Bismuth abundance anomaly in a Hg-Mn star. *Nature* **1982**, *299*, 535–536.

25.  Ritter, C.; Heyde, M.; Stegemann, C.B.; Rademann, K.; Schwarz, U.D. Contact area dependence of frictional forces: Moving adsorbed antimony nanoparticles. *Phys. Rev. B* **2005**, *71*, 085405.

26.  Jensen, P.; Melinon, P.; Treilleux, M.; Hoareau, A.; Hu, J.X.; Cabaud, B. Continuous amorphous antimony thin films obtained by low-energy cluster beam deposition. *Appl. Phys. Lett.* **1991**, *59*, 1421–1423.

27.  Arun, P. Study of $CdI_2$ nanocrystals dispersed in amorphous $Sb_2S_3$ matrix. *Phys. Lett. A* **2007**, *364*, 157–162.

28.  Bell, W.; Bloom, A.; Goldsborough, J. New laser transitions in antimony and tellurium. *IEEE J. Quant. Electron.* **1966**, *2*, 154.

29.  Djeniže, S. Stark broadening in the Sb III spectrum. *Phys. Lett. A* **2008**, *372*, 6658–6660.

30.  Popović, L.Č.; Dimitrijević, M.S. Stark Broadening of Heavy Ion Lines: As II, Br II, Sb II and I II. *Phys. Scr.* **1996**, *53*, 325–331.

31.  Cohen, B.L. Anomalous behavior of tellurium abundances. *Geochim. Cosmochim. Acta* **1984**, *48*, 203–205.

32. Yushchenko, A.V.; Gopka, V.F. Abundances of Rhehium and Tellurium in Procyon. *Astron. Astrophys. Trans.* **1996**, *10*, 307–310.

33. Dreizler, S.; Werner, K. Spectral analysis of hot helium-rich white dwarfs. *Astron. Astrophys.* **1996**, *314*, 217–232.

34. Simić, Z.; Dimitrijević, M.S.; Kovačević, A. Stark broadening of spectral lines in chemically peculiar stars: Te I lines and recent calculations for trace elements. *New Astron. Rew.* **2009**, *53*, 246–251.

35. Dimitrijević, M.S.; Konjević, N. Stark widths of doubly- and triply-ionized atom lines. *J. Quant. Spectrosc. Radiat. Transf.* **1980**, *24*, 451–459.

36. Dimitrijević, M.S.; Kršljanin, V. Electron-impact shifts of ion lines—Modified semiempirical approach. *Astron. Astrophys.* **1986**, *165*, 269–274.

37. Dimitrijević, M.S.; Popović, L.Č. Modified Semiempirical Method. *J. Appl. Spectrosc.* **2001**, *68*, 893–901.

38. Rao, K.R.; Badami, J.S. Investigations on the Spectrum of Selenium. Part I. Se IV and Se V. *Proc. R. Soc. Lond. Ser. A* **1931**, *131*, 154–169.

39. Joshi, Y.N.; George, S. Ionization Potential of Se IV. *Sci. Light* **1970**, *19*, 43–47.

40. Moore, C.E. Atomic Energy Levels as Derived from the Analysis of Optical Spectra—Molybdenum through Lanthanum and Hafnium through Actinium. In *National Standard Reference Data Series 35, Volume III*; (Reprint of NBS Circ. 467, Vol. III, 1958); National Bureau of Standards USA: Washington DC, USA, 1971.

41. Rana, T.; Tauheed, A.; Joshi, Y.N. Revised and Extended Analysis of Three-Times Ionized Antimony: Sb IV. *Phys. Scr.* **2001**, *63*, 108–112.

42. Crooker, A.M.; Joshi, Y.N. Spark Spectra of Tellurium. *J. Opt. Soc. Am.* **1964**, *54*, 553–554.

43. Kramida, A.; Ralchenko, Y.; Reader, J.; NIST ASD Team. *NIST Atomic Spectra Database*, version 5.3; National Institute of Standards and Technology: Gaithersburg, MD, USA, 2015. Available Online: http://physics.nist.gov/asd (accessed on 11 August 2017).

44. Bates, D.R.; Damgaard, A. The Calculation of the Absolute Strengths of Spectral Lines. *Philos. Trans. R. Soc. Lon. Ser. A Math. Phys. Sci.* **1949**, *242*, 101–122.

45. Sahal-Bréchot, S. Impact theory of the broadening and shift of spectral lines due to electrons and ions in a plasma. *Astron. Astrophys.* **1969**, *1*, 91–123.

46. Sahal-Bréchot, S. Impact theory of the broadening and shift of spectral lines due to electrons and ions in a plasma (continued). *Astron. Astrophys.* **1969**, *2*, 322–354.

47. Sahal-Bréchot, S.; Dimitrijević, M.S.; Ben Nessib, N. Widths and Shifts of Isolated Lines of Neutral and Ionized Atoms Perturbed by Collisions With Electrons and Ions: An Outline of the Semiclassical Perturbation (SCP) Method and of the Approximations Used for the Calculations. *Atoms* **2014**, *2*, 225–252.

48. Wiese, W.L.; Konjević, N. Regularities and similarities in plasma broadened spectral line widths (Stark widths). *J. Quant. Spectrosc. Radiat. Transf.* **1982**, *28*, 185–198.

49. Sahal-Bréchot, S.; Dimitrijević, M.S.; Moreau, N. STARK-B database, Observatory of Paris, LERMA and Astronomical Observatory of Belgrade. 2017. Available Online: http://stark-b.obspm.fr (accessed on 11 August 2017).

50. Sahal-Bréchot, S.; Dimitrijević, M.S.; Moreau, N.; Ben Nessib, N. The STARK-B database VAMDC node: A repository for spectral line broadening and shifts due to collisions with charged particles. *Phys. Scr.* **2015**, *50*, 054008.

51. Dubernet, M.L.; Boudon, V.; Culhane, J.L.; Dimitrijevic, M.S.; Fazliev, A.Z.; Joblin, C.; Kupka, F.; Leto, G.; Le Sidaner, P.; Loboda, P.A.; et al. Virtual atomic and molecular data centre. *J. Quant. Spectrosc. Radiat. Transf.* **2010**, *111*, 2151–2159.

52. Dubernet, M.L.; Antony, B.K.; Ba, Y.A.; Babikov, Y.L.; Bartschat, K.; Boudon, V.; Braams, B.J.; Chung, H.-K.; Daniel, F.; Delahaye, F.; et al. The virtual atomic and molecular data centre (VAMDC) consortium. *J. Phys. B* **2016**, *49*, 074003.

# 16

# Symmetric Atom–Atom and Ion–Atom Processes in Stellar Atmospheres

**Vladimir A. Srećković [1,*], Ljubinko M. Ignjatović [1] and Milan S. Dimitrijević [2,3]**

[1]  Institute of Physics, University of Belgrade, Pregrevica 118, Zemun, 11080 Belgrade, Serbia; ljuba@ipb.ac.rs
[2]  Astronomical Observatory, Volgina 7, 11060 Belgrade, Serbia; mdimitrijevic@aob.rs
[3]  LERMA, Observatoire de Paris, UMR CNRS 8112, UPMC, 92195 Meudon CEDEX, France
[*]  Correspondence: vlada@ipb.ac.rs

**Abstract:** We present the results of the influence of two groups of collisional processes (atom–atom and ion–atom) on the optical and kinetic properties of weakly ionized stellar atmospheres layers. The first type includes radiative processes of the photodissociation/association and radiative charge exchange, the second one the chemi-ionisation/recombination processes with participation of only hydrogen and helium atoms and ions. The quantitative estimation of the rate coefficients of the mentioned processes were made. The effect of the radiative processes is estimated by comparing their intensities with those of the known concurrent processes in application to the solar photosphere and to the photospheres of DB white dwarfs. The investigated chemi-ionisation/recombination processes are considered from the viewpoint of their influence on the populations of the excited states of the hydrogen atom (the Sun and an M-type red dwarf) and helium atom (DB white dwarfs). The effect of these processes on the populations of the excited states of the hydrogen atom has been studied using the general stellar atmosphere code, which generates the model. The presented results demonstrate the undoubted influence of the considered radiative and chemi- ionisation/recombination processes on the optical properties and on the kinetics of the weakly ionized layers in stellar atmospheres.

**Keywords:** atomic processes; molecular processes; radiative transfer; absorption quasi-molecular bands; sun:atmosphere; sun:photosphere; stars:atmospheres; white dwarfs

## 1. Introduction

Atomic and molecular data play a key role in many areas of science like atomic and molecular physics, astrophysics, nuclear physics, industry, etc. [1–9]. The interpretation of interstellar line spectra with radiative transfer calculations usually requires spectroscopic data and collision data (e.g., atomic parameters, cross sections, etc.) [10,11]. Determination of accurate fundamental stellar parameters is one of the most important of today's tasks and this area of fundamental science is very important and still current. For example, the atomic and molecular data are important for development of atmosphere models of solar and near solar type stars and for radiative transport investigations as well as an understanding of the kinetics of stellar and other astrophysical plasmas [12]. Available LTE codes for stellar atmosphere modelling like ATLAS [13,14], MARCS [15] as well as NLTE codes like e.g., PHOENIX (see e.g., [16,17]) and TLUSTY [18,19] require the knowledge of atomic and molecular data. In addition, spectrum synthesis codes (e.g., SYNTHE, SYNSPEC) for radiative transfer and spectra depend on these as input parameters [20]. Such atomic data and processes are also important in modelling early Universe chemistry (see [21]). Evaluation of chemical abundances in the standard Big Bang model are calculated from a set of chemical reactions for the early universe and among them are very important reactions with species like H, $H_2^+$ and also different Rydberg atoms [22]. The highly excited atomic states are named 'Rydberg states', and the atoms in such states are called 'Rydberg atoms'. Strictly speaking, only highly excited states, should be counted among the Rydberg states. However, in practice, an atom $A^*(n,l)$ is being treated as a Rydberg atom if $(n - n_0) \geq 4$, where $n_0$ is the principal quantum number of the outer shell of the atom $A$ in its ground state, and in the case of an atom $He^*(n)$—for any $n \geq 3$. It should be noted that, at the present time, even in the laboratory experiments, Rydberg atoms with $n$ close to $10^2$ are being explored, while the astrophysicists observe the radiation of atoms from the states with $n$ close to $10^3$. With a change of $n$, the parameters characterizing Rydberg states may change by orders of magnitude.

The content of this article is distributed in four sections. The first is devoted to the detailed description of the processes, and the corresponding methods of the determination of such processes coefficients and parameters. Sections 2 and 3 show the existing theoretical results concerning the investigated processes, their role in the low temperature layers of stellar atmospheres, as well as the methods of the investigation of such processes. Finally, at the end of this article, the current research and directions of further research are summarized.

In a series of papers [23–27], two groups of atom–Rydberg atom and ion–atom collisional processes have been studied from the point of view of their effect on the optical and kinetic properties of weakly ionized laboratory and astrophysical plasmas.

The first group of them includes chemi-ionization and chemi-recombination processes of the type

$$A^*(n) + A \Longleftrightarrow e + A + A^+, \tag{1a}$$

$$A^*(n) + A \Longleftrightarrow e + A_2^+, \tag{1b}$$

where $A^*(n)$ is an atom in a highly excited (Rydberg) state with a principal quantum number $n \gg 1$, and $e$ is a free electron. The processes caused by the action of the resonant energy exchange mechanism inside the electron component corresponding to the atomic–atomic or electron–ion–atomic system are considered [28]. These processes are illustrated in the Figure 1, where Figure 1a schematically shows the geometry of the collision $A^*(n) + A$, and Figure 1b—the essence of the resonant mechanism is that the transition of an external weakly bound electron of the system $(e_n)$ with energy $\varepsilon_n < 0$ to a free state with energy $\varepsilon_k > 0$ is accompanied by a transition of the subsystem $A + A^+$ from the first excited molecular state with energy $U_2(R)$ to the ground state with energy $U_1(R)$.

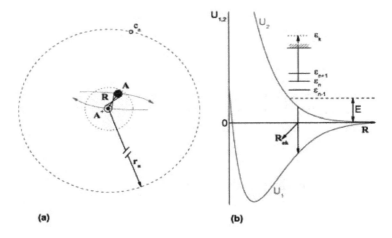

(a)   (b)

**Figure 1.** (a) schematic illustration of $A^*(n,l) + A$ collision (1) within the domain of internuclear distances $R \ll r_n$, where $r_n$ is the characteristic radius of Rydberg atom $A^*(n,l)$; (b) schematic illustration of the resonance mechanism.

The second group of processes includes radiation processes of the type:

$$\varepsilon_\lambda + A_2^+ \Longleftrightarrow A + A^+ \quad \text{photodissociation/association,} \tag{2a}$$

$$\varepsilon_\lambda + A^+ + A \Longleftrightarrow A + A^+ \quad \text{radiation charge exchange,} \tag{2b}$$

where $\varepsilon_\lambda$ stands for the energy of a photon with a wavelength $\lambda$, $A$ and $A^+$—an atom and its positive ion in the ground states, and $A_2^+$—a molecular ion in the ground electronic state. These processes are illustrated in the Figure 2, where $U_1(R)$ and $U_2(R)$ represent the adiabatic potential curves of the ground and first excited electronic state of the ion, and $R$ stands for the internuclear distance in atomic units. This figure shows that the radiative processes under study represent the result of transitions, with the emission or absorption of a photon, between the mentioned molecular electronic states.

The effect of chemi-ionization and chemi-recombination processes (1) can be estimated by comparing their intensities with the intensities of known concurrent ionization and recombination processes, namely:

$$A^+ + e \Longrightarrow \varepsilon_\lambda + A^*(n), \tag{3}$$

$$A^+ + e + e \Longleftrightarrow A^*(n) + e. \tag{4}$$

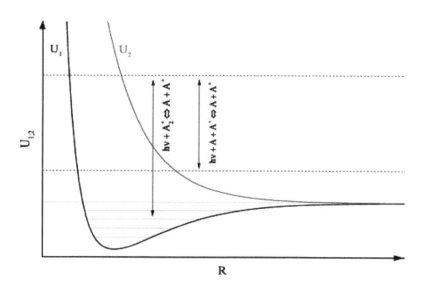

**Figure 2.** The schematic presentation of the photo-dissociation/association processes Equation (2a) and free–free processes Equation (2b): $R$ is the internuclear distance, $U_1(R)$ and $U_2(R)$ are the potential energy curves of the initial (lower) and final (upper) electronic state of molecular ion $A_2^+$, and $h\nu$ is the photon energy.

The influence of radiation processes (2) can be analysed by comparing with the intensities of known concurrent radiative processes, namely:

$$A^+ + e \Longleftrightarrow \varepsilon_\lambda + A^+ + e, \tag{5a}$$

$$A^+ + e \Longleftrightarrow \varepsilon_\lambda + A^*, \tag{5b}$$

$$A + e \Longleftrightarrow \varepsilon_\lambda + A + e, \tag{6}$$

$$A + e \Longleftrightarrow \varepsilon_\lambda + A^-, \tag{7}$$

where $A^-$ is stable negative ion.

In connection with astrophysical plasmas, two cases are considered:

- the case of hydrogen, when $A = H(1s)$ and $A^+ = H^+$,
- the case of helium, when $A = He(1s^2)$ and $A^+ = He^+(1s)$.

For the solar atmosphere, $A$ usually denotes atom H(1s) and $A^+ = H^+$, and, for the case of helium-rich white dwarf atmospheres, $A$ denotes $He(1s^2)$ and $A^+ = He^+(1s)$.

Chemi-ionization/recombination processes are very important for the evolution of Universe in the early epochs. The most important process is hydrogen recombination, and Rydberg states can play important roles in this process [22].

## 2. Chemi-Ionization and Chemi-Recombination Processes

The region of importance of chemi-ionization processes and production of Rydberg atoms i.e., chemi-recombination is in the cool dwarf stars and, especially, cool white dwarfs. Spectroscopic observations of cool white dwarfs [29] have demonstrated that white dwarfs with temperatures less than 6100 K are found to display significant flux deficits that are not predicted by the current WD model. One can suggest that such process may be absorption by atoms and molecules in highly excited Rydberg states.

Recent research of the atmospheres of cooling stars such as white dwarfs pointed out an anomaly in light emission of Rydberg atom with $n = 10$ and tabular lifetime $\tau \sim 10^{-6}$ s. The lines of the corresponding infra-red transitions have not been observed [30]. Let us note that it is just these states that correspond to the maximal values of chemi-ionization rate coefficients. According to the observational data, we have that, under such conditions $N_0 \geq 10^{17}$ cm$^{-3}$ and $N^* \geq 10^{13}$ cm$^{-3}$, where $N_0$ and $N^*$ are the densities of the ground state and Rydberg atoms. It is not difficult to estimate that the probability of a Rydberg atom being extinguished through the chemi-ionization channel is comparable to the probability of its radiative decay.

Let us not forget the importance of chemi-ionization processes and production of Rydberg atoms in chemistry, physics and related branches of science [28,31,32].

In this article, the current state of research of the processes in atom–Rydberg atom collisions is presented. The principal assumptions of the model of such processes are based on the dipole resonance mechanism.

The chemi-ionization and inverse chemi-recombination processes (1) can be considered from the point of view of their effect on the population of the excited states of the hydrogen atoms in solar and cold-star atmospheres, as well as on the population of excited states of the helium atoms in DB white dwarfs. Comparative analysis of the influence of these and concurrent processes (3) and (4) can be presented by the values of the following parameters:

$$F_{phr}^{(ab)}(2,8) = \frac{\sum\limits_{n=2}^{8} I_r^{(ab)}(n,T)}{\sum\limits_{n=2}^{8} I_{phr}^{(ab)}(n,T)}, \quad F_{eei}^{(ab)}(2,8) = \frac{\sum\limits_{n=2}^{8} I_r^{(ab)}(n,T)}{\sum\limits_{n=2}^{8} I_r^{(eei)}(n,T)}, \quad F_i(n,T) = \frac{I_{ci}(n,T)}{I_{i;ea}(n,T),} \tag{8}$$

where $I_{ci}(n,T)$, $I_{i;ea}(n,T)$, $I_r^{(ab)}(n,T)$, $I_{phr}^{(ab)}(n,T)$ and $I_r^{(eei)}(n,T)$ are the fluxes caused by ionization and recombination processes (1), (3) and (4).

### 2.1. Solar Atmosphere

As a necessary step to improve the modeling of the solar photosphere, as well as to model atmospheres of other similar and cooler stars where the main constituent is also hydrogen, it is required to take into account the influence of all the relevant collisional processes on the excited-atom populations in weakly ionized hydrogen plasmas. This is important since a strong connection between the changes in atom excited-state populations and the electron density exists in weakly ionized plasmas.

It is a fact that, with an increase of the electron density, caused by a growth of the excited hydrogen atom population, the rate of thermalization by electron–atom collisions in the stellar atmosphere will become higher. A consequence will be that the radiative source function of the line center will be more closely coupled to the Planck function, making the synthesized spectral lines stronger for a given model structure, affecting the accuracy of plasma diagnostics and determination of the atmospheric pressure.

The theoretical investigation of the processes (1) started in [33] for the hydrogen symmetric case $A = H$. Although some of the chemi-ionization processes in atom–Rydberg atom collisions had already been described in [33], their intensive astrophysical research began somewhat later. From the astrophysical point of view, in Ref. [34], an investigation was started for the chemi-recombination processes of the photosphere and the lower chromosphere of the Sun, where $4 \leq n \leq 8$. Further research of Mihajlov and coworkers continued in [35] on the chemi-ionisation processes in $H^*(n \geq 2) + H(1s)$ collisions and inverse recombination in the photosphere and the lower chromosphere of the Sun.

The partial rate coefficients for the chemi-ionization processes (1a,b) are determined by expressions

$$K_{ci}^{(a,b)}(n, T) = \int_{E_{min}(n)}^{\infty} v\sigma_{ci}^{(a,b)}(n, E)f(v; T)dv, \tag{9}$$

where $\sigma_{ci}^{(a,b)}(n, E)$ is cross section, $v$ is the atom–Rydberg–atom impact velocity, $f(v; T)$ is the velocity distribution function for the given temperature $T$, and $E_{min}(n)$ is determined by the behavior of the potential curve $U_2(R)$ and the splitting term as presented in [35].

Under the conditions that exist in the solar atmosphere, the chemi-recombination rate coefficients can be presented over chemi-ionization rate coefficient

$$K_{ci}^{(a)}(n, T) \cdot N_n N_1 = K_{cr}^{(a)}(n, T) \cdot N_1 N_{ai} N_e, \tag{10}$$

where $N_1$ and $N_n$ denote the densities of ground- and excited-state of atoms, respectively, while $N_{ai}$ is the densities of ions.

Using partial rate coefficients $K_{ci,cr}^{(a,b)}(n, T)$, we can determine the total one,

$$K_{ci,cr}(n, T) = K_{ci,cr}^{(a)}(n, T) + K_{ci,cr}^{(b)}(n, T), \tag{11}$$

which characterizes the efficiency of the chemi-ionization/recombination processes (1a,b) together. The total chemi-ionization and chemi-recombination fluxes are $I_{ci}(n, T) = K_{ci}(n, T) \cdot N_n N_1$, $I_{cr}(n, T) = K_{cr}(n, T) \cdot N_1 N_i N_e$.

Figure 3a,b show the values of the total chemi-ionization and recombination rate coefficients $Kci(n, T)$ and $Kcr(n, T)$ in the region $2 \leq n \leq 8$ and $5000\,K \leq T \leq 10,000\,K$ obtained in [35]. From the figure, it follows that the maximum values of the hydrogen $Kci(n, T)$ lies at $n = 5$ for all temperatures (except for $T = 5000\,K$, where maximum value of $Kci(n, T)$ are at $n = 4$), and maximum values for the chemi recombination rate coefficient $Kcr(n, T)$ shifts to higher $n$ (from 3 to 5) when temperature increases. The relative importance of a particular channel ('a' and 'b') for the chemi-ionization and chemi-recombination processes (1a,b) is shown in Figure 4. Branch coefficient $X(n; T) = K_{ci,cr}^{(b)}/Kci(n, T)$ present the ratio of rate coefficient for processes (1b) and total ((1a) + (1b)). From Figure 4, one can see that, for lower temperatures, i.e., 5000 K, processes (1b) are dominant while importance of processes (1a) increase with the increase of temperature as expected.

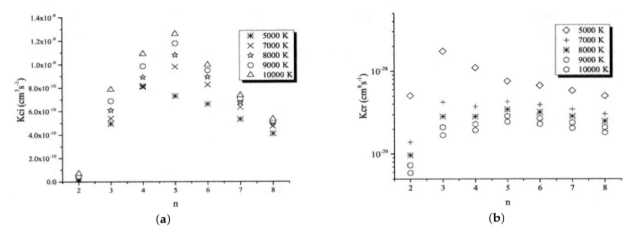

**Figure 3.** (**a**) total chemi-ionization rate coefficients $Kci(n; T; H)$ with $5000\,\text{K} \le \text{T} \le 10,000\,\text{K}$ and for principal quantum numbers $n = 2$–$10$; (**b**) same as in (**a**) but for the inverse recombination coefficients $Kr(n; T; H)$ (data taken from [35]).

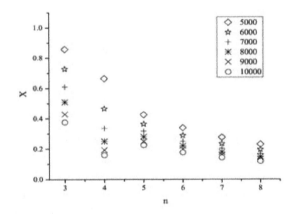

**Figure 4.** Same as in Figure 3 but for the branch coefficient $X(n; T; H)$, characterizing the relative importance of the particular channel ('a' and 'b') for the chemi-ionization and chemi-recombination processes (1a,b).

From the astrophysical point of view, the results have shown that, in some parts of the solar atmosphere, chemi-recombination processes (1) can dominate with respect to the photo-recombination process (3) and their intensity can be close to the intensity of triple electron–electron–ion recombination processes (4). The behavior of quantity $F_{phr}^{(ab)}$ as a function of $h$ is shown in Figure 5a. The behavior of the quantity $F_{eei}^{(ab)}$ as a function of height $h$ is shown in Figure 5b from [35]. One can see that the considered chemi-recombination processes dominate with respect to the concurrent electron–electron–ion recombination processes within the region $100\,\text{km} \le h \le 650\,\text{km}$ and significantly influence the optical properties of the solar photosphere. Namely, if we take the considered processes into account, we will improve the modeling of the solar photosphere, as the model atmospheres of other similar and cooler stars. If we do not take into account all relevant collisional processes on the excited-atom populations in weakly ionized plasmas, the corresponding electron density will be less accurate. For example, as stated in Reference [26], an increase of the electron density, caused by a growth of the excited hydrogen atom population, will result in a higher rate of thermalization by electron–atom collisions in the stellar atmosphere. As a consequence, the radiative source function of the line center will be more closely coupled to the Planck function, and the synthesized spectral lines will be stronger for a given model structure. This will influence on the accuracy of plasma diagnostics and determination of the atmospheric pressure.

Domination of the chemi–recombination processes with $2 \leq n \leq 8$ over the electron–ion photo–recombination processes is confirmed in a significant part of the photosphere ($-50$ km $\leq h \leq 600$ km). Thus, it is proofed that these processes are important for non-LTE modeling of solar atmosphere.

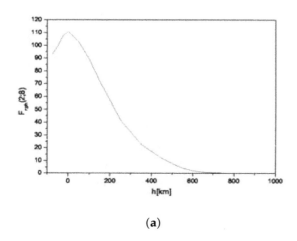 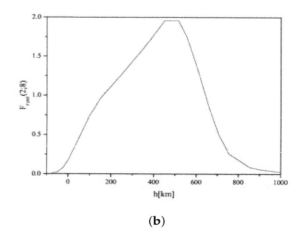

(a)                                                                        (b)

**Figure 5.** (a) parameter $F_{phr}^{(ab)}(n, T)$ (8) as functions of the height $h$ of the solar atmosphere; (b) same as in (a) but for for the parameter $F_{eei}^{(ab)}(n, T)$ (8).

*2.2. Atmospheres of the DB White Dwarfs.*

The chemi-ionization and chemi-recombination processes and consequently Rydberg atoms are of importance for cool stars and, first of all, for cool white dwarfs. Recently, a new effect has been noticed from Spitzer observations of cool white dwarfs [29]. Namely, these observations have demonstrated that some white dwarfs with $T < 6100$ K are found to display significant flux deficits in Spitzer observations (see [29]). These mid-IR flux deficits are not predicted by the current white dwarf models including collision induced absorption due to molecular hydrogen. This fact implies that the source of this flux deficit is not standard molecular absorption but some other physical process. It is possible that such process may be absorption by atoms and molecules in highly excited Rydberg states.

In the helium case of chemi-ionization and chemi-recombination processes, the situation turns out to be similar to the hydrogen case. This conclusion is based on the results obtained by Mihajlov and coworkers in [36]. The influence of symmetrical chemi-ionization and chemi-recombination processes on the helium atom Rydberg states population in weakly ionized layers of helium-rich DB white dwarfs has been investigated in [36].

Figure 6a,b show the obtained values of the total chemi-ionization and recombination rate coefficients $Kci(n, T)$ and $Kcr(n, T)$ in the region $3 \leq n \leq 10$ and 10,000 K $\leq T \leq 30,000$ K for helium plasma. The dependence of the both coefficients $Kci(n, T)$ and $Kcr(n, T)$ on the quantum number decreases with the increase of the temperature. Regarding the importance of the particular channel ('a' and 'b') for the chemi-ionization and chemi-recombination processes (1a,b), as can be seen in Figure 6b, conclusions are the same as for the chemi-ionization rate coefficients.

From the astrophysical viewpoint, chemi-ionization and chemi-recombination contribution to the Rydberg state populations have been compared with electron–electron–ion recombination, electron-excited atom ionization, and electron–ion photorecombination processes for $n$ from 3 to 10, in helium-rich DB white dwarf atmosphere layers with logarithm of gravity $\log g = 7$ and 8 and effective temperature $T_{eff} \leq 20,000$ K.

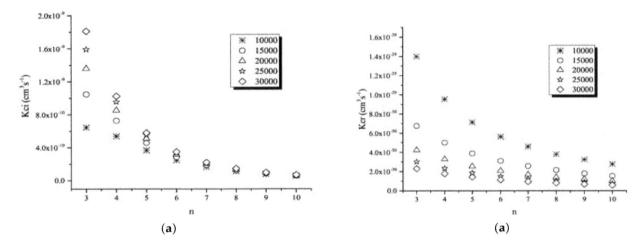

**Figure 6.** (**a**) total chemi-ionization rate coefficients $Kci(n; T; He)$ with $10,000\,\text{K} \le T \le 30,000\,\text{K}$ and for principal quantum numbers $n = 3$–$10$; (**b**) same as in (a) but for the inverse recombination coefficients $Kr(n; T; He)$ (data taken from [36]).

Some of the results obtained for the helium case [36] are illustrated by the Figure 7a,b, which are related to the recombination processes in the DB white dwarf atmosphere with an effective temperature $T_{eff} = 12,000\,\text{K}$. The values of the parameters $F_{phr}^{(ab)}(n, T)$ and $F_{eei}^{(ab)}(n, T)$ are given as functions of the logarithm of the Rosseland optical depth.

The results from [36] undoubtfully show that, for the lower temperatures, the chemi-ionization/recombination processes (1) are absolutely dominant over electron-excited atom ionization (3) and electron–electron–atom recombination (4) processes, for $n = 3, 4$, and $5$ for almost all $\log \tau < 0$ values. For $n = 6, 7$, and $8$ and in the same $\log \tau$ range, processes (1a,b) are comparable with processes (3) and (4). It is concluded that processes (1a,b) can be dominant ionization/recombination mechanisms in helium-rich DB white dwarf atmosphere layers for $\log g = 7$ and $8$ and $T_{eff} \le 20,000\,\text{K}$ and have to be implemented in relevant models of weakly ionized helium plasmas.

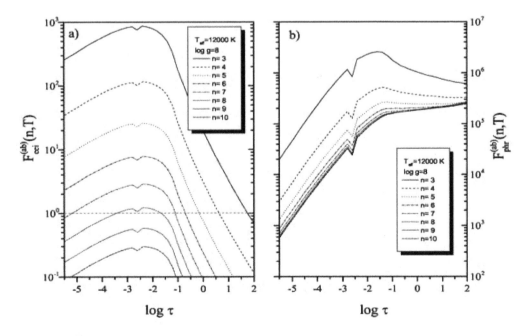

**Figure 7.** (**a**) parameter $F_{eei}^{(ab)}(n, T)$ (8) as a function of the logarithm of the Rosseland optical depth $\log \tau$; (**b**) as in (**a**) but for the parameter $F_{phr}^{(ab)}(n, T)$ (8).

*2.3. Chemi-Ionization Processes in Solar and DB White-Dwarf Atmospheres in the Presence of Mixing Channels*

In the recent paper of Mihajlov et al. [37], two kinds of atomic collision processes have been considered that simultaneously occur in the stellar atmospheres and influence each other (see Figure 8). This is about the chemi-ionization processes (1) and (n-n′)-mixing i.e., excitation-deexcitation processes

$$A^*(n,l) + A \Longrightarrow A + A^*(n',l), \quad \text{(n–n')-mixing,}  \tag{12}$$

where $A$ are atoms in their ground states, $A^*(n,l)$ is an atom in a highly excited (Rydberg) state with the principal quantum number $n \gg 1$ and orbital quantum number $l$ and $A =$ H or He.

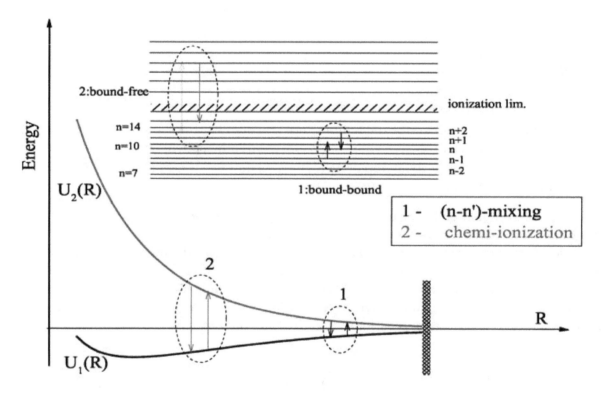

**Figure 8.** The schematic presentation of the chemi-ionization and (n-n′)-mixing processes.

It was expected that process (12) would reduce the impact of chemi-ionization processes (1), as can be seen from the data of Mihajlov et al. (see Figure 9). The arrows in Figure 9a,b illustratively present that reduction of rate coefficients.

Mihajlov and coworkers have shown that, for the lower temperatures, the chemi-ionization processes are still dominant over electron-excited atom ionization processes ($A^*(n) + e \to A^+ + e + e$), for $n = 3$, 4, and 6 almost in the whole observed atmosphere which is illustrated by Figure 10 from [37]. For $n = 6$, 7, and 8 in the whole observed atmosphere, chemi-ionization processes are comparable with electron-excited atom ionization processes. This is illustrated in Figure 10a, for WD with $T_{eff} = 12,000$ K (helium case) and in Figure 10b for solar photosphere (hydrogen case).

From the results [37,38], it follows that processes (1) are significant for such hydrogen and helium plasmas with the ratio $Ne/Na < 10^{-3}$, where $Ne$ and $Na$ are the free electron and ground state atom density. Accordingly, these processes are significant for the stellar atmospheres that contain the corresponding weakly ionized layers.

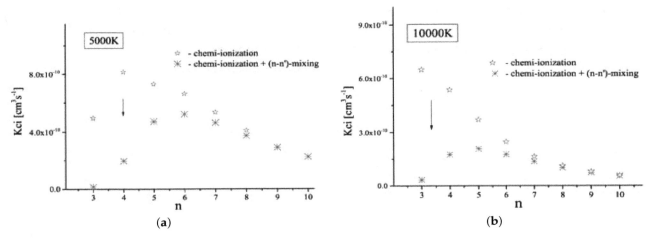

**Figure 9.** (a) comparison of the calculated values of rate coefficients of the chemi-ionization processes (1a,b) with and without inclusion of (n-n')-mixing process, case $A = H$; (b) same as in (a) but for case $A = He$.

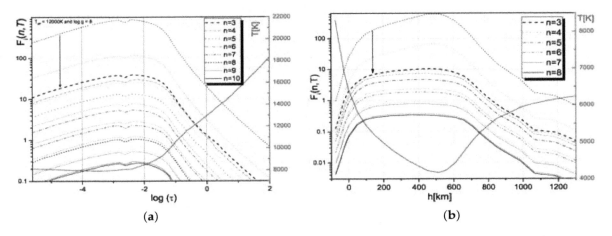

**Figure 10.** (a) parameter $F_i(n, T; He^*(n))$ (ratio of fluxes generated in atom–Rydberg–atom and electron-excited atom impact ionization) as a function of the logarithm of Rosseland optical depth $\log(\tau)$, for principal quantum numbers $n = 3$–$10$, with $T_{eff} = 12,000$ K and $\log g = 8$: wide line—present calculation; tiny line—calculation from [35]; (b) parameter $F_i(n, T; H^*(n))$ as a function of the height h, for principal quantum numbers $n = 3$–$8$, for model of solar photosphere [39]: wide line—present calculation; tiny line—calculation from [36].

## 2.4. The Atmospheres of Late Type Dwarfs (M Red Dwarfs)

The general stellar atmosphere code PHOENIX [10] has the advantage that, apart from solving the atmospheric structure, it also calculates output spectra. A good example of testing and application of PHOENIX code are processes (1) that influence the excited state populations and the free electron density, and also influence the atomic spectral line shapes.

In Reference [25], processes (1) are also considered in the case of an atmosphere of a M red dwarf with an effective temperature $T_{eff} = 3800$ K. The influence of these processes on the population of excited states of the hydrogen atom in this case was investigated using the PHOENIX code [10,16], which as a result generates a model of the considered atmosphere. In the framework of this work, the PHOENIX code includes processes (1) for $n \geq 4$. Some of the results obtained in [25] are illustrated by the Figure 11, where the values of the parameter $\zeta$ are presented, i.e., the ratio of the populations of the excited states of the hydrogen atom, calculated with and without taking into account processes (1). The presented figures show that, at least in the region $n \leq 15$, processes (1) equally affect the populations of the excited states of the hydrogen atom.

Then, Reference [25] shows that the processes (1) in the whole region of $n > 1$ also influence the free electron density (see Figure 12). This figure shows the behavior of free electron density calculated with these processes (solid curve) and without them (dashed curve).

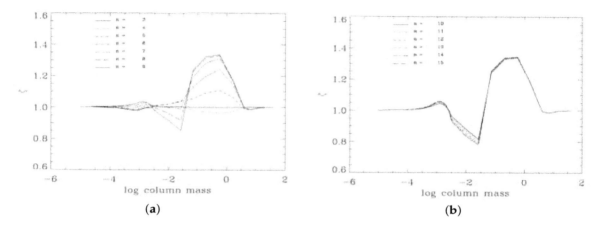

**Figure 11.** (a) the behavior of the population ratio $\zeta$ for $3 \le n \le 9$ as a function of the column mass; (b) same as in (a) but for $10 \le n \le 15$ (from [25]).

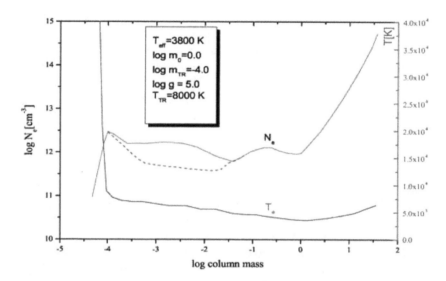

**Figure 12.** Structure of model atmosphere–electron density $Ne$ and temperature $Te$ as a function of column mass.

The presented results suggested that processes (1), due to their influence on the excited state populations and the free electron density, also should influence the atomic spectral line shapes in the atmospheres of late type dwarfs.

*2.5. Influences of Chemi-Ionization/Recombination Processes on the Hydrogen Spectral Lines in the M Red Dwarf Atmosphere*

In connection with this problem, in Reference [26], the atmospheres of a M red dwarf with an effective temperature $T_{eff} = 3800$ K was also examined. In contrast to the previous case, processes (1) with $n = 2$ and 3 were included here.

The results presented in [26] show that the chemi-ionization/recombination processes (1), are directly affecting the population of the excited states of the hydrogen atom and the electron concentration, and thus have a very strong effect on the shape of the spectral lines of the atom. For the given atmosphere, the profiles of a number of spectral lines of the hydrogen atom were calculated.

Figure 13 (from [26]) show the line profiles of $H_\alpha$, $H_\delta$, $H_\epsilon$ $Pa_\epsilon$ with and without inclusion of processes (1). Profiles are synthesized with PHOENIX code with Stark broadening contribution calculated using tables from [40] for Stark broadening of hydrogen lines (linear Stark effect). Lineshape changes, especially in the wings, show the influence of the electron density change having a direct influence on the Stark broadening of hydrogen lines.

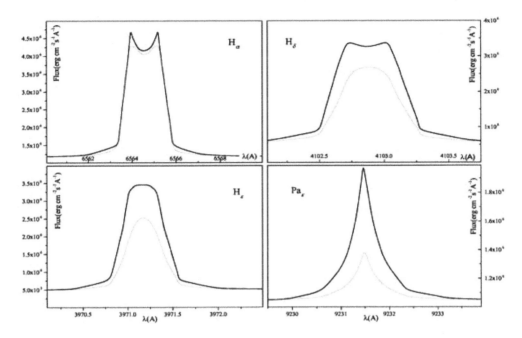

**Figure 13.** Line profiles with (full) and without (red tiny) inclusion of chemi-ionization and chemi-recombination processes for H lines.

## 3. Symmetric Ion–Atom Processes

It is well known [22] that the chemical composition of the primordial gas consists of electrons and species such as: helium—He, $He^+$, $He^{2+}$ and $HeH^+$; hydrogen—H, $H^-$, $H^+$, $H_2^+$ and $H_2$; deuterium—$D$, $D^+$, $HD$, $HD^+$ and $HD^-$; lithium—Li, $Li^+$, $Li^-$, $LiH^-$ and $LiH^+$. Evaluation of chemical abundances in the standard Big Bang model are calculated from a set of chemical reactions for the early universe [22]. Among them are very important reactions (2) that involve species like H, $H^+$, $H_2^+$, $He_2^+$, He, $He^+$, whose role in the primordial star formation is crucial.

Recently, References [41,42] have pointed out that the photodissociation of the diatomic molecular ion in the symmetric cases (2a) are of astrophysical relevance and could be important in modeling of specific stellar atmosphere layers, and they should be included in some chemical models. The data that involves reactions (2) are also useful in hydrogen and helium theoretical and laboratory plasmas research [43].

### 3.1. Solar Atmosphere: Visible Wavelength Region

The theoretical investigation of processes (2) started in [44] for the hydrogen symmetric case $A = H$. Then, in [44,45], the processes (2) were considered in relation to the photosphere and the lower solar chromosphere by characterizing their spectral emissivity $\varepsilon_{ia}(\lambda)$, (the spectral density of the radiation energy, which these processes generate from a unit volume per unit time, into an angle $4\pi$). Their contributions have been collated with concurrent processes (5)–(7) by comparative analysis of their influence using the values of the following parameters: $F_{ei}(\lambda) = \varepsilon_{ia}(\lambda)/\varepsilon_{ei}(\lambda)$, $F_{ea}^{ff}(\lambda) = \varepsilon_{ia}(\lambda)/\varepsilon_{ea}^{ff}(\lambda)$ and $F_{ea}^{fb}(\lambda) = \varepsilon_{ia}(\lambda)/\varepsilon_{ea}^{fb}(\lambda)$, where processes $\varepsilon_{ei}(\lambda)$, $\varepsilon_{ea}^{ff}(\lambda)$, and $\varepsilon_{ea}^{fb}(\lambda)$ are their emissivity.

One of the major results obtained in papers of Mihajlov et al. [44,45] for the visible spectral region of the spectrum are illustrated by the Figure 14 for the considered layer of the solar atmosphere on the basis of standard models [39,46]. It was found that in this region processes (2) give a contribution of 10–12% in comparison with dominant processes (7). This fact alone demonstrated that considered ion–atom radiative processes must be taken into account for solar atmosphere modeling. Later estimates showed, however, that the relative influence of the absorption processes (2) on the solar atmosphere opacity should significantly increase at the transition from the considered wavelength region $\lambda \geq 365$ nm to the region $\lambda_H \leq \lambda < 365$ nm, where $\lambda_H = 91.1262$ nm is the wavelength that corresponds to the ionization threshold of the H(1s) atom.

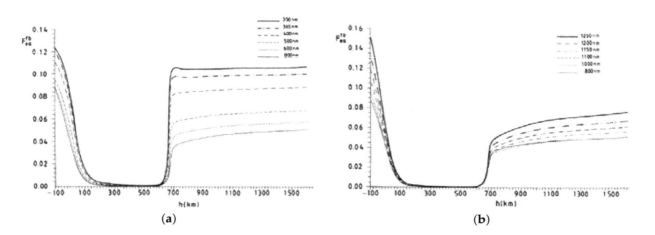

**Figure 14.** (a) the behaviour of the parameter $F_{ea}^{fb}(\lambda)$ as a functions of height $h$ in the solar atmosphere for 350 nm $\leq \lambda \leq$ 800 nm; (b) the behaviour of the parameter $F_{ea}^{fb}(\lambda)$ as a function of height $h$ of the solar atmosphere for 800 nm $\leq \lambda \leq$ 1250 nm (from [44]).

### 3.2. Solar Atmosphere: UV and VUV Wavelength Region

In the far UV region, the emission channels of processes (2) cease to play a role, while the role of absorption channels in this region grows very rapidly. The total absorption rate coefficient $K_{ia} = K_{ia}^a + K_{ia}^b$ for processes (2a,b) as well as the branch coefficient $X = K_{ia}^b / K_{ia}$ which shows the influence of channels 'a' and 'b' in reaction (2) are presented in Figure 15a,b, respectively. The quantities $K_{ia}$ and $X$ are described in detail in [47] as well as their relations with cross sections. From Figure 15a one can see that $K_{ia}$ strongly depends on temperature and wavelength and, from Figure 15b, it follows that, for lower temperatures, processes (2a) are dominant and importance of processes (2b) increases with the increase of temperature as expected.

In accordance with [47], in the Ultraviolet and Vacuum Ultraviolet regions, the influence of the absorption processes in processes (2) is estimated by the parameter, $F_k(\lambda)$ which is the ratio of the absorption coefficient of these processes and the absorption coefficient determined by the concurrent processes (5)–(7) taken together. The results obtained for the parameters $F_K(h)$ in the wavelength region 92 nm $\leq \lambda \leq$ 350 nm are presented by Figure 16.

This figure shows that in the significant part of the considered region of altitudes ($-75$ km $\leq h \leq 1065$ km) the absorption processes (2) together give the contribution that varies from about 10% to about 90% of the contribution of the absorption processes (7), which are considered as the main absorption processes in the solar photosphere. In connection with the other known concurrent absorption processes, it is shown that, in the considered region of altitude, there are significant parts where the symmetric processes completely dominate (see also [48]).

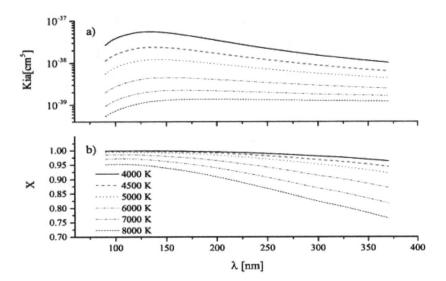

**Figure 15.** (a) the total absorption coefficient $Kia$ as a function of $\lambda$ and $T$; (b) the branch coefficient $X$ as a function of $\lambda$ and $T$. Calculations from [47] case $A = \text{H}$.

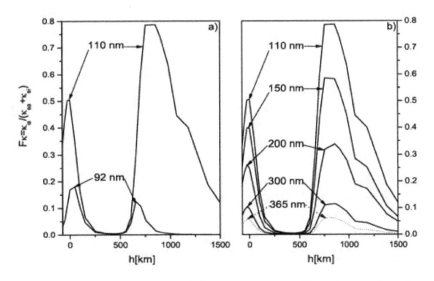

**Figure 16.** The behaviour of the parameter $F_k(\lambda)$ as a functions of height $h$ in the solar atmosphere.

### 3.3. DB White Dwarf Atmospheres: Visible Wavelength Region

The results from [23], obtained for one Koester model ($\log g = 8$ and $T_{eff} = 12{,}000$ K), have already provided a more realistic picture of the relative importance of $\text{He}^-$ (6) and $\text{He}_2^+$ (2a) total absorption processes, at least in the region $\lambda \leq 300$ nm. In a following paper [24], the relative importance of $\text{He}_2^+$ and other relevant absorption processes in the region $\lambda \geq 200$ nm was examined for several of Koester's (1980) models ($T_{eff} = 12{,}000$, $14{,}000$, $16{,}000$ K, $\log g = 7$, 8). It was shown that, in all considered cases, the contribution to opacity of the processes of $\text{He}_2^+$ molecular ion photodissociation and $\text{He} + \text{He}^+$ collisional absorption charge exchange combined is close to or at least comparable with the contribution of the absorption processes (6) and atomic absorption processes (5).

### 3.4. DB White Dwarf Atmospheres: UV and VUV Wavelength Region

Let us note in this context that, in the case of white dwarf atmospheres with dominant helium component, among all possible symmetric ion–atom absorbtion processes that are allowed by their composition, only the processes (2a,b) have to be taken into account [23,27].

Figure 17a presents the total absorption rate coefficient $K_{ia} = K_{ia}^a + K_{ia}^b$ for the processes (2a,b) as well as the branch coefficient $X = K_{ia}^b / K_{ia}$ that shows the influence of channels 'a' and 'b' in reaction (2). The quantities $K_{ia}$ and $X$ are described in detail in [27]. Regarding the importance of the particular channel ('a' and 'b') for the processes (2a,b), conclusions are the same as for the case of hydrogen.

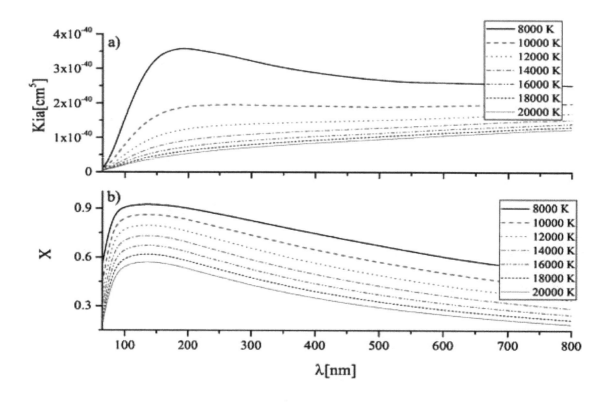

**Figure 17.** (**a**) the total absorption coefficient $Kia$ as a function of $\lambda$ and $T$; (**b**) the branch coefficient $X$ as a function of $\lambda$ and $T$. Data from [27], case $A = $ He.

For the calculation of spectral properties, the data from the corresponding DB white-dwarf atmosphere models [49] have been used as well as the data for coefficients given in Figure 17. It was established that the processes (2a,b) significantly influence the opacity of the considered DB white dwarf atmospheres, with an effective temperature $T_{eff} \geq 12,000$ K, which fully justifies their inclusion in one of the models of such atmospheres [50]. However, the same comparison demonstrated also that the dominant role in those atmospheres generally still belongs to the concurrent absorbtion process (6), while the processes (2a,b) can be treated as dominant (with respect to this concurrent process) only in some layers of those atmospheres, and only within the part 50 nm $< \lambda < 250$ nm of the far UV and EUV region (see Figure 18). The results obtained in research allow for the possibility of estimating which absorption processes give the main contribution to the opacity in DB white dwarf atmospheres in different spectral regions. Therefore, from [23,27,51,52] results, it follows that the helium absorption processes (2a,b) are dominant in the region 70 nm $\leq \lambda \leq 200$ nm, while, in the region $\lambda \geq 200$ nm, the absorption processes (6) have an important role.

From the presented material, it follows that the considered symmetric ion–atom absorption processes cannot be treated only as one channel among many equal channels with influence on the opacity of the solar atmosphere. Namely, these symmetric processes around the temperature minimum increase the absorption of the EM radiation, so that this absorption becomes almost uniform in the whole solar photosphere

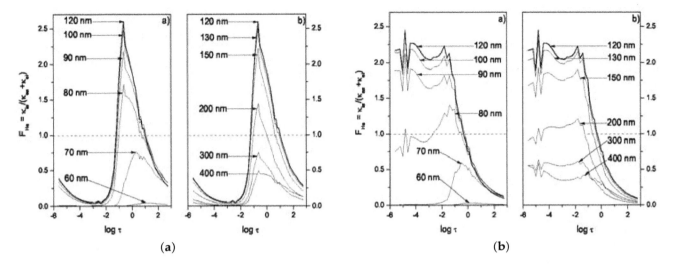

**Figure 18.** (a) behaviour of the quantity $F_{He} = \kappa_{ia}/(\kappa_{ea} + \kappa_{ei})$ (ratio of the absorption coefficients of processes (2), (5) and (6)) within the atmosphere of a DB white dwarf in the case $\log g = 8$ and $T_{eff} = 12,000$ K; (b) as in (a), but for the case $\log g = 8$ and $T_{eff} = 14,000$ K.

## 4. Conclusions

All the foregoing shows the undoubted influence of the radiation ion–atom processes and chemi-ionization/recombination processes on the optical properties and on the kinetics of weakly ionized layers of stellar atmospheres, and they should be studied from the spectroscopic aspect. In addition, it can be expected that the reported results will be a sufficient reason for including these processes in the models of stellar atmospheres. The further development of research in a new direction must be connected with the investigations of described processes but in the field of modelling in early Universe chemistry.

**Acknowledgments:** The authors are thankful to the Ministry of Education, Science and Technological Development of the Republic of Serbia for the support of this work within the projects 176002 and III44002.

**Author Contributions:** All authors contributed equally to this work.

## References

1. Marinković, B.; Pejčev, V.; Filipović, D.; Šević, D.; Milosavljević, A.; Milisavljević, S.; Rabasović, M.; Pavlović, D.; Maljković, J. Cross section data for electron collisions in plasma physics. *J. Phys. Conf. Ser.* **2007**, *86*, 012006.
2. Marinković, B.P.; Jevremović, D.; Srećković, V.A.; Vujčić, V.; Ignjatović, L.M.; Dimitrijević, M.S.; Mason, N.J. BEAMDB and MolD—Databases for atomic and molecular collisional and radiative processes: Belgrade nodes of VAMDC. *Eur. Phys. J. D* **2017**, *71*, 158.
3. Srećković, V.A.; Ignjatović, L.M.; Jevremović, D.; Vujčić, V.; Dimitrijević, M.S. Radiative and Collisional Molecular Data and Virtual Laboratory Astrophysics. *Atoms* **2017**, *5*, 31.
4. Bezuglov, N.; Borodin, V.; Eckers, A.; Klyucharev, A. A quasi-classical description of the stochastic dynamics of a Rydberg electron in a diatomic quasi-molecular complex. *Opt. Spectrosc.* **2002**, *93*, 661–669.
5. Bezuglov, N.; Borodin, V.; Klyucharev, A.; Matveev, A. Stochastic dynamics of a Rydberg electron during a single atom–atom ionizing collision. *Russ. J. Phys. Chem.* **2002**, *76*, S27–S42.
6. Boyd, T.; Sanderson, J. *The Physics of Plasmas*; Cambridge University Press: Cambridge, UK, 2003.
7. Mason, N. The status of the database for plasma processing. *J. Phys. D* **2009**, *42*, 194003.
8. Campbell, L.; Brunger, M. Modelling of plasma processes in cometary and planetary atmospheres. *Plasma Sources Sci. Technol.* **2012**, *22*, 013002.

9.   Larimian, S.; Lemell, C.; Stummer, V.; Geng, J.W.; Roither, S.; Kartashov, D.; Zhang, L.; Wang, M.X.; Gong, Q.; Peng, L.Y.; et al. Localizing high-lying Rydberg wave packets with two-color laser fields. *Phys. Rev. A* **2017**, *96*, 021403.

10.  Hauschildt, P.; Baron, E. Cool stellar atmospheres with PHOENIX. *Mem. Soc. Astron. Ital.* **2005**, *7*, 140.

11.  Christensen-Dalsgaard, J.; Dappen, W.; Ajukov, S.; Anderson, E.; Antia, H.M.; Basu, S.; Baturin, V.A.; Berthomieu, G.; Chaboyer, B.; Chitre, S.M.; et al. The current state of solar modeling. *Science* **1996**, *272*, 1286.

12.  Fontenla, J.; Curdt, W.; Haberreiter, M.; Harder, J.; Tian, H. Semiempirical models of the solar atmosphere. III. Set of non-LTE models for far-ultraviolet/extreme-ultraviolet irradiance computation. *Astrophys. J.* **2009**, *707*, 482.

13.  Kurucz, R.L. *Atlas: A Computer Program for Calculating Model Stellar Atmospheres*; SAO Special Report; Smithsonian Astrophysical Observatory: Cambridge, MA, USA, 1970; Volume 309.

14.  Kurucz, R. *ATLAS9 Stellar Atmosphere Programs and 2 km/s Grid*; Kurucz CD-ROM No. 13; Smithsonian Astrophysical Observatory: Cambridge, MA, USA, 1993; Volume 13.

15.  Gustafsson, B.; Bell, R.; Eriksson, K.; Nordlund, Å. A grid of model atmospheres for metal-deficient giant stars. I. *Astron. Astrophys.* **1975**, *42*, 407–432.

16.  Hauschildt, P.H.; Baron, E. A 3D radiative transfer framework-VI. PHOENIX/3D example applications. *Astron. Astrophys.* **2010**, *509*, A36.

17.  Husser, T.O.; Wende-von Berg, S.; Dreizler, S.; Homeier, D.; Reiners, A.; Barman, T.; Hauschildt, P.H. A new extensive library of PHOENIX stellar atmospheres and synthetic spectra. *Astron. Astrophys.* **2013**, *553*, A6.

18.  Hubeny, I.; Hummer, D.; Lanz, T. NLTE model stellar atmospheres with line blanketing near the series limits. *Astron. Astrophys.* **1994**, *282*, 151–167.

19.  Hubeny, I.; Lanz, T. Non-LTE line-blanketed model atmospheres of hot stars. 1: Hybrid complete linearization/accelerated lambda iteration method. *Astrophys. J.* **1995**, *439*, 875–904.

20.  Hubeny, I.; Lanz, T. A brief introductory guide to TLUSTY and SYNSPEC. *arXiv* **2017**, arXiv:1706.01859.

21.  Coppola, C.M.; Galli, D.; Palla, F.; Longo, S.; Chluba, J. Non-thermal photons and H2 formation in the early Universe. *Mon. Not. R. Astron. Soc.* **2013**, *434*, 114–122.

22.  Puy, D.; Dubrovich, V.; Lipovka, A.; Talbi, D.; Vonlanthen, P. Molecular fluorine chemistry in the early Universe. *Astron. Astrophys.* **2007**, *476*, 685–689.

23.  Mihajlov, A.A.; Dimitrijevic, M.S.; Ignjatovic, L.M. The influence of ion–atom radiative collisions on the continuous optical spectra in helium-rich DB white-dwarf atmospheres. *Astron. Astrophys.* **1994**, *287*, 1026–1028.

24.  Mihajlov, A.A.; Dimitrijević, M.S.; Ignjatović, L.M.; Djurić, Z. Radiative $He^+(1s) + He(1s^2)$ Processes as the Source of the DB White Dwarf Atmosphere Electromagnetic Continuous Spectra. *Astrophys. J.* **1995**, *454*, 420.

25.  Mihajlov, A.A.; Jevremović, D.; Hauschildt, P.; Dimitrijević, M.S.; Ignjatović, L.M.; Alard, F. Influence of chemi-ionization and chemi-recombination processes on the population of hydrogen Rydberg states in atmospheres of late type dwarfs. *Astron. Astrophys.* **2003**, *403*, 787–791.

26.  Mihajlov, A.A.; Jevremović, D.; Hauschildt, P.; Dimitrijević, M.S.; Ignjatović, L.M.; Alard, F. Influence of chemi-ionization and chemi-recombination processes on hydrogen line shapes in M dwarfs. *Astron. Astrophys.* **2007**, *471*, 671–673.

27.  Ignjatović, L.M.; Mihajlov, A.A.; Sakan, N.M.; Dimitrijević, M.S.; Metropoulos, A. The total and relative contribution of the relevant absorption processes to the opacity of DB white dwarf atmospheres in the UV and VUV regions. *Mon. Not. R. Astron. Soc.* **2009**, *396*, 2201–2210.

28.  Mihajlov, A.A.; Srećković, V.A.; Ignjatović, L.M.; Klyucharev, A.N. The Chemi-Ionization Processes in Slow Collisions of Rydberg Atoms with Ground State Atoms: Mechanism and Applications. *J. Clust. Sci.* **2012**, *23*, 47–75.

29.  Kilić, M.; von Hippel, T.; Mullally, F.; Reach, W.; Kuchner, M.; Winget, D.; Burrows, A. The mistery deepens: Spitzer observations of cool white dwarfs. *Astrophys. J.* **2006**, *642*, 1051.

30.  Gnedin, Y.N.; Mihajlov, A.A.; Ignjatović, L.M.; Sakan, N.M.; Srećković, V.A.; Zakharov, M.Y.; Bezuglov, N.N.; Klycharev, A.N. Rydberg atoms in astrophysics. *New Astron. Rev.* **2009**, *53*, 259–265.

31.  O'Keeffe, P.; Bolognesi, P.; Avaldi, L.; Moise, A.; Richter, R.; Mihajlov, A.A.; Srećković, V.A.; Ignjatović, L.M. Experimental and theoretical study of the chemi-ionization in thermal collisions of Ne Rydberg atoms. *Phys. Rev. A* **2012**, *85*, 052705.

32.   Lin, C.; Gocke, C.; Röpke, G.; Reinholz, H. Transition rates for a Rydberg atom surrounded by a plasma. *Phys. Rev. A* **2016**, *93*, 042711.

33.   Mihajlov, A.A.; Dimitrijević, M.S.; Djurić, Z. Rate coefficients of collisional H-H*$(n)$ ionization and H-H$^+$-e and H$_2{}^+$-e recombination. *Phys. Scr.* **1996**, *53*, 159–166.

34.   Mihajlov, A.A.; Ignjatović, L.M.; Vasilijević, M.M.; Dimitrijević, M.S. Processes of H-H$^+$-e and H$_2^+$-e recombination in the weakly-ionized layers of the solar atmosphere. *Astron. Astrophys.* **1997**, *324*, 1206–1210.

35.   Mihajlov, A.A.; Ignjatović, L.M.; Srećković, V.A.; Dimitrijević, M.S. Chemi-ionization in Solar Photosphere: Influence on the Hydrogen Atom Excited States Population. *Astrophys. J. Suppl. Ser.* **2011**, *193*, 2.

36.   Mihajlov, A.; Ignjatović, L.M.; Dimitrijević, M.; Djurić, Z. Symmetrical chemi-ionization and chemi-recombination processes in low-temperature layers of helium-rich DB white dwarf atmospheres. *Astrophys. J. Suppl. Ser.* **2003**, *147*, 369.

37.   Mihajlov, A.A.; Srećković, V.A.; Ignjatović, L.M.; Dimitrijević, M.S. Atom-Rydberg-atom chemi-ionization processes in solar and DB white-dwarf atmospheres in the presence of (n-n')-mixing channels. *Mon. Not. R. Astron. Soc.* **2016**, *458*, 2215–2220.

38.   Mihajlov, A.A.; Srećković, V.A.; Ignjatović, L.M.; Klyucharev, A.N.; Dimitrijević, M.S.; Sakan, N.M. Non-Elastic Processes in Atom Rydberg-Atom Collisions: Review of State of Art and Problems. *J. Astrophys. Astron.* **2015**, *36*, 623–634.

39.   Vernazza, J.E.; Avrett, E.H.; Loeser, R. Structure of the solar chromosphere. III - Models of the EUV brightness components of the quiet-sun. *Astrophys. J. Suppl. Ser.* **1981**, *45*, 635–725.

40.   Vidal, C.R.; Cooper, J.; Smith, E.W. Unified theory calculations of Stark broadened hydrogen lines including lower state interactions. *J. Quant. Spectrosc. Radiat. Transf.* **1971**, *11*, 263–281.

41.   Babb, J.F. State resolved data for radiative association of H and H$^+$ and for Photodissociation of H$_2^+$. *Astrophys. J. Suppl. Ser.* **2015**, *216*, 21.

42.   Heays, A.; Bosman, A.; van Dishoeck, E. Photodissociation and photoionisation of atoms and molecules of astrophysical interest. *Astron. Astrophys.* **2017**, *602*, A105.

43.   Dubernet, M.; Antony, B.; Ba, Y.; Babikov, Y.L.; Bartschat, K.; Boudon, V.; Braams, B.; Chung, H.K.; Daniel, F.; Delahaye, F.; et al. The virtual atomic and molecular data centre (VAMDC) consortium. *J. Phys. B* **2016**, *49*, 074003.

44.   Mihajlov, A.A.; Dimitrijević, M.S.; Ignjatović, L.M. The contribution of ion–atom radiative collisions to the opacity of the solar atmosphere. *Astron. Astrophys.* **1993**, *276*, 187.

45.   Mihajlov, A.A.; Dimitrijević, M.; Ignjatović, L.; Djurić, Z. Spectral coefficients of emission and absorption due to ion–atom radiation collisions in the solar atmosphere. *Astron. Astrophys. Suppl. Ser.* **1994**, *103*.

46.   Maltby, P.; Avrett, E.; Carlsson, M.; Kjeldseth-Moe, O.; Kurucz, R.; Loeser, R. A new sunspot umbral model and its variation with the solar cycle. *Astrophys. J.* **1986**, *306*, 284–303.

47.   Mihajlov, A.; Ignjatović, L.M.; Sakan, N.; Dimitrijević, M. The influence of H$_2^+$-photo-dissociation and (H + H$^+$)-radiative collisions on the solar atmosphere opacity in UV and VUV regions. *Astron. Astrophys.* **2007**, *469*, 749–754.

48.   Srećković, V.A.; Mihajlov, A.A.; Ignjatović, L.M.; Dimitrijević, M.S. Ion-atom radiative processes in the solar atmosphere: Quiet Sun and sunspots. *Adv. Space Res.* **2014**, *54*, 1264–1271.

49.   Koester, D. Model atmospheres for DB white dwarfs. *Astron. Astrophys. Suppl. Ser.* **1980**, *39*, 401–409.

50.   Bergeron, P.; Wesemael, F.; Beauchamp, A. Photometric calibration of hydrogen and helium rich white dwarf models. *Publ. Astron. Soc. Pac.* **1995**, *107*, 1047.

51.   Ignjatović, L.M.; Mihajlov, A.A.; Srećković, V.A.; Dimitrijević, M.S. Absorption non-symmetric ion–atom processes in helium-rich white dwarf atmospheres. *Mon. Not. R. Astron. Soc.* **2014**, *439*, 2342–2350.

52.   Mihajlov, A.A.; Ignjatović, L.M.; Srećković, V.A.; Dimitrijević, M.S.; Metropoulos, A. The non-symmetric ion–atom radiative processes in the stellar atmospheres. *Mon. Not. R. Astron. Soc.* **2013**, *431*, 589–599.

# 17

# Rosetta Mission: Electron Scattering Cross Sections—Data Needs and Coverage in BEAMDB Database

**Bratislav P. Marinković** [1,*], **Jan Hendrik Bredehöft** [2] , **Veljko Vujčić** [3], **Darko Jevremović** [3] and **Nigel J. Mason** [4]

[1]   Institute of Physics Belgrade, University of Belgrade, Pregrevica 118, Belgrade 11080, Serbia

[2]   Institute for Applied and Physical Chemistry, Fachbereich 2 (Biologie/Chemie), Universität Bremen, Leobener Straße 5, Bremen 28359, Germany; jhbredehoeft@uni-bremen.de

[3]   Astronomical Observatory Belgrade, Volgina 7, Belgrade 11000, Serbia; veljko@aob.rs (V.V.); darko@aob.rs (D.J.)

[4]   Department of Physical Sciences, The Open University, Milton Keynes MK7 6AA, UK; N.J.Mason@open.ac.uk

*   Correspondence: braislav.marinkovic@ipb.ac.rs

**Abstract:** The emission of [O I] lines in the coma of Comet 67P/Churyumov-Gerasimenko during the Rosetta mission have been explained by electron impact dissociation of water rather than the process of photodissociation. This is the direct evidence for the role of electron induced processing has been seen on such a body. Analysis of other emission features is handicapped by a lack of detailed knowledge of electron impact cross sections which highlights the need for a broad range of electron scattering data from the molecular systems detected on the comet. In this paper, we present an overview of the needs for electron scattering data relevant for the understanding of observations in coma, the tenuous atmosphere and on the surface of 67P/Churyumov-Gerasimenko during the Rosetta mission. The relevant observations for elucidating the role of electrons come from optical spectra, particle analysis using the ion and electron sensors and mass spectrometry measurements. To model these processes electron impact data should be collated and reviewed in an electron scattering database and an example is given in the BEAMD, which is a part of a larger consortium of Virtual Atomic and Molecular Data Centre—VAMDC.

**Keywords:** electron scattering; cross sections; Rosetta mission; atomic and molecular databases

## 1. Introduction

The Rosetta spacecraft was launched in 2004 as a part of the European Space Agency (ESA) space program, with the mission to rendez-vous with, orbit and place a lander upon periodic comet 67P/Churyumov-Gerasimenko. Rosetta was in orbit with the cometary nucleus from 2014 to September 2016 during which time it was able to closely examine how the coma of the comet and the frozen comet's surface changed relative to distance from the Sun. On November 2014 Rosetta dispatched a lander, Philae, which touched down on the comet's surface and recorded, for the first time, in situ data from the surface. This pioneering mission has provided us with new and unexpected data that are changing our understanding of the structure and chemistry of cometary systems and their role in the evolution of our solar system and possible origins of life on Earth. For example, the D to H ratio in cometary water ice is very different from that on Earth and, among the other similar findings, challenges the hypothesis that water on Earth was brought by cometary impact [1].

In this paper, we will review another intriguing and unexpected result from the Rosetta mission namely the role of electron induced dissociation in the comet's coma. The data needed to model electron processes in cometary coma and its possible relevance to the formation but also the dissociation and fragmentation of molecules observed by Rosetta instruments, will be discussed together with the current data available and the databases in which such data may be found.

## 2. Rosetta Instruments and Their Observation of Electron Scattering Processes in the Cometary Coma

Rosetta orbiter carried eleven different complex scientific instruments, while the Philae lander had ten instruments. Only those that are immediately relevant for the case study of the role of electrons in comas and the detection of more complex species that may be formed by electron induced chemistry will be reviewed here.

### 2.1. FUV Emissions Measured by the ALICE Instrument

ALICE was a far-ultraviolet (FUV) imaging spectrograph that could specially resolve spectra in the range from 70 to 205 nm. Coma emission and the reflected solar spectrum from the nucleus were recorded using ALICE throughout the Rosetta encounter. The coma was identified by a spectrum that contains several features that are weak in the solar spectrum and do not appear in the reflected light from the nucleus [2]. Besides strong hydrogen Lyman lines, lines from oxygen multiplets at 98.9, 115.2, 130.4 and 135.6 nm were observed, as were weak multiplets from carbon C I lines at 156.1 and 165.7 nm and emission bands coming from CO. The surprise was the O I line at 135.6 nm, originated from the forbidden transition $^5S^o$–$^3P$ since this is usually not seen in comas. The presence of this line and the intensity ratio of H I and O I multiplets is characteristic of the process of electron dissociative excitation of water molecules [3] and led Feldman et al. [2] to establish that electron collisions with $H_2O$ is the dominant source of these emissions. Similarly, they attributed C I emissions to electron dissociative excitation of $CO_2$. The relative contribution of the UV and electron impact to the dissociation processes are dependent on the location with respect to the nucleus and the heliocentric distance as discussed in [4,5].

### 2.2. Observations from the OSIRIS Instrument

OSIRIS (Optical, Spectroscopic and Infrared Remote Imaging System) was one of Rosetta's major imaging systems equipped with a wide-angle camera (WAC) and a series of narrow band filters covering range from 245 to 640 nm and two broad band filters, green and red, covering the spectra up to 720 nm. This instrument recorded coma emission lines and specifically targeted to the transitions of O, OH, CN, CS, NH and $NH_2$. The mapping of water distribution was possible indirectly through observations of O I and OH bandpass filters. The O I filter covers the forbidden transitions from the O I $(2p^4)$ $^1D$ state which is populated directly by photodissociation of $H_2O$ molecules, while the OH filter covers the (0–0) band of the A $^2\Sigma^+$–X $^2\Pi$ transition of OH, centred at about 308.5 nm, which is excited almost entirely by fluorescence of sunlight as pointed out in [6]. The O I $^1D$ state can be also populated from the transition from the O I $(2p^4)$ $^1S$ state. Within the CN filter lies an emission line $B^2\Sigma^+$–$X^2S^+$ (0, 0) at 388 nm, within the $NH_2$ filter there is a wide emission band Ã $^2A_1$–X $^2B_1$ (0, 10, 0) and the NH filter covers the NH $A^3\Pi_1$–$X^3\Sigma^-$ (0–0) transition [6].

From this data Bodewits et al. [6] derived column densities and calculated global production rates using the standard Haser model. They found that the water production rates derived from OH are larger than those derived from [O I], OH and [O I] photolysis. Indeed, they analysed all production rates and found a much larger drop in water production rates than diurnal variation can explain. Therefore, they concluded that the photo-dissociation and fluorescence could solely explain the processes resulting in the OH, [O I], CN and NH emission observed in the inner coma and that the fragments might emanate from different parent species and/or be formed by other processes [6]. One additional process is electron induced dissociation of water, when including this in their model a much better fit was obtained, indeed electron driven dissociation of water was found to be dominant in agreement with Feldman et al. [2].

*2.3. Detection of Organic Molecules on the Comet Surface—COSAC Mass Spectrometry*

The COSAC (COmetary SAmpling and Composition) experiment and Ptolemy were two gas analysers on the Philae lander built to monitor the chemical composition of the surface of comet 67P. Due to difficulties in landing Philae was not able to deploy all of its instruments as planned and the drill could not be deployed to collect samples for in-situ analysis. However, seven measurements were made by both COSAC and Ptolemy during Philae's hopping and at its final landing site in a so-called 'sniff mode' that had no active sampling but rather just ionized whatever molecules were present in the ionization chamber of the mass spectrometer. The sample with the richest data was acquired a few minutes after the first touchdown with subsequent decay of signal strength in the other six measurements. Both instruments measured a nearly identical decay of both the water ($m/z$ 18) and CO ($m/z$ 28) peaks. However, in the COSAC measurements the peak at $m/z$ 44 decays much slower than all the other ion species, including the water peak and, the $m/z$ 44 peak also decays much slower in the COSAC measurements than in the Ptolemy data. From these results, it has been concluded that COSAC analysed a regolith sample from the cometary nucleus in situ while Ptolemy measured cometary gas from the ambient coma [7].

The compounds detected by COSAC are listed in Table 1. All of the larger molecules can be formed from the smaller compounds carbon monoxide (CO), methane ($CH_4$), water ($H_2O$) and ammonia ($NH_3$) by simple addition reactions [8]. The $m/z$ 44 peak measured by COSAC was likely dominated by organic species, e.g., from acetaldehyde ($C_2H_4O$), formamide ($HCONH_2$) and acetamide ($CH_3CONH_2$), whereas the peak measured by Ptolemy was interpreted to be mostly due to $CO_2$. Recently, a comparison and comparative analysis of the Rosetta mass spectrometers (COSAC/Ptolemy/ROSINA) that puts some question mark on the presence of some of the nitrogen-bearing species was presented [9]. Ptolemy measurements confirmed many of the species observed by COSAC and through observation of regular peaks in the observed mass distributions indicated the presence of a sequence of compounds with additional -$CH_2$- and -O- groups (mass/charge ratios 14 and 16, respectively) which confirms COSAC's observations of acetaldehyde and may be explained by the presence of a radiation-induced polymer at the surface. Ptolemy measurements also indicated an apparent absence of aromatic compounds such as benzene and neither $H_2S$ nor $SO_2$ were observed [10]. Ammonia believed to be the precursor of N containing compounds was not unambiguously detected by either Ptolemy or COSAC, probably due to its tendency to adsorb on stainless steel surface.

**Table 1.** List of molecules identified on the comet nucleus of comet 67P by the COSAC instrument [8]. Abundances are given normalized to water, which is the most abundant compound.

| Name of Compound | Sum Formula | Abundance wrt Water |
|:---:|:---:|:---:|
| Methane | $CH_4$ | 0.5% |
| Water | $H_2O$ | 100% |
| Hydrogen cyanide | HCN | 0.9% |
| Carbon monoxide | CO | 1.2% |
| Methylamine | $CH_3NH_2$ | 0.6% |
| Acetonitrile | $CH_3CN$ | 0.3% |
| Isocyanic acid | HNCO | 0.3% |
| Acetaldehyde | $CH_3CHO$ | 0.5% |
| Formamide | $HCONH_2$ | 1.8% |
| Ethylamine | $C_2H_5NH_2$ | 0.3% |
| Methyl isocyanate | $CH_3NCO$ | 1.3% |
| Acetone | $CH_3COCH_3$ | 0.3% |
| Propionaldehyde | $C_2H_5CHO$ | 0.1% |
| Acetamide | $CH_3CONH_2$ | 0.7% |
| Glycolaldehyde | $CH_2OHCHO$ | 0.4% |
| Ethylene glycol | $HOC_2H_4OH$ | 0.2% |

Recent experiments on the irradiation of ice mixtures reveal that many of the larger molecules can be formed by electron bombardment, often at low energies and this will be discussed further in Section 5. Indeed, the bombardment and dissociation of ice species has been proposed as a route by which molecular oxygen can form. One of the most unexpected observations of Rosetta through the Rosina instrument was the detection of molecular oxygen as the fourth most abundant gas in the atmosphere of comet 67P. Oxygen is reactive so it was felt that it is unlikely to survive long periods in space. The amount of molecular oxygen detected showed a strong relationship to the amount of water measured at any given time, suggesting that their origin on the nucleus and release mechanism are linked and that irradiation of water ice leading to oxygen production and storage in the ice is a plausible mechanism for oxygen formation on a comet [11].

## 2.4. Electrons in the Cometary Coma

That there are copious amounts of electrons to induce such dissociative excitation was confirmed by ion and electron sensors (RPC-IES) on the Rosetta craft. Concentrations of particles and their time evolution in inner coma plasma was measured by Rosetta Plasma Consortium (RPC) [12] using a set of sensors developed for this purpose. The Ion and Electron Sensor (IES) was an electrostatic plasma analyser that covered an energy/charge range from 1 eV/e to 22 keV/e with a resolution of 4% [13]. The sensor provided 3D ion and electron distributions over the whole measured energy range. It was capable of simultaneously measuring electrons and positive ions with the single entrance aperture owing to two back-to-back top-hat geometry analysers. The LAP instrument (Langmuir probes) measured the plasma density in the range of ($10^0$–$10^6$ cm$^{-3}$), electron temperature ($10^2$–$10^5$ K) and plasma flow velocity (up to $10^4$ ms$^{-1}$). The LAP also measured the AC electric field up to 8 kHz [14]. The LAP was complemented by the Mutual Impedance Probe, MIP which probes the plasma and measured the natural plasma frequency which yields the electron density in the range from 2 cm$^{-3}$ to $1.5 \times 10^5$ cm$^{-3}$ and temperature from 30 K to $10^6$ K [15].

Depending on the comet distance from the Sun, both the solar wind and solar radiation interact with its nucleus and inner coma shielded by comet's own atmosphere and ionosphere. Cometary ions are created by photoionization of neutral species, mainly like $H_2O$ and $CO_2$ and their products from photodissociation, and by charge transfer with solar wind protons [16]. Solar wind electrons in interplanetary space typically have Maxwellian distribution functions with thermal energies of several eV to tens of eV [13]. This energy distribution of electrons differs from one created by photoionization of cometary neutrals by solar radiation in the cometary comas at certain distances from the Sun. Electrons of cometary origin are mainly the product of photochemistry, originating from direct photoionization and from Auger processes. They are thermalized by collisions, elastic and inelastic.

The electron density in the coma was measured complementary by the RPC Langmuir Probe (LAP) and Mutual Impedance Probe (MIP). The first findings of the spatial distribution of the plasma near comet 67P/CG showed a highly structured pattern that indicated an origin from local ionization of neutral gas. The electron density fell off with distance as $1/r$ in the range from 8 km from the nucleus up to 260 km [17]. Edberg et al. [17] concluded that this is in accord with a model in which the ionization of a neutral gas is expanding radially from the comet nucleus and when there is no significant recombination or other loss source for the plasma. However, they warned that the observed data have a large scatter around fitted $1/r$ curve and that results could be an average effect of combination of transport electric fields and solar wind.

From such data, the suprathermal energy distribution of electrons could be derived. The electron energy distribution near the comet depends on the comet distance from perihelion and mass loading process when the atoms and molecules in the cometary coma are photoionized and then interact with the solar wind flow. It spans from the energy distribution of the solar wind itself to the modified distribution where electrons are significantly decelerated as a consequence of magnetic field causing regions to pile-up. The mass loading process is connected to the outgassing rate of the comet. The formation of suprathermal electrons which are accelerated from a few eV upward to hundreds of eV, thus can play an important role in the electron driven chemistry of the comet. The Rosetta IES sensor recorded the presence of suprathermal electrons at larger distances than expected from the previous models of such weakly outgassing comets [18]. The observed electron energy distributions change by reducing a heliocentric distance from pure solar wind distribution to non-Maxwellian one that include suprathermal electrons showing maxima at energies from 10 eV to 300 eV [18]. Clark et al. [18] hypothesize that the most likely mechanisms of creating accelerated electron distributions are heating by waves generated by the pick-up ion instability and by the mixing of cometary photoelectrons, secondaries and solar wind electrons.

Further statistical analysis of Rosetta IES sensor recordings by performing fitting procedures that involve two separate sub-populations of electrons below and above 8.6 eV mean energy [19] revealed different relationships between their density and temperature and possible mechanisms of creating suprathermal electrons. Broiles et al. [19] suggested that electrons above 8.6 eV are being heated by waves driven by counter streaming solar wind protons. This conclusion arises from the observations that the population of electrons above 8.6 eV correlates well with the density of local neutrals, while the sub-population below 8.6 eV is dominated by the local magnetic field strength. Recently, Deca et al. [20] have used a fully 3D kinetic model to simulate the ion and electron dynamics of the solar wind interaction with a weakly outgassing comet 67P. They used a detailed kinetic treatment of the electron dynamics in order to cover energy distribution of electrons and to identify the origin of the warm and suprathermal electrons.

Electron energy ranges that correspond to the relevant processes in electron collisions with atoms and molecules are shown in Figure 1. The elastic cross section is dominating over low electron energies and usually is prominent even at higher energies where the ionization cross section becomes comparable in magnitude. Vibrational excitations are important at low energies but they also extend to higher energies due to resonance decay. Attachment and dissociation processes are relevant in certain domains, in water between 6 to 9 eV and 20 to 200 eV, respectively (see the summery figure of cross sections in [3]), but due to dissociative electron attachment (DEA) these processes may extend to very low energies. The production of radiation due to de-excitation depends on excitation energy levels. For water molecule production of Lyman alpha radiation has a high cross section in the electron energy range from 50 to 200 eV.

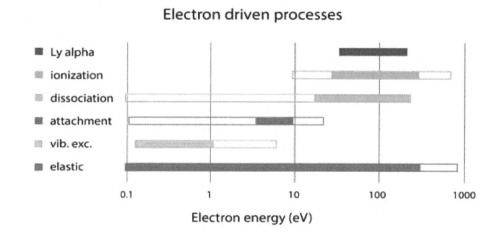

**Figure 1.** Electron energy ranges that correspond to the relevant processes in collisions with atoms and molecules. Full colour corresponds to the specific case of water (data taken from Itikawa and Mason [3]) while the open bars are extended ranges that correspond to the majority of atomic species: red, elastic scattering; green, vibrational excitations; violet, attachment; blue, dissociation; orange, ionization; darkblue, Lyman alpha.

## 3. Atomic and Molecular Data Needed for Analysing Electron Scattering Processes Relevant to Comet 67P

Despite the evidence for electron induced processes in comets many of the discrete collision processes necessary to quantify such electron driven chemistry remain uncertain. In order to develop a predictive model of cometary coma and the comet's 'atmosphere' it is necessary to assemble a 'database' of relevant electron collision processes with the different atomic and molecular species observed (or indeed inferred) from Rosetta. As discussed above Rosetta's ROSINA, as well as Ptolemy and COSAC on the Philae lander, have revealed a rich chemical inventory that would require a large atomic and molecular physics database to model all possible processes. However, if a sensitivity analysis is performed the number of important species contributing to the model may be reduced to a minimum (more manageable) number of reactions. The dominant molecules are water, CO and $CO_2$. The presence of HCN as the source of CN radicals and ammonia as source of NH and $NH_2$ is widely accepted while the simple hydrocarbons $C_2H_2$ and $C_2H_6$ are assumed to be the source of $C_2$, indeed $C_2H_6$ concentrations were unusually high in comet 67P [21]. Methane has been identified in many comets and is found in comet 67P. As discussed above several oxygen containing species were detected and apart from the ubiquitous water methanol $CH_3OH$ may be an important primary compound. The primary source of sulphur compounds may be $H_2S$ but 67P is depleted in all other sulphur bearing species ($CS_2$, OCS and $SO_2$) compared to other comets [22]. Thus, in developing an electron chemistry model of comet 67P it is necessary to have a good data base for electron interactions with $H_2O$, CO, $CO_2$, $CH_4$, $C_2H_6$, $CH_3OH$, $NH_3$, HCN and $H_2S$. The status of such a database will be discussed below but first it is necessary to understand the corollary for a 'good' database.

### 3.1. Databases

Many databases exist in order to assemble datasets and communicate them to different audiences. The NIST database collection (https://srdata.nist.gov/gateway/gateway?dblist=0) is one of the best known providing details of the structure, spectroscopy and fundamental parameters (ionization and dissociation energies) of many atoms and molecules. The need for large datasets has led to several communities investing in establishing data centres which assemble and maintain databases. For example, the fusion community has, for several decades, compiled databases in order to model plasmas in tokamak reactors and to provide data for diagnostic tools used in such plasmas. Another example of large collection of collisional data is the LXCaT database [23], which provides electron and ion scattering cross sections, swarm parameters (mobility, diffusion coefficient, etc.), reaction rates, energy distribution functions, etc. and other data required for modelling low temperature plasmas. Similarly, the astronomical community has needed large databases to interpret its observations, these include not only spectroscopic databases but also databases of chemical reaction rates (e.g., KIDA [24]) are necessary to understand the rich inventory of molecules that have been observed in the interstellar medium.

Many databases are simple collections of data but more recently the design and operation of databases has been refined. The development of IT tools has allowed data to be provided on-line, downloadable in a range of formats and allows new data to be added quickly, ensuring that the data is up to date. Previously data was reviewed and published in journal reviews which once published became gradually out of date until the next review, often a decade later (e.g., [25]). The opportunity to add new data quickly not only ensures that the latest data is adopted by the community but also reduces the likelihood of fragmentation amongst the community with different groups using different data sets in accord with their knowledge (or more commonly lack of knowledge) of the data available.

Simple assembly of data alone is not, however, the most effective form of databases. The 'user' community requires guidance as to what data to adopt. Users rarely have the necessary experience to select one dataset over another and therefore each may choose different sets, leading to systematic problems. For example, if different datasets are used in different models, cross comparison of such models is difficult and it may be hard to distinguish between the different physical and chemical hypotheses in different models from the data used in the model to explore such effects. Therefore, databases should provide 'recommended data' which is the data that the expert community providing such data believes is the optimal data reflecting state of the art measurements or calculations. These values can be updated as new data becomes available. However, when changing recommended data, it is essential to ensure that it is still 'consistent' For example in presenting a comprehensive set of electron impact cross sections the individual cross sections (elastic and inelastic (including ionization excitation etc.) should, when summed, be consistent with the recommended data for total cross sections. Databases should also present data with stated estimates of uncertainties, particularly when presenting its own composite data from several different datasets.

### 3.2. VAMDC and BEAMDB Databases

The Virtual Atomic and Molecular Data Centre—(VAMDC) and Belgrade Electron Atom and Molecule Database—(BEAMDB) are two examples of new generation of databases. The VAMDC Consortium is a worldwide consortium which federates atomic and molecular databases through an e-science infrastructure to provide easy access to data from different databases via a single portal http://portal.vamdc.eu. About 90% of the inter-connected databases are focused on data that are used for the interpretation of astronomical spectra and for modelling in many fields of astrophysics and astrochemistry, although recently the VAMDC Consortium has connected databases from the radiation damage and the plasma communities which makes it suitable for medical and industrial applications [26]. While VAMDC does not itself select and analyse data it ensures data from its component databases are accessible in a single format and ascribe to general good practices as discussed above. The VAMDC Consortium includes new databases and services on a case by case basis during annual general scientific and technical meetings.

VAMDC provides its data in a XSAMS output. XSAMS is an XML representation of an atomic and molecular data model. The system allows for distributed querying of data via the VAMDC-TAP protocol, an implementation-agnostic standard, where data providers can build their models in their own fashion and map them to the VAMDC model via a dedicated dictionary [27].

The Belgrade Electron Atom and Molecule Database—(BEAMDB) [28] is an application, database and a VAMDC node which contains data for elementary processes of electron scattering by atoms and molecules. The database covers collisional data of electron interactions with atoms and molecules in the form of differential and integrated cross sections as well as energy loss spectra. The data is stored in a relational (MySQL) database, upon a static model specifically suited to this dataset but easily extendable. There have already been several migrations of the model, the latest of which is an extension to enable storing non-neutral molecules. Currently, there are 22 species stored in the database (11 atoms and 11 molecules), presented in 71 states, involved in 59 collision processes. The web interface (http://servo.aob.rs/emol) enables on-site querying of data via an AJAX-enabled web form. The application is implemented in Django, a Python web framework and hosted on an Apache web server at the Astronomical Observatory in Belgrade.

The BEAMDB is a collisional database where several types of collisions are included: Elastic, Electronic Excitation, (Total) Inelastic, Ionization, and Total Scattering as well as electron spectroscopic data such as Energy-loss Spectra and Threshold Photoelectron Spectra. Cross sections are of several different kinds: Differential, Integral, Total, Momentum Transfer, Viscosity. Specific data that are maintained in the BEAMDB are differential cross sections (DCS) for elastic scattering and excitation of atoms and molecules. These are 3D entries since DCS depend on both electron impact energy and scattering angle. This requires two X columns while the Y column is also associated with the column representing uncertainty of data points. An example of the XSAMS output of such kind of data for He excitation is shown in Figure 2.

Producing a plot of DCS data is not available at the current stage of database development, but such data can be visualized by using either VAMDC portal or alternatively the RADAM (RADiation DAMage) database portal (http://radamdb.mbnresearch.com/). The general structure of RADAM databases covers electron/positron interactions, ionic and photonic interactions, multiscale radiation damage phenomena and radiobiological phenomena occurring at different time, spatial and energy scales in irradiated targets [29]. Examples of DCS surfaces for elastic electron scattering by helium atom and formamide ($CH_3NO$) molecule are shown in Figure 3a,b, respectively.

```xml
<?xml version="1.0" encoding="UTF-8"?>
<XSAMSData xsi:schemaLocation="http://vamdc.org/xml/xsams/1.0 http://vamdc.org/xml/xsams/1.0" xmlns:cml="http://www.xml-cml.org/schema"
xmlns:xsi="http://www.w3.org/2001/XMLSchema-instance" xmlns="http://vamdc.org/xml/xsams/1.0">
  - <Sources>
      - <Source sourceID="BIPBemol-2017-09-10-9-22-42">
          <Comments> This Source is a self-reference. It represents the database and the query that produced the xml document. The sourceID contains a
            timestamp. The full URL is given in the tag UniformResourceIdentifier but you need to unescape ampersands and angle brackets to re-use it. Query
            was: select * where (CollisionCode = 'exci') AND ((AtomSymbol = 'He')) </Comments>
          <Year>2017</Year>
          <Category>database</Category>
          <UniformResourceIdentifier>http://servo.aob.rs/emol/tap/sync?
            LANG=VSS2&REQUEST=doQuery&FORMAT=XSAMS&QUERY=select+*+where+%28CollisionCode+%3D+%27exci%27%29+AND+%28%
            28AtomSymbol+%3D+%27He%27%29%29</UniformResourceIdentifier>
          <ProductionDate>2017-09-10</ProductionDate>
          + <Authors>
      </Source>
      + <Source sourceID="BIPBemol-JPB42_145202_2009_b9_14_145202_Hoshino_etal">
  </Sources>
  - <Species>
      + <Atoms>
      + <Particles>
  </Species>
  - <Processes>
      <Radiative></Radiative>
    - <Collisions>
      - <CollisionalTransition id="PIPBemol-C56">
          <SourceRef>BIPBemol-JPB42_145202_2009_b9_14_145202_Hoshino_etal</SourceRef>
        - <ProcessClass>
            <UserDefinition>Electronic Excitation</UserDefinition>
            <Code>exci</Code>
            <IAEACode>EEX</IAEACode>
          </ProcessClass>
        - <Reactant>
            <SpeciesRef>XIPBemol-22</SpeciesRef>
            <StateRef>SIPBemol-71</StateRef>
          </Reactant>
        - <Reactant>
            <SpeciesRef>XIPBemol-XElectron</SpeciesRef>
          </Reactant>
        - <Product>
            <SpeciesRef>XIPBemol-22</SpeciesRef>
            <StateRef>SIPBemol-72</StateRef>
          </Product>
        - <DataSets>
          - <DataSet dataDescription="IOP(2009) He 3S1 (19.82) Excitation">
            - <TabulatedData>
                <SourceRef>BIPBemol-JPB42_145202_2009_b9_14_145202_Hoshino_etal</SourceRef>
              - <X units="eV" parameter="energy">
                  <DataList count="70">23.5 23.5 24 24 24 24 24 24 24 24 24 24 24 24 25.34 25.34 25.34 25.34 25.34 25.34 25.34 25.34
                    25.34 25.34 25.34 25.34 25.34 25.34 27.5 27.5 27.5 27.5 27.5 27.5 27.5 27.5 27.5 27.5 27.5 27.5 27.5 27.5 30 30 30 30
                    30 30 30 30 30 30 30 30 30 30 35 35 35 35 35 35 35 35 35 35 35 35 35 35</DataList>
                </X>
              - <X units="deg" parameter="angle">
                  <DataList count="70">60 90 30 40 50 60 70 80 90 100 110 120 130 10 15 20 30 40 50 55 60 70 80 90 100 110 120 130 10 15
                    20 30 40 50 60 70 80 90 100 110 120 130 10 15 20 30 40 50 60 70 80 90 100 110 120 130 10 15 20 30 40 50 60 70 80 90
                    100 110 120 130</DataList>
                </X>
              - <Y units="m2/sr" parameter="DCS">
                  <DataList count="70">19.1e-20 44.1e-20 5.78e-20 7.23e-20 12.6e-20 17.6e-20 25.7e-20 36e-20 39e-20 43.2e-20 33.4e-20
                    20.9e-20 12.4e-20 23.5e-20 21.2e-20 15.9e-20 14e-20 11.5e-20 11.9e-20 13.6e-20 16.1e-20 19.2e-20 23.3e-20 22.9e-20
                    19.6e-20 15.4e-20 9.77e-20 5.55e-20 22.5e-20 22.9e-20 22.9e-20 20.2e-20 17e-20 14.4e-20 19.2e-20 24.1e-20 30.9e-20
                    30.8e-20 23.4e-20 20.6e-20 13.1e-20 11.9e-20 44.9e-20 43.2e-20 32e-20 21.8e-20 13.5e-20 11.5e-20 11.8e-20 13.7e-20
                    16.4e-20 15.2e-20 12.8e-20 9.76e-20 7.94e-20 8.9e-20 44.6e-20 40e-20 34.1e-20 23.8e-20 16.7e-20 9.83e-20 8.21e-20
                    9.42e-20 9.42e-20 7.75e-20 6.57e-20 5.94e-20 7.93e-20 8.62e-20</DataList>
                - <Accuracy type="statistical" relative="false">
                    <ErrorList count="70">2.865e-20 6.615e-20 0.867e-20 1.0845e-20 1.89e-20 2.64e-20 3.855e-20 5.4e-20 5.85e-20 6.48e-
                      20 5.01e-20 3.135e-20 1.86e-20 3.525e-20 3.18e-20 2.385e-20 2.1e-20 1.725e-20 1.785e-20 2.04e-20 2.415e-20
                      2.88e-20 3.495e-20 3.435e-20 2.94e-20 2.31e-20 1.4655e-20 0.8325e-20 3.375e-20 3.435e-20 3.435e-20 3.03e-20
                      2.55e-20 2.16e-20 2.88e-20 3.615e-20 4.635e-20 4.62e-20 3.51e-20 3.09e-20 1.965e-20 1.785e-20 6.735e-20 6.48e-20
                      4.8e-20 3.27e-20 2.025e-20 1.725e-20 1.77e-20 2.055e-20 2.46e-20 2.28e-20 1.92e-20 1.464e-20 1.191e-20 1.335e-20
                      6.69e-20 6e-20 5.115e-20 3.57e-20 2.505e-20 1.4745e-20 1.2315e-20 1.413e-20 1.413e-20 1.1625e-20 0.9855e-20
                      0.891e-20 1.1895e-20 1.293e-20</ErrorList>
                  </Accuracy>
                </Y>
              </TabulatedData>
            </DataSet>
          </DataSets>
        </CollisionalTransition>
      + <CollisionalTransition id="PIPBemol-C57">
      + <CollisionalTransition id="PIPBemol-C58">
      + <CollisionalTransition id="PIPBemol-C59">
      </Collisions>
    </Processes>
  </XSAMSData>
```

**Figure 2.** XSAMS output table for differential cross section data for He electron excitation.

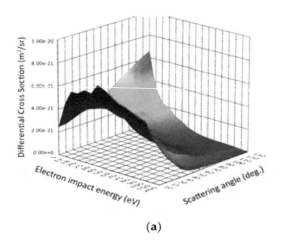

(a)

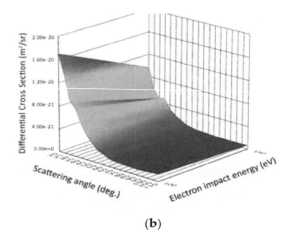

(b)

**Figure 3.** DCS surfaces as retrieved from RADAM database [29] for elastic electron scattering by: (**a**) He atom—data points are taken from ref. [30]; (**b**) Formamide molecule—data points are taken from ref. [31].

Initially BEAMDB was designed to maintain data of electron collisions with neutral species, atoms and molecules mainly in their ground state and exceptionally in excited state. However, by broadening the scope toward astrophysical applications, more specifically to the processes of comas that involve electrons, the database has been upgraded to include electron collisions with ions. The most recent data set included in BEAMDB is the one of electron ionisation cross sections of $CN^+$ cations [32]. In the next few months BEAMDB will expand to include many more molecular systems including those needed to study the electron chemistry of comet 67P.

## 4. Electron Scattering Processes and Cross Sections—Data Needs

To understand and treat by models, processes in cometary plasmas is a very challenging task since many parameters need to be taken into account in order to cover the variety of comet types and their heterogeneity. One has to consider the changing of comet distances from the Sun and hence the level of irradiation and solar wind interactions both with the comet surface and cometary plasma environment. Nevertheless, models have been developed and set of processes reviewed, including data analysis used in such models. Particles from the solar wind, secondary electrons created in plasmas and photoelectrons, produce further events of excitation, ionization and dissociation with the consequence of enhanced chemical reactions and light emission. However, comets are composed of water, silicates and carbonaceous molecules (CO, $CO_2$ and hydrocarbons) [33]. Recently, modelling of plasma processes in cometary and planetary atmospheres has been performed by Campbell and Brunger [34] with an emphasis on the role of electron-impact excitation processes. They concluded in the case of comet Hale-Bopp that electron-impact could account for 40% of the fluorescence emissions of the fourth positive bands ($A^1\Pi$–$X^1\Sigma^+$) of CO [35] and thus reducing calculated outgassing rates. Even more, their later paper [36] was focussed on electron initiated chemistry in atmospheres.

Reviews of cross section data and processes that cover electron scattering and excitations are numerous and they cover interactions with atoms [37,38], diatomic molecules [37,39], species in interstellar clouds [40] or concentrate on specific targets of triatomic molecules like water [3] or $CO_2$ and $N_2O$ [40]. Anzai et al. [41] stressed that any recommended values of cross section data currently maintained in different databases might need to be updated due to the development of new experimental techniques and theoretical methods. The number of established benchmark cross sections is rather small.

The energy of the electrons available for electron interactions with atoms and molecules in the cometary coma 'atmosphere' is such that all electron scattering processes are relevant, thus a large amount of data is required if a model of electron induced processing is to be included in a simulation model of comet 67P. In this section, the status of our knowledge of such relevant cross sections will be reviewed for the primary molecules defined in Section 3.

### 4.1. Elastic Electron Scattering and Cross Sections

Elastic scattering conserves the kinetic energy of the colliding particles. This means that quantum numbers that determine the energy are unchanged but other quantum numbers corresponding to degenerate states (e.g., helicity or spin flip) may change. In the case of many measurements not all states in the system are resolved due to the finite resolution of the electron beams used. In this case 'effective elastic cross sections' are determined which may be referred to as 'rotationally unresolved, vibrationally unresolved, electronically unresolved etc. Elastic scattering is important since, although there are no immediate changes in the target the range and hence spatial extent of the electrons is determined by such scattering.

Elastic cross sections are usually measured at specific energies and angles. These data are used to determine the 'total elastic cross section' by integrating over the entire angular range ($4\pi$). The total elastic cross section at given electron impact energy $E_i$ is given by:

$$Q_{el}(E_i) = \int_0^{4\pi} \frac{d\sigma(k; \theta, \varphi)}{d\Omega} d\Omega, \tag{1}$$

where $\frac{d\sigma(k;\theta,\varphi)}{d\Omega}$ is the elastic differential cross section, ($\theta$, $\varphi$) are the scattering angles and $k$ is the wave-vector magnitude.

Elastic scattering is one of the best-studied electron collision processes and provides one of the best tests of developing theoretical calculations of electron-molecule scattering. The development of the magnetic angle changing method [42] to extend differential cross section measurements to the full range of scattering angles from 0° to 180° has significantly improved the accuracy of total elastic cross sections, particularly for molecules with dipole moments, where elastic scattering is strongly forward peaked.

Elastic scattering cross sections have been reported for all the primary molecules $H_2O$, $CO$, $CO_2$, $O_2$, $CH_4$, $C_2H_2$, $CH_3OH$, $NH_3$, $HCN$ and $H_2S$. Elastic scattering cross sections for water have been discussed in detail as part of a wider review of electron scattering from water by Itikawa and Mason [3]. An updated review has recommended the corrected data of Khakoo et al. ([43] and erratum) for low energy scattering and Munoz et al. [44] for higher energies (where experiment and theoretical evaluations merge). The benchmarking swarm paper by de Urquijo et al. [45] on cross sections for water reproduced measured transport data in water/helium mixtures and presented the integral cross sections that are entirely self-consistent with the available total cross sections as well as the swarm data over a large range of reduced electric field, E/N.

Cross sections for elastic scattering from methane and acetylene have recently been compiled and evaluated by Song et al. [46,47]. Compilations of data for other molecules are less recent and more fragmented and should be updated. Due to its toxicity, there are few measurements of the elastic scattering cross section from HCN and therefore there is more reliance on theoretical calculations (e.g., Sanz et al. [48]).

In elastic collisions electrons do not lose energy but change the direction of motion. This is important for models where the kinetics of all particles is taken into account. In more dense plasmas the elastic momentum transfer cross section, defined as integrated DCS with the weight of $(1 - \cos\theta)$ over all scattering angles, is a more relevant quantity. Differential cross sections, although being one of the basic properties that defines electron—atom/molecule interactions, are known with relatively low accuracy. In order to illustrate the current status of the agreement of DCS amongst different experiments and theories, the case of absolute cross sections for elastic electron scattering by argon atom is presented in Figure 4. It can be seen that although at first sight all values group around the averaged values, it should be noted that the data are plotted on a logarithmic scale and that there is almost an order of magnitude disagreement for particular data points.

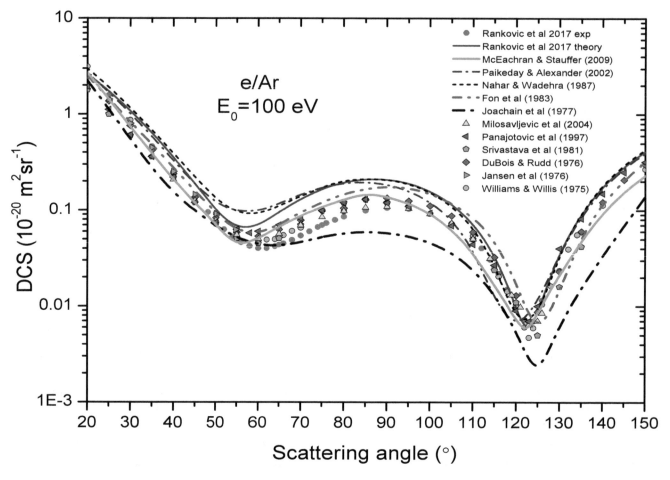

**Figure 4.** DCS for elastic electron scattering by argon atom at 100 eV impact energy. The data symbols used correspond to: (red circles) , Rankovic et al. [49] experiment; (dark red full line), Rankovic et al. [49] theory; (magenta solid line), McEachran and Stauffer [50]; (olive dash dot line), Paikeday and Alexander [51]; (dotted line), Nahar and Wadehra [52]; (violet dash dot dot line), Fon et al. [53]; (cyan dash dot line), McCarthy et al. [54]; (dashed line), Joachain et al. [55]; (cyan up triangles), Milosavljević et al. [56]; (violet left triangle), Panajotović et al. [57]; (orange pentagons), Srivastava et al. [58]; (blue diamond), DuBois and Rudd [59]; (orange right triangles), Jansen et al. [60]; (green circles), Williams and Willis [61].

## 4.2. Electron Impact Ionisation Cross Sections

Most of the ions observed in the comet ion tail are the result of photoionisation of the primary ice species and since there are fewer high energy electrons, electron induced ionisation in comets is likely to be a minor process in total ion yields. However, the mass spectrometric analysis of compounds observed on Rosetta (e.g., using the Rosina instrument) rely upon knowing the fragmentation patterns of candidate molecules which, when compared with the instrument sensitivity, can be used to calculate the relative abundances of the detected molecules. Traditionally mass spectrometers operate with electron energies of 70 eV, close to but not at, the maximum of total ionisation cross sections. Branching ratios for fragments of electron impact ionisation are available in many databases (e.g., NIST Chemistry WebBook [62]) however, whilst these ratios are often known the cross sections are not presented. These cross sections may be derived if the total ionisation cross section is known. Total ionisation cross sections may be measured to an accuracy of <10% while semi-empirical calculations provide reliable cross sections (at least above 100 eV). Data on all of the primary molecules $H_2O$, CO, $CO_2$, $CH_4$, $C_2H_2$, $CH_3OH$, $NH_3$, HCN and $H_2S$ exist with an accuracy sufficient for providing reliable data for determining their concentrations in the cometary coma and atmosphere. The recent review by

Tanaka et al. [63] covered this topic in detail employing scaled plane-wave Born models in order to provide comprehensive and absolute integral cross sections, first for ionization and then to optically allowed electronic-state excitation.

## 4.3. Anion Production

Anions have been observed in comets and may be formed by a variety of processes including radiative electron attachment, polar photodissociation, proton transfer and Dissociative Electron Attachment (DEA) where an incident electron is captured by the molecular target (AB) to form an excited state of the molecular negative ion $AB^-$. This state, a Temporary Negative Ion (TNI), generally decays by ejecting the excess electron within a finite time (a process called autodetachment) but the molecular negative ion may also decay through dissociation leading to the formation of a stable negative ion B- and a neutral (often radical) fragment (A). DEA to the list molecules have been studied, identifying the fragment channels but there are few absolute cross sections. Node of VAMDC, the IDEADB maintained by University of Innsbruck, that serves data about dissociative electron attachment to molecules, lists more than 120 different fragments resulting from this process [64]. Anion data from comet 67P is still under evaluation but earlier studies from the Giotto spacecraft of comet 1P/Halley led to a combined chemical/hydrodynamic model for the coma of comet Halley to explore various anion production mechanisms and compute the abundances of atomic and molecular anions as a function of radius in the coma [65]. The dominant anion production mechanisms are found to be polar photodissociation of water and radiative electron attachment to carbon chains in the inner coma, followed by proton transfer from $C_2H_2$ and HCN to produce $C_2H^-$ and $CN^-$, respectively. However, in the outer regions of the coma where electron temperatures reach $10^3$–$10^5$ K, dissociative electron attachment may become a dominant process. Similar effects may be understood for comet 67P. DEA to water yields $H^-$ and $OH^-$ and $O^-$ from CO and $O_2$ so there are many candidates for production of anions in comet 67P. DEA to all of the primary comet species (and most of the larger more complex species in Table 1) has been reported with DEA fragments recorded from near zero to the ionisation energy pathways. Nevertheless, very few absolute cross sections are available.

## 4.4. Electron Impact Excitation and Dissociation

Electron impact excitation and dissociative excitation of molecular systems is a critical process for a study of cometary coma and its tenuous atmosphere. As discussed above, OSIRIS and ALICE data from Rosetta shows the electron induced dissociation of water may be the source of the O I line at 135.6 nm while electron impact by CO and $CO_2$ yields C I lines at 156.1 and 165.7 nm. However, this hypothesis is handicapped by the dearth of data on electron impact excitation and electron induced neutral fragmentation for all molecules, not just those of immediate comet interest. This lack of experimental data can be attributed to difficulties in measuring neutral atoms and molecules. When an atom or molecule is in an excited state it may decay (fluoresce) with the emitted light being detected. Such experiments may identify some fragmentation or de-excitation pathways but the sensitivity of the optical detector and ability to 'capture' all of the emitted photons as well as the problem of cascades from higher lying states into the decaying state make measurements of absolute cross sections difficult. Furthermore, some excited states decay to 'dark' non-fluorescent or metastable states which makes them hard to detect. Although metastable fragments may be detected directly by surface ionization they will suffer from the same problems as photon detection i.e., cascade contributions may dominate [66,67]. Presently there are few experiments measuring electronically excited fragments by optical or metastable spectroscopy and more experiments are to be encouraged, building on the recent commissioning of electron induced fluorescence (EIF) experiment in Comenius University Bratislava. Figure 5 shows the H atom spectra recorded by electron impact of molecular hydrogen. This experiment is well equipped to study EIF of water, CO and $CO_2$ as required for cometary studies although, due to the low cross sections, data collection periods may be days or even weeks placing stringent conditions on the stability of the incident electron and gas beams.

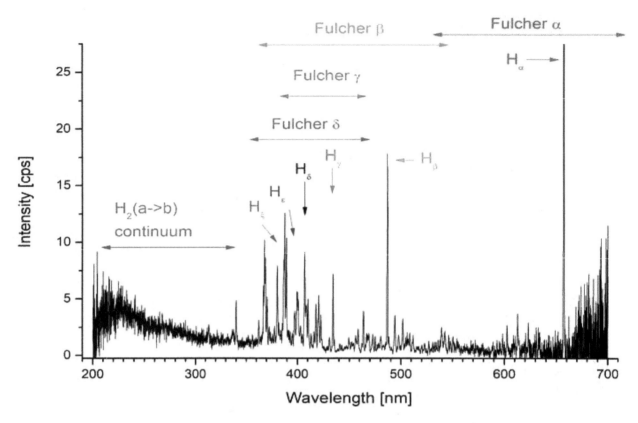

**Figure 5.** Electron induced fluorescence of the $H_2$ molecule—Balmer lines and Fulcher $\alpha$ system. Measurements performed by Danko et al. [68].

Electron Energy loss spectroscopy monitors the energy of the incident electron post collision and may also be used to probe the direct excitation cross section of the parent molecule excited states but cannot provide data on the fragmentation patterns of that excited state as it decays. Furthermore, in most molecules the electronically excited states are both close together and overlap their ro-vibrational bands, making deconvolution very difficult if discrete electronic excitation cross sections are to be derived.

Finally, the production of neutral fragments in their ground state must be considered. Photon and electron induced dissociation produces many fragments in their ground state where ground state in this case includes fragments that are ro-vibrationally excited but still in the electronic ground state. This low internal energy precludes their detection by fluorescence since IR detection has not proven possible due to IR sources in the apparatus (e.g., electron filaments). Several alternative methods have been proposed to detect ground state neutral fragments including using a second electron beam to ionize the product, or use of surfaces to 'getter' the fragments. In the current context, there are only two experiments relevant to the modelling of electron dissociation of primary comet molecules—that of Harb et al. [69] measuring OH radical production from water and C and O from CO by Cosby et al. [70] using a fast beam method.

Further experimental studies on electron impact dissociation to neutral fragments will not only benefit the cometary community but the wider electron chemistry community with applications in many plasma systems, aeronomy and radiation chemistry. However, given the experimental difficulty much of the necessary data may be provided by theoretical calculations, which require more detailed exploration.

## 5. Possibility of Electron Induced Surface Chemistry

As discussed above (Section 2.3) many of the larger more complex molecules observed by COSAC and Ptolemy on the Philae lander may be made by addition reactions from simpler molecules. How are such reactions induced? Photodissociation has been considered the primary process but electron induced chemistry within ices has been shown to be an efficient route to molecular synthesis and simple electron irradiation of primary ices has been shown to be produce most (all) of the larger molecular species. For example, Figure 6 shows the yield of formamide in an ice film composed of CO and $NH_3$ as a function of electron energy [71]. The ice was prepared with a mixing ratio of 1:8 and thickness corresponding to 12–18 monolayers and an electron exposure of 200 μC/cm$^2$. Formamide is readily formed and the resonance like feature between 6 and 12 eV is characteristic of the synthesis by reactants prepared in a dissociative electron attachment process [71]. Similar experiments have shown that as many as 15 products can be formed by electron irradiation of pure methanol ices [72] including ethylene glycol and methyl formate whilst formamide $HCONH_2$ is formed in irradiation of binary mixtures of ammonia and methanol ice and the simplest amino acid glycine from irradiation of a methylamine and carbon dioxide ice [73]. Thus, electron induced synthesis of simple cometary ices may be a route to formation of several of the organic species observed in surface material from comet 67P.

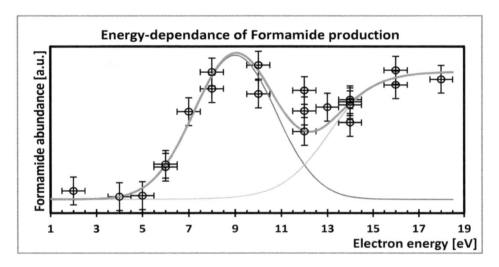

**Figure 6.** The production of formamide present in mixed multilayer films of CO and $NH_3$. The resonance like feature (red line) between 6 and 12 eV is characteristic of the synthesis of formamide by reactants prepared in a DEA process, Bredehöft et al. [71].

The route by which molecular oxygen was formed as the fourth most abundant compound in the coma observed by Rosetta, is still subject to debate. However, laboratory experiments [74,75] have shown that radiolysis of water by both electrons and photons yields molecular oxygen but also copious amounts of hydrogen peroxide. Furthermore, comparative experiments between photon and electron irradiation show that electron induced yields are higher for the same energy. Whether this is due to penetration depth of electrons or that electrons open more dissociative pathways (through dipole or spin forbidden transitions) is unknown. The role of electron induced chemistry in comets, in ice covered planetary and lunar objects and in the rich chemistry of the interstellar medium is therefore an emerging topic of modern astronomy and one that has been encouraged by the results of the Rosetta mission.

## 6. Conclusions

In this paper, we have presented a review of recent results from the Rosetta mission to comet 67P/Churyumov-Gerasimenko. The role of electron induced processes has been highlighted with the emission of [O I] lines in the coma explained by the process of electron impact dissociation of water. The role of other electron processes e.g., in the production of the unexpectedly large amounts of molecular oxygen in the coma, is handicapped by lack of detailed knowledge of electron impact cross sections We have reviewed the need for electron scattering data and discussed how such data should be collated and reviewed in electron scattering databases. The BEAMD database which is a part of a larger consortium of Virtual Atomic and Molecular Data Centre—VAMDC has been used as an example of modern generation of databases.

The importance of electron processes in comet 67P/Churyumov-Gerasimenko highlights the need for closer interactions and joint projects between the cometary and electron communities and this paper has identified some topics for joint research.

**Acknowledgments:** BPM recognizes support from MESTD-RS project #OI 171020 and the grant under 2016 VAMDC Consortium Call-#1. VV and DJ recognizes support from MESTD-RS project #III 44002. NJM recognizes support from Europlanet 2020 RI, which has received funding from the European Union's Horizon 2020 research and innovation programme under grant agreement number 654208 and ELEvaTE grant agreement number 692335, as well as the support of the UK STFC and the Leverhulme trust. We thank Juraj Országh for providing us with the unpublished figure of the hydrogen spectrum (Figure 5).

**Author Contributions:** B.P.M. wrote Section 2.1, Section 2.2, parts of Section 3 and provided Figures 1–4. J.H.B. wrote Section 2.3, provided Figure 6 and Table 1 and provided language editing. V.V. and D.J. wrote Section 3.2. N.J.M. wrote Sections 2, 4 and 5 and edited the paper. B.P.M. is responsible for maintaining electron cross section data in BEAMDB, while V.V. and D.J. are responsible for maintaining the database according to the VAMDC standards.

# References

1.  Altwegg, K.; Balsiger, H.; Bar-Nun, A.; Berthelier, J.J.; Bieler, A.; Bochsler, P.; Briois, C.; Calmonte, U.; Combi, M.; De Keyser, J.; et al. 67P/Churyumov-Gerasimenko, a Jupiter family comet with a high D/H ratio. *Science* **2015**, *347*. [CrossRef] [PubMed]

2.  Feldman, P.D.; A'Hearn, M.F.; Bertaux, J.-L.; Feaga, L.M.; Parker, J.W.; Schindhelm, E.; Steffl, A.J.; Stern, S.A.; Weaver, H.A.; Sierks, H.; et al. Measurements of the near-nucleus coma of comet 67P/Churyumov-Gerasimenko with the Alice far-ultraviolet spectrograph on Rosetta. *Astron. Astrophys.* **2015**, *583*, A8. [CrossRef]

3.  Itikawa, Y.; Mason, N.J. Cross Sections for Electron Collisions with Water Molecules. *Phys. Chem. Ref. Data* **2005**, *34*, 1–22. [CrossRef]

4.  Galand, M.; Héritier, K.L.; Odelstad, E.; Henri, P.; Broiles, T.W.; Allen, A.J.; Altwegg, K.; Beth, A.; Burch, J.L.; Carr, C.M.; et al. Ionospheric plasma of comet 67P probed by Rosetta at 3 au from the Sun. *Mon. Not. R. Astron. Soc.* **2016**, S331–S351. [CrossRef]

5.  Héritier, K.L.; Henri, P.; Vallières, X.; Galand, M.; Odelstad, E.; Eriksson, A.I.; Johansson, F.L.; Altwegg, K.; Behar, E.; Beth, A.; et al. Vertical structure of the near-surface expanding ionosphere of comet 67P probed by Rosetta. *Mon. Not. R. Astron. Soc.* **2017**, S130–S141. [CrossRef]

6.  Bodewits, D.; Lara, L.M.; A'Hearn, M.F.; La Forgia, F.; Gicquel, A.; Kovacs, G.; Knollenberg, J.; Lazzarin, M.; Lin, Z.-Y.; Shi, X.; et al. Changes in the physical environment of the inner coma of 67P/Churyumov–Gerasimenko with decreasing heliocentric distance. *Astron. J.* **2016**, *152*. [CrossRef]

7.  Krüger, H.; Goesmann, F.; Giri, C.; Wright, I.; Morse, A.; Bredehöft, J.H.; Ulamec, S.; Cozzoni, B.; Ehrenfreund, P.; Gautier, T.; et al. Decay of COSAC and Ptolemy mass spectra at comet 67P/Churyumov-Gerasimenko. *Astron. Astrophys.* **2017**, *600*, A56. [CrossRef]

8.  Goesmann, F.; Rosenbauer, H.; Bredehöft, J.H.; Cabane, M.; Ehrenfreund, P.; Gautier, T.; Giri, C.; Krüger, H.; Le Roy, L.; MacDermott, A.J.; et al. Organic compounds on comet 67P/Churyumov-Gerasimenko revealed by COSAC mass spectrometry. *Science* **2015**, *349*. [CrossRef] [PubMed]

9.  Altwegg, K.; Balsiger, H.; Berthelier, J.J.; Bieler, A.; Calmonte, U.; Fuselier, S.A.; Goesmann, F.; Gasc, S.; Gombosi, T.I.; Le Roy, L.; et al. Organics in comet 67P—A first comparative analysis of mass spectra from ROSINA-DFMS, COSAC and Ptolemy. *Mon. Not. R. Astron. Soc.* **2017**, S130–S141. [CrossRef]

10. Wright, I.P.; Sheridan, S.; Barber, S.J.; Morgan, G.H.; Andrews, D.J.; Morse, A.D. CHO-bearing organic compounds at the surface of 67P/Churyumov-Gerasimenko revealed by Ptolemy. *Science* **2015**, *349*. [CrossRef] [PubMed]

11. Bieler, A.; Altwegg, K.; Balsiger, H.; Bar-Nun, A.; Berthelier, J.-J.; Bochsler, P.; Briois, C.; Calmonte, U.; Combi, M.; De Keyser, J.; et al. Abundant molecular oxygen in the coma of comet 67P/Churyumov–Gerasimenko. *Nature* **2015**, *526*, 678–681. [CrossRef] [PubMed]

12. Carr, C.; Cupido, E.; Lee, C.G.Y.; Balogh, A.; Beek, T.; Burch, J.L.; Dunford, C.N.; Eriksson, A.I.; Gill, R.; Glassmeier, K.H.; et al. RPC: The Rosetta Plasma Consortium. *Space Sci. Rev.* **2007**, *128*, 629–647. [CrossRef]

13. Burch, J.L.; Goldstein, R.; Cravens, T.E.; Gibson, W.C.; Lundin, R.N.; Pollock, C.J.; Winningham, J.D.; Young, D.T. RPC-IES: The Ion and Electron Sensor of the Rosetta Plasma Consortium. *Space Sci. Rev.* **2007**, *128*, 697–712. [CrossRef]

14. Eriksson, A.I.; Boström, R.; Gill, R.; Åhlén, L.; Jansson, S.-E.; Wahlund, J.-E.; André, M.; Mälkki, A.; Holtet, J.A.; Lybekk, B.; et al. RPC-LAP: The Rosetta Langmuir Probe Instrument. *Space Sci. Rev.* **2007**, *128*, 723–744. [CrossRef]

15. Trotignon, J.G.; Michau, J.L.; Lagoutte, D.; Chabassière, M.; Chalumeau, G.; Colin, F.; Décréau, P.M.E.; Geiswiller, J.; Gille, P.; Grard, R.; et al. RPC-MIP: The Mutual Impedance Probe of the Rosetta Plasma Consortium. *Space Sci. Rev.* **2007**, *128*, 713–728. [CrossRef]

16. Cravens, T.E.; Kozyra, J.U.; Nagy, A.F.; Gombosi, T.I.; Kurtz, M. Electron impact ionization in the vicinity of comets. *J. Geophys. Res. Space* **1987**, *92*, 7341–7353. [CrossRef]

17. Edberg, N.J.T.; Eriksson, A.I.; Odelstad, E.; Henri, P.; Lebreton, J.P.; Gasc, S.; Rubin, M.; André, M.; Gill, R.; Johansson, E.P.G.; et al. Spatial distribution of low-energy plasma around comet 67P/CG from Rosetta measurements. *Geophys. Res. Lett.* **2015**, *42*, 4263–4269. [CrossRef]

18. Clark, G.; Broiles, T.W.; Burch, J.L.; Collinson, G.A.; Cravens, T.; Frahm, R.A.; Goldstein, J.; Goldstein, R.; Mandt, K.; Mokashi, P.; et al. Suprathermal electron environment of comet 67P/Churyumov-Gerasimenko: Observations from the Rosetta Ion and Electron Sensor. *Astron. Astrophys.* **2015**, *583*, A24. [CrossRef]

19. Broiles, T.W.; Burch, J.L.; Chae, K.; Clark, G.; Cravens, T.E.; Eriksson, A.; Fuselier, S.A.; Frahm, R.A.; Gasc, S.; Goldstein, R.; et al. Statistical analysis of suprathermal electron drivers at 67P/Churyumov–Gerasimenko. *Mon. Not. R. Astron. Soc.* **2016**, *462*, S312–S322. [CrossRef]

20. Deca, J.; Divin, A.; Henri, P.; Eriksson, A.; Markidis, S.; Olshevsky, V.; Horányi, M. Electron and Ion Dynamics of the Solar Wind Interaction with a Weakly Outgassing Comet. *Phys. Rev. Lett.* **2017**, *118*. [CrossRef] [PubMed]

21. Le Roy, L.; Wegg, K.; Balsiger, H.; Berthelier, J.-J.; Bieler, A.; Briois, C.; Calmonte, U.; Combi, M.R.; De Keyser, J.; Dhooghe, F.; et al. Rosetta mission results pre-perihelion Special feature Inventory of the volatiles on comet 67P/Churyumov-Gerasimenko from Rosetta/ROSINA. *Astron. Astrophys.* **2015**, *583*, A1. [CrossRef]

22. Calmonte, U.; Altwegg, K.; Balsiger, H.; Berthelier, J.J.; Bieler, A.; Cessateur, G.; Dhooghe, F.; van Dishoeck, E.F.; Fiethe, B.; Fuselier, S.A.; et al. Sulphur-bearing species in the coma of comet 67P/Churyumov–Gerasimenko. *Mon. Not. Roy. Astron. Soc.* **2016**, *462*, S253–S273. [CrossRef]

23. Pitchford, L.C.; Alves, L.L.; Bartschat, K.; Biagi, S.F.; Bordage, M.-C.; Bray, I.; Brion, C.E.; Brunger, M.J.; Campbell, L.; Chachereau, A.; et al. LXCat: An Open-Access, Web-Based Platformfor Data Needed for Modeling LowTemperature Plasmas. *Plasma Process. Polym.* **2016**, *14*. [CrossRef]

24. Wakelam, V.; Herbst, E.; Loison, J.-C.; Smith, I.W.M.; Chandrasekaran, V.; Pavone, B.; Adams, N.G.; Bacchus-Montabonel, M.-C.; Bergeat, A.; Béroff, K.; et al. A KInetic Database for Astrochemistry (KIDA). *Astrophys. J. Suppl. Ser.* **2012**, *199*. [CrossRef]

25. Hibbert, A. Calculation of Rates of 4p–4d Transitions in Ar II. *Atoms* **2017**, *5*, 8. [CrossRef]

26. Dubernet, M.L.; Antony, B.K.; Ba, Y.A.; Babikov, Y.L.; Bartschat, K.; Boudon, V.; Braams, B.J.; Chung, H.-K.; Daniel, F.; Delahaye, F.; et al. The virtual atomic and molecular data centre (VAMDC) consortium. *J. Phys. B* **2016**. [CrossRef]

27. VAMDC Dictionary. Available online: http://dictionary.vamdc.eu/ (accessed on 30 August 2017).

28. Marinković, B.P.; Jevremović, D.; Srećković, V.A.; Vujčić, V.; Ignjatović, L.M.; Dimitrijević, M.S.; Mason, N.J. BEAMDB and MolD—Databases for atomic and molecular collisional and radiative processes: Belgrade nodes of VAMDC. *Eur. Phys. J. D* **2017**, *71*. [CrossRef]

29. Denifl, S.; Garcia, G.; Huber, B.A.; Marinković, B.P.; Mason, N.; Postler, J.; Rabus, H.; Rixon, G.; Solov'yov, A.V.; Suraud, E.; et al. Radiation damage of biomolecules (RADAM) database development: Current status. *J. Phys. Conf. Ser.* **2013**, *438*. [CrossRef]

30. Register, D.F.; Trajmar, S.; Srivastava, S.K. Absolute elastic differential electron scattering cross sections for He: A proposed calibration standard from 5 to 200 eV. *Phys. Rev. A* **1980**, *21*, 1134–1151. [CrossRef]

31. Maljković, J.B.; Blanco, F.; García, G.; Marinković, B.P.; Milosavljević, A.R. Elastic electron scattering from formamide molecule. *Nucl. Instrum. Methods Phys. Res. B* **2012**, *279*, 124–127. [CrossRef]

32. Belić, D.S.; Urbain, X.; Cherkani-Hassani, H.; Defrance, P. Electron-impact dissociation and ionization of CN+ ions. *Phys. Rev. A* **2017**, *95*. [CrossRef]

33. Huebner, W.F. Composition of comets: Observations and models. *Earth Moon Planets* **2002**, *89*, 179–195. [CrossRef]

34. Campbell, L.; Brunger, M.J. Modelling of plasma processes in cometary and planetary atmospheres. *Plasma Sources Sci. Technol.* **2013**, *22*. [CrossRef]

35. Simmons, J.D.; Bass, A.M.; Tilford, S.G. The Fourth Positive System of Carbon Monoxide Observed in Absorption at High Resolution in the Vacuum Ultraviolet Region. *Astrophys. J.* **1969**, *155*, 345–358. [CrossRef]

36. Campbell, L.; Brunger, M.J. Electron collisions in atmospheres. *Int. Rev. Phys. Chem.* **2016**, *35*, 297–351. [CrossRef]

37. Zecca, A.; Karwasz, G.P.; Brusa, R.S. One century of experiments on electron-atom and molecule scattering. A critical review of integral cross-sections, I. Atoms and diatomic molecules. *Riv. Nuovo Cimento* **1996**, *19*, 1–146. [CrossRef]

38. Bransden, B.H.; McDowell, M.R.C. Electron scattering by atoms at intermediate energies II. Theoretical and experimental data for light atoms. *Phys. Rep.* **1978**, *46*, 249–394. [CrossRef]

39. Brunger, M.J.; Buckman, S.J. Electron–molecule scattering cross-sections. I. Experimental techniques and data for diatomic molecules. *Phys. Rep.* **2002**, *357*, 215–458. [CrossRef]

40. Flower, D.R. Molecular collision processes in interstellar clouds. *Phys. Rep.* **1989**, *174*, 1–66. [CrossRef]

41. Anzai, K.; Kato, H.; Hoshino, M.; Tanaka, H.; Itikawa, Y.; Campbell, L.; Brunger, M.J.; Buckman, S.J.; Cho, H.; Blanco, F.; et al. Cross section data sets for electron collisions with $H_2$, $O_2$, CO, $CO_2$, $N_2O$ and $H_2O$. *Eur. Phys. J. D* **2012**, *66*. [CrossRef]

42. King, G.C. The Use of the Magnetic Angle Changer in Electron Spectroscopy. In *Changer in Electron Scattering. Physics of Atoms and Molecules*; Whelan, C.T., Mason, N.J., Eds.; Springer: Boston, MA, USA, 2005; pp. 111–120.

43. Khakoo, M.A.; Silva, H.; Muse, J.; Lopes, M.C.A.; Winstead, C.; McKoy, V. Electron scattering from $H_2O$: Elastic scattering. *Phys. Rev. A* **2008**, *78*. [CrossRef]

44. Muñoz, A.; Oller, J.C.; Blanco, F.; Gorfinkiel, J.D.; Limão-Vieira, P.; Maira-Vidal, A.; Borge, M.J.G.; Tengblad, O.; Huerga, C.; Téllez, M.; et al. Energy deposition model based on electron scattering cross section data from water molecules. *J. Phys. Conf. Ser.* **2008**, *133*. [CrossRef]

45. de Urquijo, J.; Basurto, E.; Juárez, A.M.; Ness, K.F.; Robson, R.E.; Brunger, M.J.; White, R.D. Electron drift velocities in He and water mixtures: Measurements and an assessment of the water vapour cross-section sets. *J. Chem. Phys.* **2014**, *141*. [CrossRef] [PubMed]

46. Song, M.-Y.; Yoon, J.-S.; Cho, H.; Itikawa, Y.; Karwasz, G.P.; Kokoouline, V.; Nakamura, Y.; Tennyson, J. Cross Sections for Electron Collisions with Methane. *J. Phys. Chem. Ref. Data* **2015**, *44*. [CrossRef]

47. Song, M.-Y.; Yoon, J.-S.; Cho, H.; Karwasz, G.P.; Kokoouline, V.; Nakamura, Y.; Tennyson, J. Cross Sections for Electron Collisions with Acetylene. *J. Phys. Chem. Ref. Data* **2017**, *46*. [CrossRef]

48. Sanz, A.G.; Fuss, M.C.; Blanco, F.; Sebastianelli, F.; Gianturco, F.A.; García, G. Electron scattering cross sections from HCN over a broad energy range (0.1–10 000 eV): Influence of the permanent dipole moment on the scattering process. *J. Chem. Phys.* **2012**, *137*. [CrossRef] [PubMed]

49. Ranković, M.L.; Maljković, J.B.; Tökesi, K.; Marinković, B.P. Elastic electron differential cross sections for argon atom in the intermediate energy range from 40 eV to 300 eV. *Eur. Phys. J. D* **2017**, submitted.

50. McEachran, R.P.; Stauffer, A.D. An optical potential method for elastic electron and positron scattering from argon. *J. Phys. B* **2009**, *42*. [CrossRef]

51. Paikeday, J.M.; Alexander, J. Polarization Potential for e-Argon Scattering by Differential Scattering Minimization at Intermediate Energies. *Int. J. Quant. Chem.* **2002**, *90*, 778–785. [CrossRef]

52. Nahar, S.N.; Wadehra, J.M. Elastic scattering of positrons and electrons by argon. *Phys. Rev. A* **1987**, *35*, 2051–2064. [CrossRef]

53. Fon, W.C.; Berrington, K.A.; Burke, P.G.; Hibbert, A. The elastic scattering of electrons from inert gases. III. Argon. *J. Phys. B* **1983**, *16*, 307–321. [CrossRef]

54.    McCarthy, I.E.; Noble, C.J.; Phillips, B.A.; Turnbull, A.D. Optical model for electron scattering by inert gases. *Phys. Rev. A* **1977**, *15*, 2173–2185. [CrossRef]

55.    Joachain, C.J.; Vanderpoorten, R.; Winters, K.H.; Byron, F.W., Jr. Optical model theory of elastic electron- and positron-argon scattering at inter mediate energies. *J. Phys. B* **1977**, *10*, 227–238. [CrossRef]

56.    Milosavljević, A.R.; Telega, S.; Šević, D.; Sienkiewicz, J.E.; Marinković, B.P. Elastic electron scattering by argon in the vicinity of the high-energy critical minimum. *Rad. Phys. Chem.* **2004**, *70*, 669–676. [CrossRef]

57.    Panajotović, R.; Filipović, D.M.; Marinković, B.P.; Pejčev, V.; Kurepa, M.; Vušković, L. Critical minima in elastic electron scattering by argon. *J. Phys. B* **1997**, *30*, 5875–5894. [CrossRef]

58.    Srivastava, S.K.; Tanaka, H.; Chutjian, A.; Trajmar, S. Elastic scattering of intermediate-energy electrons by Ar and Kr. *Phys. Rev. A* **1981**, *23*, 2156–2166. [CrossRef]

59.    DuBois, R.D.; Rudd, M.E. Differential cross sections for elastic scattering of electrons from argon, neon, nitrogen and carbon monoxide. *J. Phys. B* **1976**, *9*, 2657–2667. [CrossRef]

60.    Jansen, R.H.J.; de Heer, F.J.; Luyken, H.J.; van Wingerden, B.; Blaauw, H.J. Absolute differential cross sections for elastic scattering of electrons by helium, neon, argon and molecular nitrogen. *J. Phys. B* **1976**, *9*, 185–212. [CrossRef]

61.    Williams, J.F.; Willis, B.A. The scattering of electrons from inert gases, I. Absolute differential elastic cross sections for argon atoms. *J. Phys. B* **1975**, *8*, 1670–1682. [CrossRef]

62.    NIST Chemistry WebBook. Available online: http://webbook.nist.gov/chemistry/ (accessed on 30 August 2017).

63.    Tanaka, H.; Brunger, M.J.; Campbell, L.; Kato, H.; Hoshino, M.; Rau, A.R.P. Scaled plane-wave Born cross sections for atoms and molecules. *Rev. Mod. Phys.* **2016**, *88*, 025004. [CrossRef]

64.    IDEADB. Available online: http://ideadb.uibk.ac.at/ (accessed on 18 October 2017).

65.    Cordiner, M.A.; Charnley, S.B. Negative ion chemistry in the coma of comet 1P/Halley. *Meteorit. Planet. Sci.* **2014**, *49*, 21–27. [CrossRef]

66.    Barnett, S.M.; Mason, N.J.; Newell, W.R. Production of the $N_2(A^1\Sigma_{u+})$ Metastable State by Electron Dissociation of $N_2O$. *Chem. Phys.* **1991**, *153*, 283–295. [CrossRef]

67.    Barnett, S.M.; Mason, N.J.; Newell, W.R. Dissociative Excitation of Metastable Fragments by Electron Impact on Carbonyl Sulphide, Carbon Dioxide and Carbon Monoxide. *J. Phys B* **1992**, *25*, 1307–1320. [CrossRef]

68.    Danko, M.; Ribar, A.; Ďurian, M.; Országh, J.; Matejčík, Š. Electron induced fluorescence of the $H_2$ molecule—Balmer lines and Fulcher α system. *Plasma Sources Sci. Technol.* **2016**, *25*. [CrossRef]

69.    Harb, T.; Kedzierski, W.; McConkey, J.W. Production of ground state OH following electron impact on $H_2O$. *J. Chem. Phys.* **2001**, *115*, 5507–5512. [CrossRef]

70.    Cosby, P.C. Electron-impact dissociation of carbon monoxide. *J. Chem. Phys.* **1993**, *98*, 7804–7818. [CrossRef]

71.    Bredehöft, J.H.; Böhler, E.; Schmidt, F.; Borrmann, T.; Swiderek, P. Electron-Induced Synthesis of Formamide in Condensed Mixtures of Carbon Monoxide and Ammonia. *ACS Earth Space Chem.* **2017**, *1*, 50–59. [CrossRef]

72.    Boyer, M.C.; Rivas, N.; Tran, A.A.; Verish, C.A.; Arumainayagam, C.R. The role of low-energy (≤20 eV) electrons in astrochemistry. *Surf. Sci.* **2016**, *652*, 26–32. [CrossRef]

73.    Mason, N.J.; Nair, B.; Jheeta, S.; Szymańska, E. Electron induced chemistry: A new frontier in astrochemistry. *Faraday Discuss.* **2014**, *168*, 235–247. [CrossRef] [PubMed]

74.    Zheng, W.; Jewitt, D.; Kaiser, R.I. Formation of Hydrogen, Oxygen and Hydrogen Peroxide in Electron-irradiated Crystalline Water Ice. *Astrophys. J.* **2006**, *639*, 534–548. [CrossRef]

75.    Zheng, W.; Jewitt, D.; Kaiser, R.I. Temperature Dependence of the Formation of Hydrogen, Oxygen and Hydrogen Peroxide in Electron-Irradiated Crystalline Water Ice. Ice. *Astrophys. J.* **2006**, *648*, 753–761. [CrossRef]

# 18

# Regularities and Systematic Trends on Zr IV Stark Widths

**Zlatko Majlinger** [1,‡], **Milan S. Dimitrijević** [1,2,*,†,‡] **and Zoran Simić** [1,‡]

[1]  Astronomical Observatory, Volgina 7, Belgrade 11060, Serbia; zlatko.majlinger@gmail.com (Z.M.); zsimic@aob.rs (Z.S.)

[2]  LERMA (Laboratoire d'Etudes du Rayonnement et de la Matière en Astrophysique et Atmosphères), Observatoire de Paris, 5 Place Jules Janssen, 92195 Meudon CEDEX, France

*  Correspondence: mdimitrijevic@aob.rs or milan.dimitrijevic@obspm.fr

†  Current address: Astronomical Observatory, Volgina 7, Belgrade 11060, Serbia.

‡  These authors contributed equally to this work.

**Abstract:** Regularities and systematic trends among the Stark widths of 18 Zr IV spectral lines obtained by modified semiempirical approach have been discussed. Also we compared those calculated Stark broadening parameters with estimates according to Cowley, Purić et al. and Purić and Šćepanović and checked the possibility to find some new estimates. It is demonstrated as well that the formula of Cowley (1971) overestimates Stark widths, obtained by using modified semiempirical method, with the increase of angular orbital momentum quantum number due to its neglection. It is also found that the results obtained by using formula for simple estimates of Purić et al. (1991) are in agreement with the modified semiempirical results within the estimated error bars of both methods, while the estimates using formula of Purić and Šćepanović (1999) are in strong disagreement which increases with the increase of angular orbital momentum quantum number.

**Keywords:** Stark broadening; line profiles; atomic data

## 1. Introduction

Stark broadening theory is very important for many laboratory and astrophysical plasma investigations, since the collisional processes between the charged particles contribute significantly to the spectral line broadening in high-temperature dense plasma. However, available numerous data on Stark widths of spectral lines are up to now still incomplete and despite of a great effort from different groups including French and Belgrade Schools and their co-workers (see e.g., [1–3] and references therein) in determination of Stark widths and shifts for spectral lines of many atoms and ions, the further work on the determination of missing Stark broadening data for applications in astrophysics and plasma physics is still needed.

When the required Stark broadening parameters are missing, and there is no data for the adequate use of more accurate theoretical methods, simple estimates seem very useful. Such quick estimates of Stark-broadening parameters may be based on different regularities and systematic trends among the Stark widths of atomic spectral lines (see e.g., [4]). We focus on present work on searching regularities and systematic trends of Stark widths for 18 Zr IV spectral lines calculated in [5] by modified semiempirical approach (MSE—Dimitrijević & Konjević [6]). This study has two directions. In the first, we compare existing calculated MSE results with estimates made in this work by formulae of Cowley [7], Purić et al. [8] and Purić & Šćepanović [9]. In the second, we try to find estimates based on systematic trends among 18 calculated MSE results. We note as well that there is no experimental Stark broadening data in the literature.

## 2. Methods

In all of our estimates in this work, the condition that temperatures are near threshold for collisional excitation or below, when Gaunt factor is constant, is satisfied.

$$3kT/2|E_j - E_{j'}| < 2, \tag{1}$$

where $k$ is Boltzmann constant and $j'$ is the perturbing level closest to initial or final level. In such a case, from MSE and the semiempirical method follows (see for further details Dimitrijević & Konjević [6] and Griem [10]) that the dependence of Stark broadening parameters with temperature is

$$f(T) = T^{-1/2}. \tag{2}$$

For Zr IV lines considered in this work, the condition (1) is in most cases satisfied up to around 20,000 K, for some cases even 30,000 K, so that for all our analysed widths calculated for temperature $T = 10,000$ K, the condition (1) is satisfied.

Estimation methods used in present research can be divided into two groups: (a) estimates derived from theory (b) estimates based on purely statistical analysis of existing data. We search the estimation formula in two versions. First, in dependence on effective ionization potential:

$$W_{E_1} = a_1 Z^{c_1} \lambda^2 N f(T) (E_{ion} - E_j)^{-b_1} \tag{3}$$

Second, in dependence on effective quantum number:

$$W_{E_2} = a_2 Z^{c_2} \lambda^2 N f(T) (n_j^*)^{-b_2}. \tag{4}$$

The relation between effective principal quantum number and upper or lower state is given by:

$$n_j^{*2} = Z^2 [E_H / (E_{ion} - E_j)], \tag{5}$$

where $Z - 1$ is ionic charge, $n^*$—effective principal quantum number, $W_E$ estimated Stark width in Å, $\lambda$ wavelength in Å, $N$-perturber density in m$^{-3}$, $T$ temperature in K, $E_H$—hydrogen atom energy, $E_{ion}$—ionization energy, $E_j$—energy of upper ($j$ = upper) and lower ($j$ = lower) level. Coefficients $a$, $b$ and $c$ in Equations (3) and (4) are independent of temperature, ionization potential and electron density for a given transition.

First we will estimate Stark widths with the simple formula of Cowley [7] which is often in use by some modern authors in original or modified form (see, for example, Killian et al. [11], Ziegler et al. [12] and Przybilla et al. [13]) and which has recently used by our team for comparison with MSE calculated values of Stark widths of Lu III spectral lines too, see Majlinger et al. [14]. In Majlinger et al. [5] we derived it from the Griem [10] semiempirical formula at low temperature limit and added the contribution to the width of the lower energy level. In such a case, the difference between original formula of Cowley and the modified formula in [5] is for a factor $2/3^{3/2}$ plus the contribution of the lower energy level. Here, we want to test the original formula of Cowley. After rearranging the original equation of Cowley [7] to express width in Å, we obtain:

$$W_{Cow} = \frac{h^2}{2\pi c} (1/2m^3 \pi k)^{1/2} \frac{\lambda^2 N}{Z^2 \sqrt{T}} (n_i^{*4}) \tag{6}$$

where $h$ is Planck's constant, $c$ speed of light and $m$ mass of perturber.

The Cowley's formula is, as we can see, of the type of Equation (4), where $j$ = upper level, and parameters are $a_2 = 2.92 \times 10^{-30}$, $b_2 = 4$ and $c_2 = -2$.

Further, we test the possibility to use the statistical estimates of [8,9] for quick prediction of unknown Stark widths. Comparing the great amount of Stark width data from STARK-B database [1,2], Purić and his co-workers found the correlation between Stark width and difference between ionization energy and energy of the upper state (which is, according to Purić, called upper effective ionization potential) and offered a set of different estimation formulae. Thus we compared our results from MSE calculations with two different approximations: so-called "generalized" estimate (Purić & Šćepanović [9], in the rest of this text PS99) and estimate based on statistical regression analysis of multiply charged heavy ions (Purić et al. [8], in the rest of this text P91). All of Purić's formulae are of type of Equation (1). Purić's works are shown to be useful in some analysis and comparisons, and investigations of systematic trends and regularities, for example in the cases of searching the regularities and trends via specific homologous sequences [15,16].

The equations for Stark width obtained from Purić's regression analysis of existing set of Stark broadening data was adapted for our purpose. After conversion of width from angular frequency units to Angstroms (see for example Hamdi [17]), coefficient values are recalculated from PS99 and P91, where these coefficients are given for Stark widths measured in $s^{-1}$. Estimation formulae from PS99 and P91 finally is:

$$W_{PS99} = 3.27 \times 10^{-28} Z^{5/2} \lambda^2 N \sqrt{T} (E_{ion} - E_j)^{-3.1} \tag{7}$$

and

$$W_{P91} = 2.52 \times 10^{-27} Z^0 \lambda^2 N \sqrt{T} (E_{ion} - E_j)^{-1.73} \tag{8}$$

respectively, if perturber density $N$ is expressed in $m^{-3}$ and energies in eV.

## 3. Results and Discussion

In Tables 1 and 2, calculated Stark widths of 18 Zr IV lines made by different estimation methods mentioned above are compared to previous MSE calculations of Stark FWHM Majlinger et al. [5]. All results presented in Tables 1 and 2 are calculated for a temperature of 10,000 K and electron density of $10^{23}$ $m^{-3}$. First we can see that widths $W_{PS99}$ estimated according to Equation (7) are systematically higher than $W_{COW}$ done by Equation (6).

Second, we can see that $W_{PS99}$ does not fit very well to WMSE widths, although this estimate should be generally applicable, the reasons for which we have already explained in detail in Majlinger et al. [5]. On the contrary, much better agreement is found between all of MSE results and results of estimate for multiple charged heavy ions obtained by Equation (8) in P91 (Equation (8) also in this paper), but only if dependence of estimate on residual charge $Z$ is neglected (e.g., parameters should be adopted to be $c = 0$ or $Z = 1$). In that case, the corresponding ratios of this comparison range from 0.4 to 2.2.

In Table 3, we are presented ranges of values of ratios of $W_{COW}$, $W_{PS99}$ and $W_{P91}$ with $W_{MSE}$ for different types of transitions. As we can see, ratios between $W_{COW}$ and $W_{MSE}$ range from 0.8 to 4, and they increase with the increase of the orbital angular momentum quantum number $\ell$, which is neglected in the Cowley's formula. For s-p transitions differences are from 0.82 up to 2.82, for p-d from 1.57 up to 1.86 and for d-f from 3.74 up to 3.97. Consequently, obtained results demonstrate that for higher $\ell$ Cowley's formula overestimates the MSE values due to the neglection of $\ell$.

For the estimates $W_{PS99}$ the considered ratio for transitions of s-p type ranges from 1.40 to 8.34 while for other considered types of transitions this ratio is from 4.43 up to 24.27. The ratio of $W_{P91}$ with $W_{MSE}$ vary from 0.48 up to 2.13 which is acceptable for a rough estimate.

In articles of Cowley [7] and Purić & Šćepanović [9] there is no discussion of the corresponding error bars. The assumed accuracy of MSE method is within the limits of 50% for simpler spectra while for complex spectra like for Zr IV might be worse. In Purić et al. [8] is stated that the agreement is within 40% error bars for the majority of cases. Consequently the ratio of $W_{P91}$ with $W_{MSE}$ is within mutual error bars.

**Table 1.** Estimated Stark widths for 18 spectral lines of Zr IV according to Cowley (1971) using Equation (6) ($W_{COW}$) for a temperature of T = 10,000 K and electron density of N = $10^{23}$ m$^{-3}$. Estimated Stark widths are compared with MSE calculated Stark widths $W_{MSE}$ [5].

| Transition | $\lambda$ (Å) | $W_{COW}$(Å) | $W_{COW}/W_{MSE}$ |
|---|---|---|---|
| 5s $^2S_{1/2}$–5p $^2P^o_{1/2}$ | 2287.38 | 0.0768 | 0.91 |
| 5s $^2S_{1/2}$–5p $^2P^o_{3/2}$ | 2164.36 | 0.0706 | 0.92 |
| 5s $^2S_{1/2}$– 6p $^2P^o_{1/2}$ | 760.16 | 0.0279 | 2.82 |
| 5s $^2S_{1/2}$–6p $^2P^o_{3/2}$ | 754.39 | 0.0280 | 2.75 |
| 5p $^2P^o_{1/2}$–5d $^2D_{3/2}$ | 1546.17 | 0.0784 | 1.86 |
| 5p $^2P^o_{3/2}$–5d $^2D_{3/2}$ | 1607.95 | 0.0847 | 1.81 |
| 5p $^2P^o_{3/2}$–5d $^2D_{5/2}$ | 1598.95 | 0.0842 | 1.82 |
| 5d $^2D_{3/2}$–5f $^2F^o_{5/2}$ | 1836.10 | 0.324 | 3.97 |
| 5d $^2D_{5/2}$–5f $^2F^o_{5/2}$ | 1848.03 | 0.328 | 3.97 |
| 5d $^2D_{5/2}$–5f $^2F^o_{7/2}$ | 1846.37 | 0.328 | 3.97 |
| 6s $^2S_{1/2}$–6p $^2P^o_{1/2}$ | 5781.45 | 1.62 | 0.82 |
| 6s $^2S_{1/2}$–6p $^2P^o_{3/2}$ | 5463.85 | 1.47 | 0.83 |
| 6p $^2P^o_{1/2}$–6d $^2D_{3/2}$ | 3577.13 | 1.13 | 1.61 |
| 6p $^2P^o_{3/2}$–6d $^2D_{3/2}$ | 3795.07 | 1.27 | 1.57 |
| 6p $^2P^o_{3/2}$–6d $^2D_{5/2}$ | 3687.95 | 1.20 | 1.59 |
| 6d $^2D_{3/2}$–6f $^2F^o_{5/2}$ | 3751.67 | 2.80 | 3.74 |
| 6d $^2D_{5/2}$–6f $^2F^o_{5/2}$ | 3775.08 | 2.83 | 3.74 |
| 6d $^2D_{5/2}$–6f $^2F^o_{7/2}$ | 3765.39 | 2.82 | 3.77 |

**Table 2.** Estimated Stark widths for 18 spectral lines of Zr IV according to Purić and Šćepanović, 1999 ($W_{PS99}$) and Purić et al., 1991 ($W_{P91}$) for a temperature of T = 10,000 K and electron density of N = $10^{23}$ m$^{-3}$. Estimated Stark widths are compared with MSE calculated Stark widths $W_{MSE}$ [5].

| Transition | $\lambda$ (Å) | $W_{PS99}$(Å) | $W_{P91}$(Å) | $W_{PS99}/W_{MSE}$ | $W_{P91}/W_{MSE}$ |
|---|---|---|---|---|---|
| 5s $^2S_{1/2}$–5p $^2P^o_{1/2}$ | 2287.38 | 0.118 | 0.0530 | 1.40 | 0.63 |
| 5s $^2S_{1/2}$–5p $^2P^o_{3/2}$ | 2164.36 | 0.110 | 0.0530 | 1.40 | 0.63 |
| 5s $^2S_{1/2}$–6p $^2P^o_{1/2}$ | 760.16 | 0.0826 | 0.0164 | 8.34 | 1.66 |
| 5s $^2S_{1/2}$–6p $^2P^o_{3/2}$ | 754.39 | 0.0837 | 0.0164 | 8.21 | 1.61 |
| 5p $^2P^o_{1/2}$–5d $^2D_{3/2}$ | 1546.17 | 0.187 | 0.0485 | 4.43 | 1.15 |
| 5p $^2P^o_{3/2}$–5d $^2D_{3/2}$ | 1607.95 | 0.202 | 0.0525 | 4.31 | 1.12 |
| 5p $^2P^o_{3/2}$–5d $^2D_{5/2}$ | 1598.95 | 0.201 | 0.0521 | 4.34 | 1.13 |
| 5d $^2D_{3/2}$–5f $^2F^o_{5/2}$ | 1836.10 | 1.39 | 0.173 | 17.03 | 2.13 |
| 5d $^2D_{5/2}$–5f $^2F^o_{5/2}$ | 1848.03 | 1.41 | 0.176 | 17.05 | 2.12 |
| 5d $^2D_{5/2}$–5f $^2F^o_{7/2}$ | 1846.37 | 1.41 | 0.176 | 17.09 | 2.13 |
| 6s $^2S_{1/2}$–6p $^2P^o_{1/2}$ | 5781.45 | 4.77 | 0.950 | 2.42 | 0.48 |
| 6s $^2S_{1/2}$–6p $^2P^o_{3/2}$ | 5463.85 | 4.39 | 0.862 | 2.47 | 0.48 |
| 6p $^2P^o_{1/2}$–6d $^2D_{3/2}$ | 3577.13 | 4.64 | 0.611 | 6.61 | 0.87 |
| 6p $^2P^o_{3/2}$–6d $^2D_{3/2}$ | 3795.07 | 5.22 | 0.688 | 6.44 | 0.85 |
| 6p $^2P^o_{3/2}$–6d $^2D_{5/2}$ | 3687.95 | 4.96 | 0.652 | 6.56 | 0.86 |
| 6d $^2D_{3/2}$–6f $^2F^o_{5/2}$ | 3751.67 | 18.0 | 1.36 | 24.10 | 1.82 |
| 6d $^2D_{5/2}$–6f $^2F^o_{5/2}$ | 3775.08 | 18.2 | 1.37 | 24.04 | 1.82 |
| 6d $^2D_{5/2}$–6f $^2F^o_{7/2}$ | 3765.39 | 18.2 | 1.37 | 24.27 | 1.83 |

Additionally, following Purić's approach, we tried without success to represent our MSE widths for Zr IV, in angular frequency units, as a polynomial of upper effective principal quantum number or to obtain a linear dependence between them in log-log scale with correlation coefficient above 90% (which proves that the dependence is reliable). However, the correlation coefficient for log-log regression with lower $n^*$ is around 88%:

$$W_{E3} = 1.37 \times 10^{-9} \lambda^2 n^*_{lower}{}^{2.47} \tag{9}$$

(in this case, $W_{MSE}$ can be estimated as $W_{MSE} = 1.54 \text{ Å} \times W_{E3} - 0.14$ with correlation coefficient of around 93%) against, for example, around 60% for upper $n^*$. But we found the polynomial dependence between Stark widths and the corresponding spectral line wavelengths instead, in a form:

$$W_{E4} = C_4\lambda^4 + C_3\lambda^3 + C_2\lambda^2 + C_1\lambda + C_0 \tag{10}$$

where $C_4 = -0.6 \times 10^{-14}, C_3 = 0.63 \times 10^{-10}, C_2 = -0.12 \times 10^{-6}, C_1 = 0.7 \times 10^{-4}$ and $C_0 = 0.44 \times 10^{-2}$ [5]. Correlation coefficient for this polynomial regression of $4^{th}$ degree is 99.9%. In all of those estimations parameters are obtained for condition of T = 10,000 K and N = $10^{23}$ m$^{-3}$.

Unfortunately there is no experimental data for Zr IV. A comparison of our results with reliable experimental data will be of interest so we encourage such measurements.

**Table 3.** Ranges of values of ratios of $W_{COW}$, $W_{PS99}$ and $W_{P91}$ with $W_{MSE}$ for different types of transitions.

| Transition | $W_{COW}/W_{MSE}$ | $W_{PS99}/W_{MSE}$ | $W_{P91}/W_{MSE}$ |
|---|---|---|---|
| s–p | 0.82–2.82 | 1.40–8.34 | 0.48–1.66 |
| p–d | 1.57–1.86 | 4.43-6.61 | 0.85–1.15 |
| d–f | 3.74–3.97 | 17.03–24.27 | 1.82–2.13 |

## 4. Conclusions

We tested Cowley's [7] formula and several expressions for quick calculation of electron-impact widths on the basis of regularity and systematic trends, comparing them with MSE results of Zr IV spectral lines calculated elsewhere Majlinger et al. [5]. We demonstrated that Cowley's formula increasingly overestimate Stark width, obtained by using the modified semiempirical method, with the increase of angular momentum quantum number because of its neglection in the Cowley's approach. It is also found that the results obtained by using formula for simple estimates of Purić et al. (1991) are in agreement with the modified semiempirical results within the estimated error bars of both methods, while the estimates using formula of Purić and Šćepanović (1999) are in strong disagreement which increases with the increase of angular orbital momentum quantum number.

**Acknowledgments:** The support of Ministry of Education, Science and Technological Development of Republic of Serbia through project 176002 is gratefully acknowledged.

**Author Contributions:** These authors contributed equally to this work.

## References

1. Sahal-Bréchot, S.; Dimitrijević, M.S.; Moreau, N.; Ben Nessib, N. The Stark-B database VAMDC node for spectral line broadening by collisions with charged particles. In *SF2A-2014, Proceedings of the Annual Meeting of the French Society of Astronomy and Astrophysics, Paris, France, 3–6 June 2014*; Ballet, J., Bournaud, F., Martins, F., Monier, R., Reyle, C., Eds.; French Society of Astronomy & Astrophysics (SF2A): Paris, France, 2014; pp. 515–521.
2. Sahal-Bréchot, S.; Dimitrijević, M.S.; Moreau, N.; Ben Nessib, N. The STARK-B database VAMDC node: A repository for spectral line broadening and shifts due to collisions with charged particles. *Phys. Scr.* **2015**, *90*, 054008.
3. Sahal-Bréchot, S.; Dimitrijević, M.S.; Moreau, N. STARK-B Database, Observatory of Paris, LERMA and Astronomical Observatory of Belgrade. 2017. Available online: http://stark-b.obspm.fr (accessed on 24 August 2017).
4. Wiese, W.L.; Konjević, N. Regularities and similarities in plasma broadened spectral line widths (Stark widths). *J. Quant. Spectrosc. Radiat. Transf.* **1982**, *28*, 185–198.

5.    Majlinger, Z.; Simić, Z.; Dimitrijević, M.S. Stark broadening of Zr iv spectral lines in the atmospheres of chemically peculiar stars. *Mon. Not. R. Astron. Soc.* **2017**, *470*, 1911–1918.

6.    Dimitrijević, M.S.; Konjević, N. Stark widths of doubly- and triply-ionized atom lines. *J. Quant. Spectrosc. Radiat. Transf.* **1980**, *24*, 451–459.

7.    Cowley, C.R. An approximate Stark broadening formula for use in spectrum synthesis. *Observatory* **1971**, *91*, 139–140.

8.    Purić, J.; Ćuk, M.; Dimitrijević, M.S.; Lessage, A. Regularities of Stark parameters along the periodic table. *Astrophys. J.* **1991**, *382*, 353–357.

9.    Purić, J.; Šćepanović, M. General Regularities of Stark Parameters for Ion Lines. *Astrophys. J.* **1999**, *521*, 490–491.

10.   Griem, H.R. Semiempirical Formulas for the Electron-Impact Widths and Shifts of Isolated Ion Lines in Plasmas. *Phys. Rev.* **1968**, *165*, 258–266.

11.   Killian, J.; Montenbruck, O.; Nissen, P.E. Chemical abundances in early B-type stars. II—Line identification and atomic data for high resolution spectra. *Astron. Astrophys.* **1991**, *88*, 101–119.

12.   Ziegler, M.; Rauch, T.; Werner, K.; Köpen, J.; Kruk, J.W. BD 22°3467, a DAO type star exciting the nebula Abell 35. *Astron. Astrophys.* **2012**, *548*, A109.

13.   Przybilla, N.; Fossati, L.; Hubrig, S.; Nieva, M.-F.; Järvinen, S.P.; Castro, N.; Schöller, M.; Ilyin, I.; Butler, K.; Schneider, F.R.N.; et al. B fields in OB stars (BOB): Detection of a magnetic field in the He-strong star CPD 57°3509. *Astron. Astrophys.* **2016**, *587*, A7.

14.   Majlinger, Z.; Simić, Z.; Dimitrijević, M.S. On the Stark Broadening of Lu III Spectral Lines. *J. Astrophys. Astron.* **2015**, *36*, 671–679.

15.   Peláez, R.J.; Djurović, S.; Ćirišan, M.; Aparacio, J.A.; Mar, S. Stark halfwidth trends along the homologous sequence of singly ionized noble gases. *Astron. Astrophys.* **2010**, *518*, A60.

16.   Elabidi, H.; Sahal-Bréchotot, S. Checking the dependence on the upper level ionization potential of electron impact widths using quantum calculations. *Eur. Phys. J. D* **2011**, *61*, 285–290.

17.   Hamdi, R.; Ben Nessib, N.; Dimitrijević, M.S.; Sahal-Bréchot, S. Stark broadening of spectral lines of Pb IV. *Mon. Not. R. Astron. Soc.* **2013**, *431*, 1039–1047.

# 19

# Line Shape Modeling for the Diagnostic of the Electron Density in a Corona Discharge

**Joël Rosato [1],\*, Nelly Bonifaci [2], Zhiling Li [3] and Roland Stamm [1]**

[1]  Laboratoire PIIM, Aix-Marseille Université and CNRS, 13397 Marseille CEDEX 20, France; roland.stamm@univ-amu.fr
[2]  Laboratoire G2Elab, CNRS and Grenoble University, 25 rue des Martyrs, 38042 Grenoble, France; nelly.bonifaci@g2elab.grenoble-inp.fr
[3]  Guizhou Institute of Technology, Caiguan Road 1, Guiyang 550003, China; lzlfrance@yahoo.fr
\*   Correspondence: joel.rosato@univ-amu.fr

Academic Editor: Luka Č. Popović

**Abstract:** We present an analysis of spectra observed in a corona discharge designed for the study of dielectrics in electrical engineering. The medium is a gas of helium and the discharge was performed at the vicinity of a tip electrode under high voltage. The shape of helium lines is dominated by the Stark broadening due to the plasma microfield. Using a computer simulation method, we examine the sensitivity of the He 492 nm line shape to the electron density. Our results indicate the possibility of a density diagnostic based on passive spectroscopy. The influence of collisional broadening due to interactions between the emitters and neutrals is discussed.

**Keywords:** line shapes; stark broadening; neutral broadening; corona discharge

## 1. Introduction

Corona discharges of helium were performed in an experiment devoted to the investigation of the dielectric properties of insulators in the context of electrical engineering. This setup consisted of a point-plane electrode system placed inside a helium cryostat; a high voltage was applied to the system and a streamer of either positively or negatively charged particles, ions or electrons, was generated and formed a plasma locally. Here, we report the analysis of the helium 492 nm line (singlet state, 1s4d–1s2p transition) observed in discharges at room temperature. This line is strongly sensitive to the electron density due to the Stark broadening generated by the plasma microfield. We show that a diagnostic of the electron density can be performed from an analysis of the line width and its forbidden component induced by the electric field. This work completes previous research on the analysis of Hβ [1], and some preliminary results were discussed at the fourth Spectral Line Shape in Plasmas code comparison workshop (SLSP4, see http://plasma-gate.weizmann.ac.il/slsp/). It will be followed with subsequent analyses of spectra observed in helium corona discharges performed at low temperatures (a few K) with liquid helium.

## 2. Presentation of the Experiment

A point-plane electrode system was placed inside a helium cryostat. The point electrode was negatively polarized by a stabilized high voltage DC power supply (Spellman RHSR/20PN60) and the current-voltage characteristics were measured using a Tektronix TDS540 oscilloscope and a Keithley 610C ammeter. The corona discharge in this geometry is axially symmetric and appears as a luminous spherical region (ionization region) localized near the point electrode against the dark background.

More details on the experimental setup can be found elsewhere, e.g., [2]. The measured intensity of the radiation was averaged over the exposure time. A liquid $N_2$ cooled 2D-CCDTKB-UV/AR detector was located directly in the exit plane of the spectrograph. The noise level of the CCD detector was determined only by the read-out noise, as the dark current of the camera was less than 1 e/pixel/h at 153 K. The wavelength and intensity response of the detection system was calibrated by using low pressure helium and tungsten ribbon lamps. The line broadening due to the instrument response (1200 g/mm grating) was estimated from the helium lines of a helium lamp as $\Delta\lambda_{ins} < 0.10$ nm.

## 3. An Analysis of the Helium 492 nm Line Observed at 300 K

In concert with previous research [1], we here report an analysis of the helium 492 nm line (singlet state, 1s4d–1s2p transition) which has been observed on spectra. Discharges were carried out at pressures from 1 to 12 bar, with a current of 100 μA. In order to simplify the interpretation of the spectra, we focus on the lowest pressure values (namely, between 1 and 2 bar) because the line overlaps with nearby molecular bands at higher pressures.

Figure 1 shows a plot of the helium 492 nm line observed at 1, 1.5, and 2 bar. The line width increases with the pressure. The bump on the blue wing denotes a forbidden transition (1s4f–1s2p) induced by the microfield. An analysis of the resonance broadening resulting from atom-atom collisions using a collision operator model [3] indicates that this effect is not the dominant line broadening one. The instrumental broadening is also not dominant. We examined the role of the Stark effect related to the plasma microfield. A computer simulation method [4] was applied to the He 492 nm line at the same pressure values as above. The He$^+$ ion microfield evolution was simulated from a quasiparticle model and the line broadening was calculated from a numerical integration of the time-dependent Schrödinger equation. The contribution of the electrons was evaluated using a collision operator. In our calculations, we used the Griem-Baranger-Kolb-Oertel model [5], assuming an electron temperature of $10^4$ K and leaving the electron density as an adjustable parameter. The ion temperature was assumed to be equal to the atomic (300 K) temperature. Our calculations indicate that the Stark broadening was mainly due to the ions. Figure 2 shows an example of the adjustment performed using the simulation method at $p = 1$ bar. A value of $1.8 \times 10^{15}$ cm$^{-3}$ was obtained for the electron density. Our calculations at $p = 1.5$ bar and 2 bar yielded higher values for the electron density, which indicates an increase trend in terms of the pressure. This result is preliminary and will be completed with further analyses.

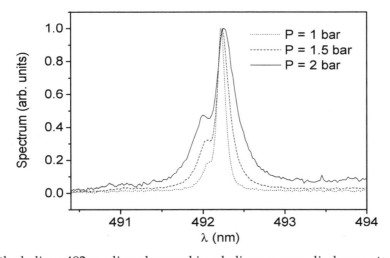

**Figure 1.** Plot of the helium 492 nm line observed in a helium corona discharge at room temperature, for different pressure values. The line width increases with the pressure. This trend is also observed on other lines. The bump on the blue wing denotes a forbidden transition induced by the microfield.

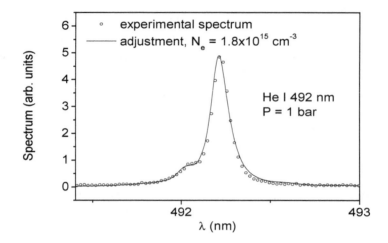

**Figure 2.** The plasma microfield yields an additional broadening, which can serve as a probe of the density. Here, the Stark broadening was evaluated using a computer simulation method [4].

## 4. Conclusions

We have analyzed spectral profiles of the helium 492 nm line in helium corona discharges by means of a computer simulation method. An application to 1, 1.5, and 2 bar gas pressures indicates that the plasma microfield yields an obvious Stark broadening, which can serve as a probe for the electron density. This result is a preliminary step in ongoing investigations of liquid corona discharges. New experiments, with liquid helium, are planned and will be analyzed using spectroscopic techniques.

**Author Contributions:** All authors contributed equally to this work.

## References

1.  Rosato, J.; Bonifaci, N.; Li, Z.; Stamm, R. A spectroscopic diagnostic of the electron density in a corona discharge. *J. Phys. Conf. Ser.* **2017**, *810*, 012057. [CrossRef]

2.  Li, Z.-L.; Bonifaci, N.; Aitken, F.; Denat, A.; von Haeften, K.; Atrazhev, V.M.; Shakhatov, V.A. Spectroscopic investigation of liquid helium excited by a corona discharge: Evidence for bubbles and "red satellites". *Eur. Phys. J. Appl. Phys.* **2009**, *47*, 2821. [CrossRef]

3.  Ali, A.W.; Griem, H.R. Theory of Resonance Broadening of Spectral Lines by Atom-Atom Impacts. *Phys. Rev.* **1965**, *140*, A1044–A1049; reprinted in *Phys. Rev.* **1966**, *144*, 366. [CrossRef]

4.  Rosato, J.; Marandet, Y.; Capes, H.; Ferri, S.; Mossé, C.; Godbert-Mouret, L.; Koubiti, M.; Stamm, R. Stark broadening of hydrogen lines in low-density magnetized plasmas. *Phys. Rev. E* **2009**, *79*, 046408. [CrossRef] [PubMed]

5.  Griem, H.R.; Baranger, M.; Kolb, A.C.; Oertel, G. Stark Broadening of Neutral Helium Lines in a Plasma. *Phys. Rev.* **1962**, *125*, 177–195. [CrossRef]

# 20

# Effect of Turbulence on Line Shapes in Astrophysical and Fusion Plasmas

**Ibtissem Hannachi** [1,2,*], **Mutia Meireni** [1], **Paul Génésio** [1], **Joël Rosato** [1], **Roland Stamm** [1] and **Yannick Marandet** [1]

[1]   Département de physique, Aix-Marseille Université, CNRS, PIIM UMR 7345, 13397 Marseille CEDEX 20, France; mutia_meireni@ymail.com (M.M.); paul.genesio@univ-amu.fr (P.G.); joel.rosato@univ-amu.fr (J.R.); roland.stamm@univ-amu.fr (R.S.); yannick.marandet@univ-amu.fr (Y.M.)

[2]   PRIMALAB, Faculty of Sciences, University of Batna 1, Batna 05000, Algeria

*   Correspondence: ibtissam.hannachi@univ-batna.dz

Academic Editors: Milan S. Dimitrijević and Luka Č. Popović

**Abstract:** We look at the effect of wave collapse turbulence on a hydrogen line shape in plasma. An atom immersed in plasma affected by strong Langmuir turbulence may be perturbed by a sequence of wave packets with a maximum electric field magnitude that is larger than the Holtsmark field. For such conditions, we propose to calculate the shape of the hydrogen Lyman $\alpha$ Lyman $\beta$ and Balmer $\alpha$ lines with a numerical integration of the Schrödinger equation coupled to a simulation of a sequence of electric fields modeling the effects of the Langmuir wave. We present and discuss several line profiles of Lyman and Balmer lines.

**Keywords:** line shape; wave collapse; electric field solitons; plasma turbulence

## 1. Introduction

The problem of plasma turbulence is of interest both from a theoretical point of view and from an experimental one for laboratory, fusion, and astrophysical plasmas. Plasma turbulence affects the transport and radiation properties of many kinds of plasma. In magnetic fusion studies, the quality of the plasma confinement is strongly dependent on the level of turbulent fluctuations. The first observations of turbulent fluctuations have been made in astrophysics on line shapes dominated by the Doppler effect [1,2]. If nonthermal movements take place on the line of sight, the line shape no longer corresponds to a Maxwellian velocity distribution at the emitter temperature. In the simplest models, a nonthermal velocity is defined as one that allows a quantitative measure of turbulence. The study of line shapes may then provide valuable information on the nature of turbulence, and this has been used in astrophysical and laboratory plasmas [3]. In this work, we are mainly interested in the contribution of Stark effect to the line shapes of hydrogen plasmas. The turbulent fluctuations of the plasma are created by the instabilities appearing in the different types of plasma studied [3].

One kind of plasma turbulence suspected to be present in astrophysical and fusion plasma is driven by plasma waves and electromagnetic waves. We studied the case of nonlinear wave collapse turbulence, a phenomenon occurring in the presence of an external source of energy, and coupling nonlinearly to the Langmuir waves with ion sound and electromagnetic waves. Due to this coupling, the density fluctuations associated with ion sound waves refracts the Langmuir waves in regions of low densities. Coherent wave packets localize in such regions, and experience a cycle driven by the ponderomotive force, which decreases them to shorter scales and enhances their intensity (wave collapse). In such conditions, numerous wave collapse sites are present in the plasma, which change its radiative properties. We proposed a model for calculating the change in the line shape of atoms submitted to the electric field of a nearby wave collapse [3,4]. Our model uses the numerical solution of the emitter Schrödinger equation submitted to an electric field taken as a sequence of envelope solitons oscillating at the plasma frequency. We used the results of numerical simulations of wave collapse [5] to sample the lifetime of each soliton as well as the probability density function for the magnitude of the electric field. We will present the changes expected on a line shape of hydrogen for plasma conditions of interest in astrophysical and fusion plasmas.

The aim of this work is to study the effect of wave collapse on spectral line shapes of Lyman α, Lyman β, and Balmer α emitted by hydrogen atoms. In the following, we consider an atom submitted to a sequence of an electric field modulated by an envelope soliton, and we calculate numerically the atomic dipole autocorrelation function in the single presence of such solitons. Plasmas submitted to an energetic beam of particles or to a strong radiation can be found in many situations. In an astrophysical context, active galactic nuclei [6], pulsar radio sources [7], planetary foreshocks [8], or solar type III radio bursts [9] are possible candidates. Relevant laboratory plasmas are laser plasmas [10], radio experiments [11], or possibly also magnetic fusion plasma, since such plasmas can be affected by the energetic beams of runaway electrons [12].

The paper is organized as follows: in Section 2, we recall the main properties of turbulent Langmuir fields, and we propose a model for computing the dipole autocorrelation function (DAF) and the line shape in Section 3. We present and discuss our results on the hydrogen lines in Section 4.

## 2. Wave Collapse and Strong Turbulence

The physics of Langmuir turbulence have been studied in detail, since it is necessary to the understanding many radiative and transport properties of plasma. Langmuir turbulence describes a plasma state affected by a high level of excited Langmuir waves [13]. Using the Zakharov equations [14], it is possible to distinguish between weak and strong Langmuir turbulence. A useful quantity for making such a distinction is the ratio $W$ of the wave energy density to the thermal energy density:

$$W = \varepsilon_0 E_L^2 / 4 N_e k_B T, \tag{1}$$

with $T$ and $N_e$ representing the hydrogen plasma temperature and density, $E_L$ the magnitude of the wave, $k_B$ the Boltzmann constant, and $\varepsilon_0$ the permittivity of free space. Beyond a critical value of $W$, which depends on the plasma conditions, the Langmuir waves couple with ion sound and electromagnetic waves, resulting in a strong turbulence regime. Strong Langmuir turbulence occurs in plasma submitted to an external source of energy, which may be coupled to the plasma waves, thus increasing their intensity, and allowing the start of nonlinear processes such as wave-wave interactions [15]. In this strong turbulence regime, the energy density of the waves can exceed the plasma energy density, and a large amount of energy is available. The physical process at work is the creation of low density regions by the coupling of the density fluctuations associated to the ion sound wave with the Langmuir wave. Wave packets are refracted in regions of low density, which are also regions of high refractive index. The nonlinear ponderomotive force then moves part of the plasma out of the region of maximum field value, thus starting a dynamic process where coherent wave packets evolve to shorter scales and higher intensities reaching several hundred times the average microfield $E_0 = 1/(4\pi\varepsilon_0 r_0^2)$, where $r_0$ is the average distance between particles defined by $r_0^3 = 3/(4\pi N_e)$. Plasma computer simulations reveal the existence of a wave packet cycle with a collapse arrested by dissipation, and a nucleation mechanism allowing the creation of new wave packets [5]. The electric field of such wave packets oscillates at a frequency close to the plasma frequency $\omega_p = \sqrt{N_e e^2 / m \varepsilon_0}$, where $m$ is the electron mass, and the wave packet is modulated by an envelope soliton with a Gaussian or Lorentzian shape. The average duration of a cycle is an estimate for the characteristic time of strong turbulence. It can be obtained from plasma simulations such as particle in cells codes and scales as much as 40 times the inverse of the average of $W$ [5], using units of the inverse plasma frequency. Taking account of this relation and of the expression of $W$ given by Equation (1), the choice of a value of $W$ also determines the values of the electric field modulus and of the average duration of a cycle for given background plasma conditions. If energy is supplied from an external source, localized wave packets may be created at a high rate, and the plasma will contain many of those coexisting wave packets. The localized wave packets appear to be densely packed, so that a large number of emitters experience the field of a wave packet.

## 3. Line Shape Model for Wave Collapse

In this work, we are interested in studying the effect of nonlinear wave collapse on a Stark spectral line shape of hydrogen atoms. Following our description of wave packet collapse, we assume that a large number of hydrogen atoms are submitted to the electric field of a wave packet. We propose to model the electric field of the wave by a sequence of solitons using a renewal process. The maximum magnitude of the electric field is sampled with a probability density function (PDF) that we assume to be half-normal. We jump from one soliton to the next using an exponential waiting time distribution $\nu$ exp(-$\nu t$), with the jumping frequency $\nu$ chosen as the inverse of the average duration of a cycle [3,4]. We call such a sequence of envelope solitons a single electric field history with a duration of the order of the line shape time of interest (inverse of the line width). This time is also the decorrelation time of dipole autocorrelation function (DAF), a quantity $C(t)$ defined by [16]:

$$C(t) = Tr\langle \vec{D} \cdot U^+(t)\vec{D}U(t)\rho\rangle, \qquad (2)$$

where the trace is over the atomic states, $\vec{D}$ and $U$ are the atomic dipole and evolution operators, $\rho$ is the density matrix, and the angle brackets imply an average over the configurations of the perturbation. The atom perturbation dynamics are obtained by numerically solving the Schrödinger equation for the evolution operator $U(t)$ for each history. The DAF is obtained as an average over a large number ($10^4$) of independent electric field histories. The sampling of the stochastic variables may be done on the computer with pseudorandom number algorithms, associated to numerical techniques such as transformation or rejection methods [17,18]. The line shape is also obtained numerically using a Fourier transform of the DAF:

$$I(\omega) = \frac{1}{\pi}Re\int_0^\infty \exp(i\omega t)C(t)dt. \qquad (3)$$

The calculation presented in the following concerns the hydrogen Lyman $\alpha$ ($L_\alpha$), Lyman $\beta$ ($L_\beta$), and Balmer $\alpha$ ($H_\alpha$) lines, neglecting fine structure in order to obtain a fast numerical evaluation, and using the spherical quantum number n, l, m. For our calculations of the Balmer lines, we neglect the broadening of the lower states of the transitions. Using our simulation, it is possible to calculate the effect of Langmuir solitons alone. It is possible to compare to Stark broadening (impact approximation or ab initio) in a hydrogen plasma in equilibrium [16]. Using a convolution, it is also possible to calculate a profile taking into account both equilibrium Stark broadening and Langmuir solitons.

## 4. Results

We first compute the solution of the Schrödinger equation in the presence of the sequence of solitons alone. We calculate the dipole autocorrelation function $C(t)$ and the line profile for $L_\alpha$, $L_\beta$, and $H_\alpha$, for a density of $10^{19}$ m$^{-3}$ and for a temperature of $10^5$ K, conditions which can be found in the edge of a tokamak plasma. Solitons are generated for an average value of $W \approx 1.1$, resulting in a jumping frequency of $\nu \approx \omega_p/37$ and an average peak value of $150E_0$ for the electric field. After a study of the shape effect of the shape of the envelope soliton [3], and with the help of plasma simulations, we chose a soliton shape with a width equal to 20% of the wave cycle duration, a value which minimizes the broadening effect of the soliton sequence [3].

Figures 1–3 plot the DAF of $L_\alpha$, $L_\beta$, and $H_\alpha$ for the pure Stark effect using an impact approximation (dashed line), pure soliton effect (solid line), and the product of the two preceding DAFs (dotted line). In Figure 1, a strong decay is observed on the soliton DAF of $L_\alpha$ (solid line) for times shorter than the average duration of a wave packet cycle. This strong initial decay can be attributed to the large magnitude of the soliton electric field. A similar behavior of the DAF for short times is also observed for $L_\beta$ (Figure 2, solid line) and $H_\alpha$ (Figure 3, solid line). For intermediate times, the soliton DAF of $L_\alpha$ has a weaker decay than the Stark DAF, but the two curves are similar for long times. For times longer than the wave packet cycle, we observe for $L_\beta$ and $H_\alpha$ a weaker decay of the soliton than the impact DAF. Small amplitude oscillations are seen on these DAF, and are in phase with the plasma frequency. In all cases, the decay of the product DAF (dotted line) is significantly larger than for the Stark DAF.

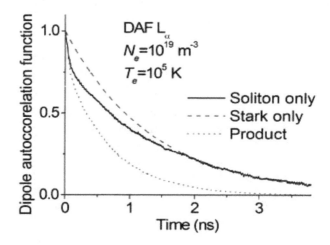

**Figure 1.** Dipole autocorrelation functions (DAF) for $L_\alpha$ submitted to $10^4$ solitons sequences calculated for $W = 1.1$ (solid line) for 10,000 histories, and compared to the Stark DAF (dashed line) calculated with an impact approximation, and the product DAF (dotted line) in plasma with a density of $N_e = 10^{19}$ m$^{-3}$ and temperature of $T_e = 10^5$ K.

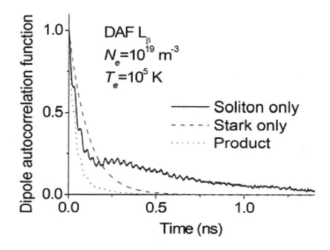

**Figure 2.** Same as Figure 1 for $L_\beta$.

In Figures 4–6, we present the line shapes of $L_\alpha$, $L_\beta$, and $H_\alpha$ for the same plasma conditions. Although being especially studied for high density plasmas, detailed Stark line shapes are also needed for low density plasmas, since they enter in the modeling of radiative transfer together with Doppler broadening [19]. We calculated the line shape profile of $L_\alpha$, $L_\beta$, and $H_\alpha$ by a Fourier transform of the dipole autocorrelation functions already discussed. In Figure 4, we found that the line profile which is only affected by solitons (solid line) is similar to the Stark profile (dashed line) for $L_\alpha$. The convolution profile (dotted line) is, however, about 2.1 broader than the Stark profile. For $L_\beta$ (Figure 5) and $H_\alpha$ (Figure 6), the profile only affected by solitons (solid line) is narrower than the pure Stark profile.

The convolution profile (dotted line) is broader by a factor 2 than the Stark profile of $L_\beta$, and broader by a factor 1.7 for $H_\alpha$.

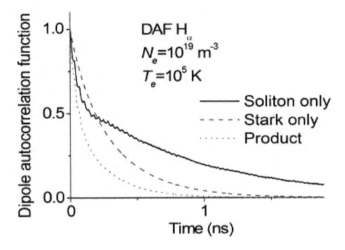

**Figure 3.** Same as Figure 1 for $H_\alpha$.

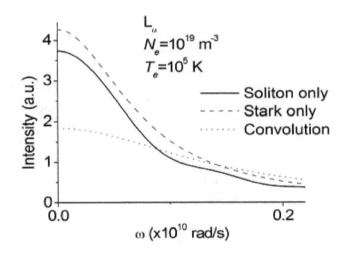

**Figure 4.** Line shape of $L_\alpha$ for soliton only with $W = 1.1$ (solid line), for Stark only in the impact approximation (dashed line), and compared to a convolution (dotted line) of the latter two for the plasma condition of Figure 1.

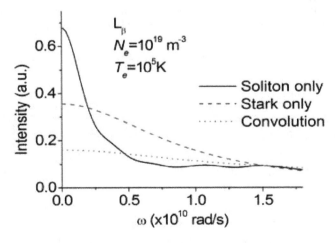

**Figure 5.** Same as Figure 4 for $L_\beta$.

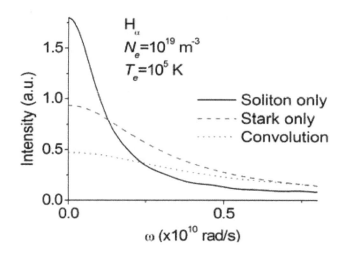

**Figure 6.** Same as Figure 4 for $H_\alpha$.

## 5. Conclusions

The study of Zhakarov equations and numerous numerical simulations has revealed the complex behavior of plasma affected by strong Langmuir turbulence. The nonlinear coupling of the plasma waves creates numerous localized wave packets, subject to collapse. Each wave packet experiences a cycle during which the electric field magnitude grows to values of more than hundred $E_0$ ($E_0 \approx$ Holtsmark field). Using the main properties of strong Langmuir turbulence obtained from simulation calculations, we proposed a simple stochastic renewal model for the electric field of the wave packets. This model field is well-suited to study the effect of Langmuir turbulence on the line shape emitted in plasma. We used a simulation to generate field histories with a prescribed probability density function and waiting time distribution. The dipole autocorrelation function was obtained by a numerical integration of the Schrödinger equation and an average over $10^4$ histories. Our calculations concern the hydrogen $L_\alpha$, $L_\beta$, and $H_\alpha$ lines for plasma conditions for a density equal to $10^{19}$ m$^{-3}$ and a temperature of $10^5$ K. Strong turbulence brings a significant additional broadening to the pure Stark profile for all three lines.

In the future, we will look for other lines, plasma conditions, and wave collapse conditions, and make comparisons with line shapes observed in turbulent plasma.

**Acknowledgments:** This work is supported by the funding agency Campus France (Pavle Savic PHC project 36237PE). This work has also been carried out within the framework of the EUROfusion Consortium and has received funding from the Euratom research and training program 2014–2018 under grant agreement no. 633053. The views and opinions expressed herein do not necessarily reflect those of the European Commission.

**Author Contributions:** This work is based on the numerous contributions of all the authors.

## References

1.    Rosseland, S. Viscosity in the Stars. *Mon. Not. R. Astron. Soc.* **1928**, *89*, 49–53. [CrossRef]
2.    Struve, O. Thermal Doppler effect and turbulence in stellar spectra of early class. *Proc. Natl. Acad. Sci. USA* **1932**, *18*, 585–589. [CrossRef] [PubMed]
3.    Hannachi, I.; Stamm, R.; Rosato, J.; Marandet, Y. Effect of nonlinear wave collapse on line shapes in a plasma. *Europhys. Lett.* **2016**, *114*, 23002. [CrossRef]
4.    Stamm, R.; Hannachi, I.; Meireni, M.; Capes, H.; Godbert-Mouret, L.; Koubiti, M.; Rosato, J.; Marandet, Y.; Dimitrijević, M.; Simić, Z. Line shapes in turbulent plasmas. *Eur. Phys. J. D* **2017**, *71*, 68. [CrossRef]
5.    Robinson, P.A. Nonlinear wave collapse and strong turbulence. *Rev. Mod. Phys.* **1997**, *69*, 507–574. [CrossRef]
6.    Miller, H.; Wiita, P. *Active Galactic Nuclei*; Springer: Berlin, Germany, 1987.

7.   Asseo, E.; Porzio, A. Strong Langmuir turbulence in a pulsar emission region: Statistical analysis. *Mon. Not. R. Astron. Soc.* **2006**, *369*, 1469–1490. [CrossRef]
8.   Sigsbee, K.; Kletzing, C.A.; Gurnett, D.A.; Pickett, J.S.; Balogh, A.; Lucek, E. Statistical behavior of foreshock Langmuir waves observed by the cluster wideband data plasma wave receiver. *Ann. Geophys.* **2004**, *22*, 2337–2344. [CrossRef]
9.   Ratcliffe, H.; Kontar, E.P.; Reid, A.S. Large-scale simulations of solar type III radio bursts: Flux density, drift rate, duration, and bandwidth. *Astron. Astrophys.* **2014**, *572*, A111. [CrossRef]
10.  Kruer, W.L. *The Physics of Laser-Plasma Interactions*; Addison-Wesley: Redwood City, CA, USA, 1988.
11.  Bauer, B.; Wong, A.; Scurry, L.; Decyk, V. Efficiency of caviton Formation as a function of plasma density gradient. *Phys. Fluids B* **1990**, *2*, 1941. [CrossRef]
12.  Paz-Soldan, C.; Eidietis, N.; Granetz, R.; Hollmann, E.; Moyer, R.; Wesley, J.; Zhang, J.; Austin, M.; Crocker, N.; Winger, A.; et al. Growth and decay of runaway electrons above the critical electric field under quiescent conditions. *Phys. Plasmas* **2014**, *21*, 022514. [CrossRef]
13.  Zakharov, V.; Musher, S.; Rubenchik, A. Weak Langmuir turbulence of an isothermal plasma. *Sov. Phys. JETP* **1975**, *42*, 80–86.
14.  Zakharov, V. Collapse of Langmuir waves. *Sov. Phys. JETP* **1972**, *35*, 908–914.
15.  Bellan, P.M. *Fundamental of Plasma Physics*; Cambridge University Press: Cambridge, UK, 2006.
16.  Griem, H.R. *Spectral Line Broadening by Plasmas*; McGraw-Hill: New York, NY, USA, 1964.
17.  Vesely, F. *Computational Physics, an Introduction*; Plenum Press: New York, NY, USA, 1994.
18.  IMSL. 2012. Available online: http://www.roguewave.com (accessed on 12 August 2017).
19.  Rosato, J.; Reiter, D.; Kotov, V.; Marandet, Y.; Capes, H.; Godbert-Mouret, L.; Koubiti, M.; Stamm, R. Progress Radiative Transfer modelling in Optically Thick Divertor plasmas. *Contrib. Plasma Phys.* **2010**, *50*, 398–403. [CrossRef]

# 21

# Nonlinear Spectroscopy of Alkali Atoms in Cold Medium of Astrophysical Relevance

**Dmitry K. Efimov** [1,2], **Martins Bruvelis** [3], **Nikolai N. Bezuglov** [1], **Milan S. Dimitrijević** [4,5,6], **Andrey N. Klyucharev** [1], **Vladimir A. Srećković** [7,*], **Yurij N. Gnedin** [8] and **Francesco Fuso** [9]

[1]   Saint Petersburg State University, St. Petersburg State University, 7/9 Universitetskaya nab., St. Petersburg 199034, Russia; dmitry.efimov@uj.edu.pl (D.K.E.); bezuglov50@mail.ru (N.N.B.); anklyuch@gmail.com (A.N.K.)

[2]   Instytut Fizyki im. Mariana Smoluchowskiego, Uniwersytet Jagielloński, 30-348 Kraków, Poland

[3]   Laser Centre, University of Latvia, LV-1002 Riga, Latvia; martins.bruvelis@gmail.com

[4]   Astronomical Observatory, Volgina 7, 11060 Belgrad, Serbia; mdimitrijevic@aob.rs

[5]   IHIS-Technoexperts, Bezanijska 23, 11080 Zemun, Serbia

[6]   Observatoire de Paris, 92195 Meudon CEDEX, France

[7]   Institute of Physics, University of Belgrad, P.O. Box 57, 11001 Belgrad, Serbia

[8]   Pulkovo Observatory, Russian Academy of Sciences, St. Petersburg 196140, Russia; gnedin@gao.spb.ru

[9]   Dipartimento di Fisica Enrico Fermi and CNISM, Università di Pisa, I-56127 Pisa, Italy; francesco.fuso@unipi.it

*   Correspondence: vlada@ipb.ac.rs

**Abstract:** The time-dependent population dynamics of hyperfine (HF) sublevels of $n^2p_{3/2}$ atomic states upon laser excitation in a cold medium of alkali atoms is examined. We demonstrate some peculiarities of the absorption HF multiplet formation in $D2$-line resulting from a long interaction time ($\sim$200 μs) interaction between light and Na ($n = 3$) and Cs ($n = 6$) atoms in a cold and slow sub-thermal ($T \sim$ 1K) beam. We analytically describe a number of $D2$-line-shape effects that are of interest in spectroscopic studies of cold dusty white dwarfs: broadening by optical pumping, intensity redistribution within components of $D2$-line HF multiplet for partially closed transitions and asymmetry of absorption lines induced by AC Stark shifts for cyclic transitions.

**Keywords:** spectroscopy; alkali atoms; astrophysics

---

## 1. Introduction

Investigation of the fluorescence spectrum of sodium atoms is an important data source for astrophysics and, especially, for understanding the physical processes in cool stars, namely, for brown and white dwarfs. For example, the measured precise atmospheric parameters for shortest period binary white dwarfs confirm the existence of metal-rich envelopes around extremely low-mass white dwarfs and allow us to examine the distribution of the abundance of non-hydrogen elements, including Na, as a function of effective temperature and mass [1]. It is worth emphasizing that we can expect the gravitational wave strain for such systems.

Other interesting astrophysical objects are metal polluted white dwarfs and dusty white dwarfs. It has long been suspected that metal polluted white dwarfs (types DAZ, DBZ and DZ) and white dwarfs with dusty disks can possess planetary systems [2]. Therefore, the spectroscopic observations of sodium atoms in these objects can confirm the validity of this hypothesis. It is in fact known that all dusty white dwarfs show evidence for alkali atoms accretion onto their dusty disk.

In reference [3], the observational constraints on the origin of metals and alkali atoms in cool white dwarfs are discussed. The presence of absorption lines of Na, Mg, Fe etc in the photospheres of cool hydrogen atmosphere DA-type white dwarfs has been an unexplained problem for a long time. However, in [4], it was shown that the metal abundances in the atmospheres of white dwarfs can be explained by episodic accretion events whenever the white dwarf travels through relatively overdense clouds.

Furthermore, Na 8183.27, 8192.81 Å absorption doublet was discovered in astrophysical systems of white dwarf–main sequence binaries. It allows us to investigate in detail the process of mass transfer interaction in these complex astrophysical systems.

In astrophysics, there is a wide interest in the search for dusty white dwarfs with powerful infrared excess. This excess is produced by orbiting dust disk that contains planetary systems and planetesimals. All these materials get accreted onto the white dwarf and enrich its pure hydrogen or/and helium atmosphere. Studying these heavy elements enriched white dwarfs becomes an effective way to measure directly the bulk compositions of extrasolar planetesimals [5]. Na atoms can be the essential part of these heavy-element clouds; therefore, detailed knowledge of Na emission features can be extremely relevant for investigating such astrophysical problems.

Recently, combined Spitzer and ground-based Korea Microlensing Telescope Network observations identified and precisely measured an Earth-mass planet orbiting ultra cool dwarf [6]. Observations of Na atoms in ultra cool dwarfs can become the effective method for determining planetesimals and Earth-mass planets in dusty disks surrounding the central ultra cool dwarf.

We note as well that, already in 1974, Brown [7] detected the first neutral sodium cloud near Jovian satellite Io (see also [8]). The investigation of these sodium clouds is needed for better understanding of the interaction between Io's atmosphere and Jovian magnetosphere [9] and of processes in the Jovian environment [10].

We are concerned here with important particularities of cold alkali atoms absorption spectra in D2-lines (see Figure 1) that result from a long interaction time of light and matter and, as a consequence, from a strong involvement of optical pumping phenomena within hyperfine (HF) components of the ground and excited states. The described novel theoretical predictions are compared with experimental data obtained in sub-thermal (cold) atomic beams [11,12] operated with two different alkali species, namely Na and Cs. Atomic units are used unless stated otherwise.

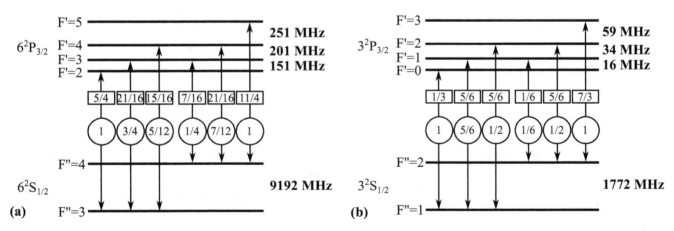

**Figure 1.** Transition strength $\tilde{S}_{F''F'}$ (boxed) and branching $\Pi$ (circled) factors for the hyperfine components of D2-lines for **(a)** cesium (transition $6^2s_{1/2} \rightarrow 6^2p_{3/2}$) and **(b)** sodium (transition $3^2s_{1/2} \rightarrow 3^2p_{3/2}$) atoms. For cyclic (closed) transitions, the factor $\Pi = 1$. The natural lifetime $\tau$ and saturation threshold $\Omega_{st} = 1/(\tau\sqrt{2})$ of the resonance states are $\tau = 30.5$ ns and $\Omega_{st} = 3.7$ MHz for Cs while $\tau = 16.4$ ns and $\Omega_{st} = 6.9$ MHz for Na, respectively.

## 2. Formation of HF Absorption Multiplet for Partially Open Transitions

Absorption spectra of alkali atoms have a multiplet form due to HF structure of atomic states (see Figure 1). There are two different classes of multiplet components corresponding to cyclic (or closed) and partially open (or non-cyclic) transitions between HF sublevels $F'$ and $F''$ of the resonant excited (e) and the ground (g) states, respectively. In the partially open case, each spontaneously emitted photon induces two transitions $F' \rightarrow F'' = 2,1$ in Na and $F' \rightarrow F'' = 4,3$ in Cs with probabilities $\Pi$ (see Figure 1). The dimensionless parameter $\Pi$ is called the branching ratio. The cyclic transitions correspond to $\Pi = 1$ and they are particularly important in cooling and trapping techniques of neutral atoms, for example, in magneto-optical traps (MOTs) [13].

Excitation of the partially open individual transition $F'' \rightarrow F'$ is accompanied by optical pumping phenomenon that is usually associated with redistribution of population within HF sublevels of the ground state because of interaction with resonant light fields [14]. Line-shape effects due to optical pumping (the so-called depletion broadening) in the weak excitation limit and long interaction time $\tau_{tr}$ were examined in detail in [15]. A convenient approximation in describing the time-dependent population dynamics of HF sublevels $F'', F'$ is a two-level model. Atoms are excited on the atomic transition by monochromatic laser radiation with frequency $\omega_L$, amplitude $A_0$, and detuning $\delta \equiv \omega_L - \omega_{eg}$. If the laser intensity is insufficient to saturate the transition, which corresponds to Rabi frequency $\Omega$ ($\Omega = \langle e| A_0 \hat{d} | g \rangle$, where $\hat{d}$ is the atomic dipole operator) being below the natural linewidth $\Gamma_e$ of the excited state, then the rate of spontaneous transitions per one atom is equal to [14]:

$$\Gamma_g = \frac{\Omega^2 \Gamma_e}{4\delta^2 + \Gamma_e^2}; \quad \Omega \ll \Gamma_e. \tag{1}$$

In a partially open system, each spontaneously emitted photon returns to g-state the $\Pi$-portion of the population, while the remaining $(1 - \Pi)$-portion is transferred to states outside the two-level system. It means the depletion rate $\Gamma_{pum}$ is equal to $\Gamma_{pum} = (1 - \Pi)\Gamma_g$ and the lower state population $n_g$ of the atom decays exponentially $n_g(\tau) \approx \exp(-\tau/\tau_{pum})$ at the time $\tau$ upon transit through the laser beam. The corresponding pumping time is $\tau_{pum} = 1/\Gamma_{pum}$, i.e.,

$$\tau_{pum} = \frac{1}{\Gamma_{pum}} = \frac{1}{1 - \Pi} \frac{4\delta^2 + \Gamma_e^2}{\Omega^2 \Gamma_e}. \tag{2}$$

In experiments involving atom beams crossed by the laser radiation [11,12], we can identify a transit time $\tau_{tr}$. In such conditions, the total population after the atom beam has crossed the laser radiation, $n_g(\infty)$, becomes $n_g(\infty) \approx \exp(-\tau_{tr}/\tau_{pum})$. Thus, the criterion for the development of optical pumping is $\tau_{tr} > \tau_{pum}$, or in terms of Rabi frequencies

$$\Omega > \Omega_{cr}; \quad \Omega_{cr} = \Gamma_e \sqrt{\frac{1 + 4\delta^2/\Gamma_e^2}{1 - \Pi} \cdot \frac{\tau_e}{\tau_{tr}}}, \tag{3}$$

where $\tau_e = 1/\Gamma_e$ is the radiative lifetime of the upper e-state. The ratio $\tau_e/\tau_{tr}$ for cold atomic beams acquires values of $\sim 10^{-4}$ ($\tau_e \approx 20$ ns, $\tau_{tr} \approx 200$ μs) [3]. Therefore, the depletion manifestation can be observed at Rabi frequencies well below (by two orders of magnitude) the saturation frequency $\Omega_{st} \equiv \Gamma_e/\sqrt{2}$; exceeding this value results in the development of nonlinear (power broadening) and quantum optics effects [14].

Important spectroscopic features of optical pumping emerge when the Rabi frequency is set close, or above, the critical value $\Omega_{cr}$ (3) [15]. Conventional experiments would usually register the fluorescence signal $J$ (absorption line) from the entire excitation volume, which is proportional to the total number $\Gamma_e \bar{n}_e$ of photons emitted by a single atom. On the other hand, the number $(1 - \Pi)\Gamma_e \bar{n}_e$ of spontaneously emitted photons on transitions outside the g-e system is equal to the total loss $1 - n_g(\infty)$ of the ground state population, so that the signal $J$ can be written as

$$J \equiv \Gamma_e \bar{n}_e = \frac{1}{(1 - \Pi)} \left(1 - \exp(-\tau_{tr}/\tau_{pum})\right). \tag{4}$$

Equation (4) yields a dependence of the fluorescence signal on the laser detuning $\delta$ in the following explicit form:

$$J(\delta) = \frac{\Omega^2 \tau_{tr}}{\Gamma_e} \frac{1}{P_{pum}} \left[ 1 - \exp\left( -\frac{P_{pum}}{1 + 4\delta^2/\Gamma_e^2} \right) \right],$$ (5)

$$P_{pum} = \frac{\tau_{tr}}{\tau_{pum}^0}; \quad \tau_{pum}^0 = \frac{1}{1 - \Pi} \frac{\Gamma_e}{\Omega^2}.$$ (6)

The above equation and data presented in Figure 2 show that the absorption spectrum strongly depends on the pumping parameter $P_{pum}$ defined as in Equation (6), which is given by the ratio of transit time $\tau_{tr}$ and pumping time $\tau_{pum}^0$. The latter has the meaning of optical pumping time at resonant excitation, when $\delta = 0$. Importantly, the parameter $P_{pum}$ can be large even at laser intensities well below the saturation limit: $P_{pum} >> 1$ when $\tau_{tr} >> \tau_{pum}^0$ and $\Omega << \Omega_{cr}$.

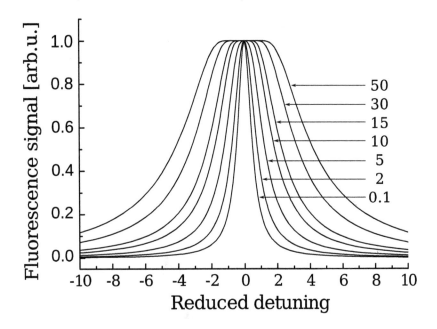

**Figure 2.** Dependence of the absorption signal $J$ Equation (5) on the reduced detuning $\delta/\Gamma_e$ for different values of the pumping parameter $P_{pum}$ Equation (6) (shown as labels of the curves). All curves are unity-normalized at the line center $\delta = 0$.

According to Equation (4), optical pumping leads to saturation of the fluorescence signal at the value of $1/(1 - \Pi)$. This fact along with redistribution of intensities within $D2$-line multiplet have been predicted and experimentally demonstrated in [15]. Another phenomenon described by Equations (4) and (5) is related to depletion broadening of the absorption line. As an estimation of the characteristic full width $\Delta_{pm}$ of the spectral absorption profile induced by optical pumping, one can write

$$\Delta_{pm} = \sqrt{\frac{\tau_{tr}}{\tau_e}(1 - \Pi) \cdot \Omega^2 - \Gamma_e^2}.$$ (7)

In the case of cold beams of alkali atoms, with $\tau_{tr}/\tau_e \sim 10^4$ and $\Omega > \Omega_{cr}$ for $\delta = 0$ (see Equation (3)), the width $\Delta_{pm}$ acquires the form $\Delta_{pm} \approx \Omega\sqrt{(1 - \Pi)\tau_{tr}/\tau_e} \sim 100 \cdot \Omega$, which may essentially exceed the power broadening effects ($\Delta_{pw} \sim \Omega$) even in the limit of weak excitation. The corresponding experimental investigation of spectral broadening due to optical pumping (the parameter $\tau_{tr}/\tau_e \sim 100$) in partially open HF level systems in Na has been reported in [15]. We note that, since the experiments were performed at low number densities ($\sim 10^{10}$ cm$^{-3}$) of Na atoms, line-shape modifications by radiation trapping in collimated beams [16] can be disregarded.

## 3. Experimental Setup and Spectroscopic Data

Common experimental practice foresees the use of two types of thermal beams: effusive beam with longitudinal velocity $v_{lg} \approx 400$ m/s [17], and crossed beams with $v_{lg} \approx 600$ m/s [18]. The second type represents a supersonic beam with $v_{lg} \approx 1100$ m/s [15] and a rather specific velocity distribution function [19]. An additional experimental possibility is represented by laser cooled atomic beams [20,21], which are characterized by small $v_{lg}$ around 12 m/s, which corresponds to the sub-thermal temperature interval lying below 1 K and practically negligible transversal velocity.

At the core of the relevant setup is a hollow pyramid with reflective inner surfaces and a hole at its vertex. By shining a single, large diameter (35 mm) laser beam along the pyramid axis, the optical configuration of an MOT is achieved. Imbalance of the radiation pressure along the pyramid axis pushes Cs atoms out of the pyramid hole. Further collimation of the atoms is achieved by a transverse optical molasses right after the pyramid hole (Figure 3). The atom beam is then excited by a diode laser tunable over the hyperfine transitions belonging to the $D2$-line of Cs; blue laser radiation is superposed in order to ionize the excited atoms. The production rate of ions is proportional to the total number of excited atoms in the excitation volume. Ions are effectively collected and detected, providing a sensitive probe of the excited population [12]. Note that line-shape modifications by radiation trapping in a cold medium [22,23] can be disregarded for the above experimental conditions.

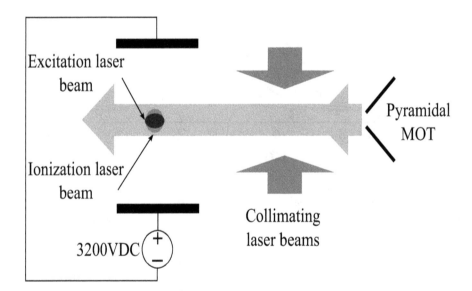

**Figure 3.** Schematic diagram of the experimental setup for the production and excitation of a cold Cs atom beam.

Figure 4 shows an experimentally obtained absorption profile (dots) upon excitation of the cyclic $F'' = 4 \rightarrow F' = 5$ transition for Cs atoms. We underline two main points: (i) although the main line-profile results from symmetric power broadening; (ii) there is a slight asymmetry in the line-shape.

This asymmetry is induced by the other HF components of $n^2 p_{3/2}$ sublevels and, as it will be shown in the next section, is strongly affected by the relationships between laser Rabi frequency and values of HF splitting. For comparison, we show in Figure 5 a situation occurring for Na atoms in similar experimental conditions to demonstrate the occurrence of a quite non-standard line shape, whose explanation is presented in the following.

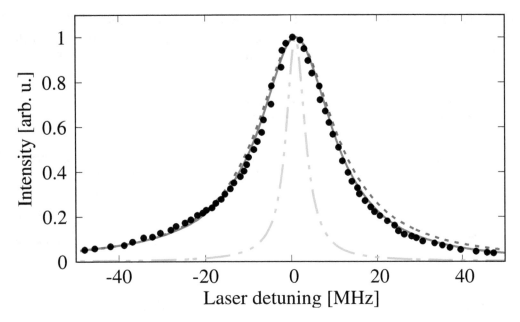

**Figure 4.** Theoretical (solid curve, Equation (19)) and experimental [11] (dots) absorption profile of the $D2$-line of Cs upon an excitation of the closed $F'' = 4 \to F' = 5$ transition in a cold sub-thermal Cs beam for laser power of 1.0 mW (the corresponding Rabi frequency $\Omega = 21.2$ MHz). The dashed curve corresponds to the power broadening profile. The bar-dashed curve exhibits the natural broadening profile.

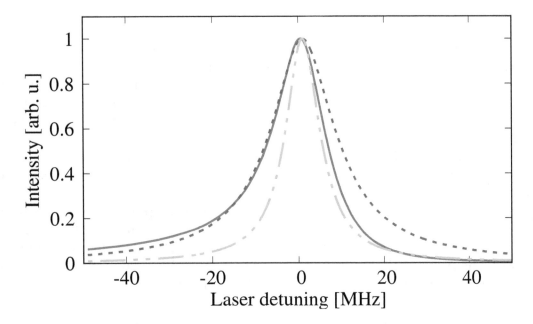

**Figure 5.** The same as in Figure 4 in the case of $D2$-line of sub-thermal Na atoms and the excitation of the closed $F'' = 2 \to F' = 3$ transition with laser Rabi frequency $\Omega = 21.2$ MHz.

## 4. Cyclic Transitions Treatment: Modeling and Discussion

The description of cyclic transitions can rarely be reduced to the two-level system model: even upon resonant excitation ($\delta = 0$) in a cold beam, the presence of other HF levels may result in the appearance of fundamentally new effects. Let us add the third level $|\,3\rangle$ to a two-level closed system $|\,4\rangle,|\,2\rangle$ (see Figure 6a), which can decay to the passive state $|\,1\rangle$ due to spontaneous emission.

We associate states $|1\rangle$, $|2\rangle$ and $|3\rangle$, $|4\rangle$ of Figure 6 with HF g-sublevels $F'' = 1$, $F'' = 2$ and e-sublevels $F' = 2$, $F' = 3$, accordingly, in the case of Na atoms (see Figure 1), while, for Cs HF structure, we choose HF g-sublevels $F'' = 3$, $F'' = 4$ and e-sublevels $F' = 4$. $F' = 5$.

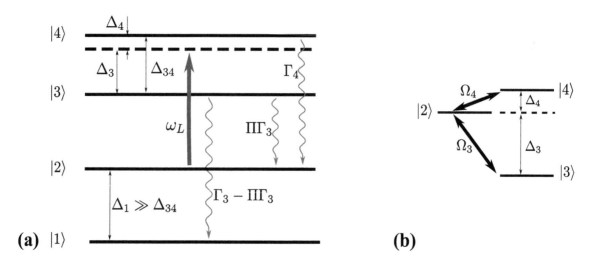

**Figure 6.** (a) schematic illustration of a three-level system having a cyclic transition $|4\rangle \rightarrow |2\rangle$ and a partially open transition $|3\rangle \rightarrow |2\rangle$ with the branching factor $\Pi$. The additional passive state $|1\rangle$ accumulates the population due to the spontaneous transition $|3\rangle \rightarrow |1\rangle$; **(b)** the same scheme in the rotating wave approximation. The bare ground state $|2\rangle$ is selected as the position of zero energy (dashed horizontal line).

We are concerned here with the resonant excitation of cyclic HF transitions between HF sublevels. As a consequence, we set the following constrains for laser detuning $\delta = \omega_L - \omega_{42}$, as indicated in Figure 6: $\delta = -\Delta_4 \sim \Gamma << \Delta_3 \approx \Delta_{34}$, where $\Gamma = \Gamma_3 = \Gamma_4$ is the unique natural linewidth of the upper $n^2 p_{3/2}$ levels. The lower levels $|1\rangle, |2\rangle$ correspond to the HF sublevels of the ground state $n^2 s_{3/2}$. Its large HF splitting ($\Delta_1$), compared to the one ($\Delta_{34}$) of resonant states $n^2 p_{3/2}$, transforms the state $|1\rangle$ into a dark state that is not interacting with the pump laser. Figure 6b shows the dressed state configuration obtained using the rotating wave approximation [14]. If the energy $\varepsilon_2$ of the state 2 is chosen as zero, then the energies $\varepsilon_{3,4}$ of the dressed states $|3\rangle, |4\rangle$ turn out to be determined by the laser detuning $\delta$: $\varepsilon_4 = \Delta_4 = -\delta$, $\varepsilon_3 = -\Delta_3$.

It is convenient to represent the Rabi frequencies $\Omega_{F''F'}$ of the pump laser in terms of a single reduced frequency $\Omega_{red}$ [15], which is the product of the amplitude $A_0$ of the laser field and the reduced dipole matrix element $D_{S,P} = (n^2 S_{1/2}||D||n^2 P_{3/2})$ [24] for the respective non HF resolved fine transition: $\Omega_{red} = A_0 D_{S,P}$. The partial Rabi frequency values $\Omega_{F''F'}$ for HF transitions $S_{1/2}, F'' \rightarrow P_{3/2}, F'$ are determined by the corresponding, so-called, line strengths $S_{F''F'}$ [24]:

$$\Omega_{F''F'} = \Omega_{red}\sqrt{\tilde{S}_{F''F'}}; \quad \tilde{S}_{F''F'} = S_{F''F'}/D_{S,P}^2. \tag{8}$$

The values of non-dimensional parameters $\tilde{S}_{F''F'}$ are reported within rectangular frames in Figure 1. In Figure 6b, and subsequent discussions, we are using the abbreviations $\Omega_i = \Omega_{2''i'}$ ($i = 3, 4$).

A fundamentally new aspect for the three-level system is the appearance of the dynamic (AC) Stark shifts [13,14] of the state 2, due to its laser induced mixing with the state 3. As a result, the energy defect (detuning) between states 4 and 2 is also undergoing a shift (see also below, Equation (16))

$$\Delta_4 \rightarrow \delta_4 = \Delta_4 - \Omega_3^2/(4\Delta_{34}). \tag{9}$$

In the case of a resonance ($\Delta_4 = 0$), the increase of the pump laser intensity leads to an increasing shift of the actual detuning $\delta_4$. It is worth noting that, in order to significantly affect the light-induced asymmetry in the line profile, the absolute value $|\delta_4|$ of detuning must be larger than the natural linewidth $\Gamma$, i.e., according to Equation (9), it is necessary that $\Omega_3$ exceeds the saturation threshold: $\Omega_3^2 > 2\Gamma\Delta_3 > \Gamma^2$.

In the weak excitation limit ($\Omega_4 < \Gamma$), the induced transitions $|2\rangle \rightarrow |4\rangle$ and $|2\rangle \rightarrow |3\rangle$ are independent from each other, as they are relatively weak compared to spontaneous transitions. This means that, due to the partially open $|2\rangle \rightarrow |3\rangle$ transition, the ground state depletion should take place and, consequently, the cyclic transition $|2\rangle \rightarrow |4\rangle$ should be affected by the depletion broadening. The results of Section 2 are applicable for the partially open transition $|2\rangle \rightarrow |3\rangle$. In particular, using the notation of Figures 1 and 3, the relation (2) can be rewritten as

$$\frac{\tau_{pum}}{\tau_{tr}} \approx \frac{1}{\tau_{tr}\Gamma_{pum}} = \frac{\tau/\tau_{tr}}{1-\Pi}\frac{4\Delta_3^2}{\Omega_3^2}. \tag{10}$$

The inequality $\tau_{tr}/\tau_{pum} > 1$ characterizing the onset of optical pumping is fulfilled at $\Omega_3 > \Omega_{cr} = 1.50$ MHz and $\Omega_3 > \Omega_{cr} = 9.6$ MHz for Na and Cs, respectively, for the experimental ratio $\tau_{tr}/\tau \sim 10^4$. It is well seen that, in the case of Cs, its critical value $\Omega_{cr}$ lies beyond the saturation threshold. At such Rabi frequencies, the linear approximation is no longer applicable and a more accurate approach is required to describe the light–matter interaction.

### 4.1. Adiabatic Approach

The large value of the ratio $\tau_{tr}/\tau \sim 10^4$ allows one to use the method of adiabatic elimination [14,25] to obtain explicit qualitative description of the above effects for cyclic transitions. The exact analysis of the dynamics of optical pumping should be carried out within the framework of the density matrix [14] for the three-level system model shown in Figure 6b. Equations describing the interaction of a single atom (from the atomic beam) with a classical exciting radiation are the optical Bloch Equations [14]:

$$i\dot{\rho}_{ij} = \omega_{ij}\rho_{ij} + \sum_k \left(H_{ik}\,\rho_{kj} - \rho_{ik}\,H_{kj}\right) + il_{ij}, \tag{11}$$

where non-diagonal elements $\rho_{ij}$ ($i \neq j$) are associated with the so-called coherence while diagonal elements give the level population $n_i{:}n_i = \rho_{ii}$. The matrix $l_{ij}$ describes relaxation processes due to spontaneous radiative transitions. The matrix $H_{ik}$ determines the interaction of atoms with the laser light. If the electric field of the excitation laser in Figure 3 has a Gaussian distribution, $E(z) = A_0\exp(-z^2/(2d^2))$, with width $d$ along the atomic beam axis (coordinate $z$), then in the rotating wave approximation (RWA) [14] $H_{ik}$ has the following representation (see also Equation (8)):

$$\Omega_{F''F'}(t) = 0.5 \cdot \Omega_{red}\sqrt{\tilde{S}_{F''F'}}\exp(-2t^2/\tau_{tr}^2). \tag{12}$$

The time of flight of an atom through the excitation zone is determined by the laser beam waist size $d$: $\tau_{tr} = 2\,d/v$. Under the experimental conditions of [12], the respective values are $\tau_{tr} \approx 1.1$ mm and $\tau_{tr} \approx 200$ μs.

It is worth noting that all individual members of Equation (12) may be rewritten in the universal form:

$$\dot{X}_k = -\left(\Gamma_k + i\omega_k\right)X_k + F_k(t). \tag{13}$$

Equation (13) can be interpreted as the equation of a linear oscillator with complex coordinates $X_k$, which are subject to a slowly varying external force $F_k(t)$. One can see that the frequency detuning $\omega_k$ plays the role of rigidity, while the width $\Gamma_k$ is associated with the dissipation constant. From classical mechanics, it is well known that characteristic time $\tau_{rel}$ of the evolution of a forced oscillator to a steady-state condition is determined as $\tau_{rel} = 1/\sqrt{\Gamma_k^2 + \omega_k^2}$ [26]. If $\tau_{tr} >> \tau_{rel}$, the left-hand side of Equation (13) becomes negligible, so that Equation (13) has the following solutions:

$$X_k = F_k(t) / (\Gamma_k + i\omega_k) ; \tau_{tr} >> 1/\sqrt{\Gamma_k^2 + \omega_k^2}. \tag{14}$$

*4.2. Strict Results*

Adiabatic elimination for Bloch equations (11) was implemented in [27]. At the first stage, using Equation (14), we express non-diagonal elements $\rho_{ij}$ ($i \neq j$) via diagonal ones, i.e., via populations $n_i$. As a result, the following closed equation systems describe the population dynamics:

$$\dot{n}_2 = \Gamma n_4 + \Gamma \Pi n_3 + r_4(n_4 - n_2) + r_3(n_3 - n_2), \tag{15a}$$

$$\dot{n}_i = -\Gamma n_i - r_i(n_i - n_2); \quad i = 3, 4, \tag{15b}$$

$$r_i = \Omega_{i2}^2 \frac{\Gamma_i}{4\delta_i^2 + \Gamma_i^2}, \quad \delta_i = \omega_{i2} - \frac{\Omega_{j2}^2}{4\omega_{43}}, \quad \Gamma_i = \Gamma \left(1 + \frac{\Omega_{j2}^2}{2\omega_{43}^2}\right), \quad j \neq i, 2, \tag{16}$$

where $\omega_{43}$ is the difference between energies $\varepsilon_4$ and $\varepsilon_3$. These relations have the structure of balance equations for population transfer between the levels due to the spontaneous, with rate $\Gamma$, and laser stimulated, with rate $r_i$, transitions. An important result is the appearance of the AC Stark shifts $\Omega_{j2}^2 / (4\omega_{34})$ (see also Equation (9)) in the pumping rate constant $r_i$.

Owing to the adiabatic elimination, system (15) can be further reduced to a single equation for the total population $N(t)$ ($N = n_2 + n_4 + n_3$) [25] of a single atom:

$$\dot{N} = -\frac{\Gamma \cdot (1 - \Pi) \cdot r_3}{\Gamma + 2r_3 + r_4 (\Gamma + r_3) / (\Gamma + r_4)} N, \tag{17}$$

$$n_2 = \frac{N}{1 + r_4/ (\Gamma + r_4) + r_3/ (\Gamma + r_3)}; \quad n_i = \frac{r_i}{\Gamma + r_i} n_2 \quad i = 3, 4. \tag{18}$$

This reduction allows us to obtain an explicit representation for the fluorescence signal $J$, which is proportional to the total number $\Gamma(\bar{n}_4 + \bar{n}_3)$ of photons emitted by a single atom from the excitation volume

$$J = J(t) \mid_{t=\infty} \equiv \Gamma \int_{-\infty}^{t} d\tau \cdot (n_4(\tau) + n_3(\tau)) \mid_{t=\infty}. \tag{19}$$

*4.3. Discussion of the Line–Shape Structure*

Solid lines in Figures 4 and 5 are plotted using Formula (19). Note that the theoretical results well describe the experimental data. Both profiles for Na and Cs atoms have an asymmetry, manifested in a rapid drop of the right wings. This asymmetry is due to the upwards AC Stark energy shift (see Equations (9) and (16)) of level $\mid 2 \rangle$ in Figure 6, which corresponds to the HF sublevel $F'' = 2$ for Na and $F'' = 4$ for Cs. As a result, the value of actual detuning $\delta_4 = \Delta_4 - \Omega_3^2/\Delta_{43}$ depends on the sign of $\Delta_4 = -\delta$, i.e., the actual detuning $\delta_4$ for the same absolute value of laser detuning $\delta$ is larger for the right wing side compared to the left wing side.

There is another point that complicates the situation. The structure of Equation (17) implies a slow decay of the population resulting from a weak laser stimulated mixture between states $\mid 2 \rangle$ and $\mid 3 \rangle$. It means the cyclic transition $\mid 2 \rangle \rightarrow \mid 3 \rangle$ ceases to be closed. The decay factor should manifest itself in a depletion broadening. The actual profile, thus, is formed via interplay of space dependent (see Equations (16) and (17)) optical pumping and AC Stark effects.

# 5. Conclusions

One of the characteristic features of a cold medium is the long interaction time $\tau_{tr}$ between light and atoms. As a result, a variety of nonlinear optical effects may take place even for moderate values of light power. We have experimentally observed and theoretically modeled such nonlinear effects by studying HF-selective laser interaction with cold atomic beams consisting of alkali atoms.

Significant modifications of the optical features are found in closed transitions. In particular, we have predicted and experimentally demonstrated the appearance of an asymmetry in the corresponding absorption lines and have explained this occurrence through AC Stark shifts of the involved states. The long transit time ($\sim$0.2 ms) through the excitation zone, combined with a relatively small mixing of HF sublevels of the resonant $n^2p_{3/2}$ state due to the laser coupling results in opening a decay channel for cyclic transitions. The particularities of the line shape formation are a result of the strong interplay between time dependent optical pumping and AC Stark effects.

The results discussed here are of potential interest for interpretation of spectroscopic data obtained from fluorescence spectra of a cold medium of astrophysical relevance such as different modifications of cold white dwarfs or neutral sodium clouds near Jovian moon Io.

**Acknowledgments:** This work was supported with partial funding from the European Commission projects EU FP7 IAPP COLDBEAM and EU FP7 Centre of Excellence FOTONIKA-LV (REGPOT-CT-2011–285912- FOTONIKA), and the US Office of Naval Research Grant No. N00014-12-1-0514. In addition, the authors are thankful to the MESTD of the Republic of Serbia Grant 176002. We wish to thank Arturs Cinins and Aigars Ekers for stimulating discussions at the early stage of the present work.

**Author Contributions:** F.F. conceived, designed and performed the experiments; A.N.K. contributed analysis tools; V.A.S. analyzed the data, D.K.E. and M.B. performed the calculations; N.N.B., M.S.D., M.B. and Y.N.G. wrote the paper.

# References

1.  Debes, J.H.; Walsh, K.J.; Stark, C. The link between planetary systems, dusty white dwarfs, and metal-polluted white dwarfs. *Astrophys. J.* **2012**, *747*, 148–156. [CrossRef]
2.  Gianninas, A.; Dufour, P.; Kilic, M.; Brown, W.R.; Bergeron, P.; Hermes, J. Precise atmospheric parameters for the shortest-period binary white dwarfs: gravitational waves, metals, and pulsations. *Astrophys. J.* **2014**, *794*, 35–52. [CrossRef]
3.  Chary, R.; Zuckerman, B..; Becklin, E.E. *Observational Constraints on the Origin of Metals in Cool DA-Type White Dwarfs*; The Universe as Seen by ISO; Cox, P., Kessler, M., Eds.; ESA Special Publication: Noordwijk, The Netherlands, 1999, Volume 427, pp. 289–291. [astro-ph/9812090].
4.  Dupuis, J.; Fontaine, G.; Wesemael, F. A study of metal abundance patterns in cool white dwarfs. III-Comparison of the predictions of the two-phase accretion model with the observations. *Astrophys. J. Suppl. Ser.* **1993**, *87*, 345–365. [CrossRef]
5.  Xu, S.; Jura, M. The Drop during Less than 300 Days of a Dusty White Dwarf's Infrared Luminosity. *Astrophys. J. Lett.* **2014**, *792*, L39. [CrossRef]
6.  Shvartzvald, Y.; Yee, J.; Novati, S.C.; Gould, A.; Lee, C.U.; Beichman, C.; Bryden, G.; Carey, S.; Gaudi, B.; Henderson, C.; et al. An Earth-mass Planet in a 1 au Orbit around an Ultracool Dwarf. *Astrophys. J. Lett.* **2017**, *840*, L3. [CrossRef]
7.  Brown, R.A. Optical Line Emission from Io. In *Exploration of the Planetary System*; Woszczyk, A., Iwaniszewska, C., Eds.; IAU Symposium: Paris, France 1974; Volume 65, pp. 527–531. [CrossRef]
8.  Brown, R.A.; Chaffee, F.H., Jr. High-resolution spectra of sodium emission from Io. *Astrophys. J.* **1974**, *187*, L125–L126. [CrossRef]
9.  Wilson, J.K.; Mendillo, M.; Baumgardner, J.; Schneider, N.M.; Trauger, J.T.; Flynn, B. The dual sources of Io's sodium clouds. *Icarus* **2002**, *157*, 476–489. [CrossRef]
10. Mendillo, M.; Baumgardner, J.; Flynn, B.; Hughes, W.J. The extended sodium nebula of Jupiter. *Nature* **1990**, *348*, 312–314. [CrossRef]

11. Porfido, N. A Slow and Cold Particle Beam for Nanotechnological Purposes. Ph.D. Thesis, Pisa University, Pisa, Italy, 2012.

12. Porfido, N.; Bezuglov, N.; Bruvelis, M.; Shayeganrad, G.; Birindelli, S.; Tantussi, F.; Guerri, I.; Viteau, M.; Fioretti, A.; Ciampini, D.; et al. Nonlinear effects in optical pumping of a cold and slow atomic beam. *Phys. Rev. A* **2015**, *92*, 043408. [CrossRef]

13. Metcalf, H.J.; Van der Straten, P. *Laser Cooling and Trapping*; Springer: New York, NY, USA, 1999.

14. Shore, B.W. *Manipulating Quantum Structures Using Laser Pulses*; Cambridge University Press: Cambridge, UK, 2011.

15. Sydoryk, I.; Bezuglov, N.; Beterov, I.; Miculis, K.; Saks, E.; Janovs, A.; Spels, P.; Ekers, A. Broadening and intensity redistribution in the Na (3p) hyperfine excitation spectra due to optical pumping in the weak excitation limit. *Phys. Rev. A* **2008**, *77*, 042511. [CrossRef]

16. Bezuglov, N.; Ekers, A.; Kaufmann, O.; Bergmann, K.; Fuso, F.; Allegrini, M. Velocity redistribution of excited atoms by radiative excitation transfer. II. Theory of radiation trapping in collimated beams. *J. Chem. Phys.* **2003**, *119*, 7094–7110. [CrossRef]

17. Beterov, I.; Tretyakov, D.; Ryabtsev, I.; Bezuglov, N.; Miculis, K.; Ekers, A.; Klucharev, A. Collisional and thermal ionization of sodium Rydberg atoms III. Experiment and theory for nS and nD states with $n = 8$–20 in crossed atomic beams. *J. Phys. B* **2005**, *38*, 4349–4361. [CrossRef]

18. Miculis, K.; Beterov, I.; Bezuglov, N.; Ryabtsev, I.; Tretyakov, D.; Ekers, A.; Klucharev, A. Collisional and thermal ionization of sodium Rydberg atoms: II. Theory for nS, nP and nD states with $n = 5$–25. *J. Phys. B* **2005**, *38*, 1811–1831. [CrossRef]

19. Zakharov, M.Y.; Bezuglov, N.; Lisenkov, N.; Klyucharev, A.; Beterov, I.; Michulis, K.; Ékers, A.; Fuso, F.; Allegrini, M. Optimization of sub-Doppler absorption contour in gas-dynamic beams. *Optic. Spectrosc.* **2010**, *108*, 877–882. [CrossRef]

20. O'Dwyer, C.; Gay, G.; de Lesegno, B.V.; Weiner, J.; Camposeo, A.; Tantussi, F.; Fuso, F.; Allegrini, M.; Arimondo, E. Atomic nanolithography patterning of submicron features: writing an organic self-assembled monolayer with cold, bright Cs atom beams. *Nanotechnology* **2005**, *16*, 1536–1541. [CrossRef]

21. Camposeo, A.; Piombini, A.; Cervelli, F.; Tantussi, F.; Fuso, F.; Arimondo, E. A cold cesium atomic beam produced out of a pyramidal funnel. *Opt. Commun.* **2001**, *200*, 231–239. [CrossRef]

22. Bezuglov, N.; Molisch, A.; Fioretti, A.; Gabbanini, C.; Fuso, F.; Allegrini, M. Time-dependent radiative transfer in magneto-optical traps. *Phys. Rev. A* **2003**, *68*, 063415. [CrossRef]

23. Bezuglov, N.; Molisch, A.; Fuso, F.; Allegrini, M.; Ekers, A. Nonlinear radiation imprisonment in magneto-optical vapor traps. *Phys. Rev. A* **2008**, *77*, 063414. [CrossRef]

24. Sobelman, I.I. Radiative Transitions. In *Atomic Spectra and Radiative Transitions*; Springer: Berlin, German, 1992; pp. 200–302.

25. Stenholm, S. *Foundations of Laser Spectroscopy*; Courier Dover: New York, NY, USA, 2005.

26. Landau, L.; Lifshits, E. *Mechanics*; Pergamon Press: New York, NY, USA, 1969.

27. Bruvelis, M.; Cinins, A.; Leitis, A.; Efimov, D.; Bezuglov, N.; Chirtsov, A.; Fuso, F.; Ekers, A. Particularities of optical pumping effects in cold and ultra-slow beams of Na and Cs in the case of cyclic transitions. *Optic. Spectrosc.* **2015**, *119*, 1038–1048. [CrossRef]

# 22

# Quantum and Semiclassical Stark Widths for Ar VII Spectral Lines

**Rihab Aloui [1], Haykel Elabidi [2,*], Sylvie Sahal-Bréchot [3] and Milan S. Dimitrijević [3,4†]**

[1] Laboratoire Dynamique Moléculaire et Matériaux Photoniques, GRePAA, École Nationale Supérieure d'ingénieurs de Tunis, University of Tunis, 1008 Tunis, Tunisia; rihabaloui88@gmail.com

[2] LDMMP, GRePAA, Faculté des Sciences de Bizerte, University of Carthage, 7021 Bizerte, Tunisia

[3] Sorbonne Université, Observatoire de Paris, Université PSL, CNRS, LERMA, F-92190 Meudon, France; sylvie.sahal-brechot@obspm.fr (S.S.-B.); mdimitrijevic@aob.rs (M.S.D.)

[4] Astronomical Observatory, Volgina 7, 11060 Belgrade, Serbia

\* Correspondence: haelabidi@uqu.edu.sa

† Current address: Astronomical Observatory, Volgina 7, 11060 Belgrade, Serbia.

**Abstract:** We present in this paper the results of a theoretical study of electron impact broadening for several lines of the Ar VII ion. The results have been obtained using our quantum mechanical method and the semiclassical perturbation one. Results are presented for electron density $10^{18}$ cm$^{-3}$ and for electron temperatures ranging from $2 \times 10^4$ to $5 \times 10^5$ K required for plasma modeling. Our results have been compared to other semiclassical ones obtained using different sources of atomic data. A study of the strong collisions contributions to line broadening has been performed. The atomic structure and collision data used for the calculations of line broadening are also calculated by our codes and compared to available theoretical results. The agreement found between the two calculations ensures that our line broadening procedure uses adequate structure and collision data.

**Keywords:** Stark broadening; Ar VII line profiles; stars; white dwarfs; atomic data; scattering

## 1. Introduction

Atomic and line broadening data for many elements and their ions are very useful for solving many astrophysical problems, such as the calculations of opacity and radiative transfer [1]. Especially, accurate Stark broadening parameters are important to obtain a reliable modelization of stellar interiors. The Stark broadening mechanism is also important for the investigation, analysis, and modeling of B-type and particularly A-type stellar atmospheres, as well as white dwarf atmospheres [2,3]. Furthermore, the development of computers and instruments, such as the new X-ray space telescope *Chandra*, has motivated the calculations of line broadening of trace elements in the X-ray wavelength range. It has been shown that analysis of white dwarf atmospheres, where Stark broadening is dominant compared to the thermal Doppler broadening, needs models taking into account heavy element opacity.

In Rauch et al. [4], the authors reported problems encountered in their determination of element abundances: the line cores of the S VI resonance doublet appear too deep to match the observation and they are not well suited for an abundance determination, and the same problem exists in relation to the N V and O VI. This is due to the lack of line broadening data for these ions. Some other data exist, but the required temperatures and electron densities are lacking, and it is necessary to extrapolate such data to obtain the temperatures and densities at the line-forming regions, especially the line cores. This procedure of extrapolation can provide inaccurate results especially in the case of extrapolating to obtain temperatures, since the temperature dependence of line widths may be very different. This lack of data represents an inconvenience for the development of spectral analysis by means of the NLTE model atmosphere techniques. We quote here the conclusion in Rauch et al. [4]: "spectral analysis by means of NLTE model atmospheres has presently arrived at a high level of sophistication, which is now hampered largely by the lack of reliable atomic data and accurate line-broadening tables."

Astrophysical interest of Ar VII illustrates for example recent discovery of far UV lines of this ion in the spectra of very hot central stars of planetary nebulae and white dwarfs [5]. In this article, the authors have also shown the importance of the line broadening data for this element in its various ionization stages. Argon also has an important role in plasma technological applications and devices [6]. It produces favorable conditions for very stable discharges and is also very often used as a carrier gas in plasma, which contains a mixture of other gases. Thus, the knowledge of the Stark broadening parameters of neutral and ionized argon lines is an important tool for plasma electron density diagnostic.

The Stark broadening calculations in the present work are based on two approaches: the quantum mechanical approach and the semiclassical perturbation one. The quantum mechanical expression for electron impact broadening calculations for intermediate coupling was obtained in Elabidi et al. [7]. The first applications were performed for the 2s3s−2s3p transitions in Be-like ions from nitrogen to neon [8] and for the 3s−3p transitions in Li-like ions from carbon to phosphor [9]. This approach was also used in Elabidi & Sahal-Bréchot [10] to check the dependence on the upper level ionization potential of electron impact widths and in Elabidi et al. [11] to investigate the influence of strong collisions and quadrupolar potential contributions on line broadening. Our quantum approach is an ab initio method; i.e., all the parameters required for the calculations of the line broadening such as radiative atomic data (energy levels, oscillator strengths ...) or collisional data (collision strengths or cross sections, scattering matrices ...) are evaluated during the calculation and not taken from other data sources. We used the sequence of the University College London (UCL) atomic codes SUPERSTRUCTURE/DW/JAJOM that have been used for many years to provide fine energy levels, wavelengths, radiative probability rates, and electron impact collision strengths. Recently, they have been adapted to line broadening calculations [8].

In the present paper, we continue the effort to provide atomic and line broadening data for argon ions. Quantum Stark broadening of 12 lines of the Ar VII ion have been calculated using 9 configurations ($1s^22s^22p^6$: $3s^2$, 3s3p, $3p^2$, 3s3d, 3p3d, 3s4s, 3s4p, 3s4d, and 3s5s). Our calculations have been made for a set of temperatures ranging from $2 \times 10^4$ to $5 \times 10^5$ K. These parameters will be useful for a more accurate determination of photospheric properties. We perform also a semiclassical calculations for these lines using our atomic data from the code SUPERSTRUCTURE. We compare these results to the semiclassical ones [12], for which the atomic structure has been calculated with the Bates and Damgaard approximation [13].

## 2. Outline of the Theory and Computational Procedure

### 2.1. Quantum Mechanical Formalism

We present here an outline of our quantum formalism for electron impact broadening. More details can be found elsewhere [7,8]. The calculations are made within the frame of the impact approximation, which means that the time interval between collisions is much longer than the duration of a collision. The expression of the Full Width at Half Maximum (FWHM) $W$ obtained in Elabidi et al. [8] is:

$$
\begin{aligned}
W \;=\; & 2N_e \left(\frac{\hbar}{m}\right)^2 \left(\frac{2m\pi}{k_B T}\right)^{\frac{1}{2}} \\
& \times \int_0^\infty \Gamma_w(\varepsilon) \exp\left(-\frac{\varepsilon}{k_B T}\right) d\left(\frac{\varepsilon}{k_B T}\right)
\end{aligned}
\tag{1}
$$

where $k_B$ is the Boltzmann constant, $N_e$ the electron density, $T$ the electron temperature, and

$$\Gamma_w(\varepsilon) \;=\; \sum_{J_i^T J_f^T l K_i K_f} \frac{\left[K_i, K_f, J_i^T, J_f^T\right]}{2}$$

$$\times \left\{ \begin{matrix} J_i K_i l \\ K_f J_f 1 \end{matrix} \right\}^2 \left\{ \begin{matrix} K_i J_i^T s \\ J_f^T K_f 1 \end{matrix} \right\}^2$$

$$\times \left[1 - (\mathrm{Re}\,(S_I)\mathrm{Re}\,(S_F) + \mathrm{Im}\,(S_I)\mathrm{Im}\,(S_F))\right] \tag{2}$$

where $L_i + S_i = J_i$, $J_i + l = K_i$ and $K_i + s = J_i^T$. $L$ and $S$ represent the atomic orbital angular momentum and spin of the target, $l$ is the electron orbital momentum, and the superscript $T$ denotes the quantum numbers of the total electron+ion system. $S_I$ ($S_F$) are the scattering matrix elements for the initial (final) levels, expressed in the intermediate coupling approximation, $\mathrm{Re}\,(S)$ and $\mathrm{Im}\,(S)$ are respectively the real and the imaginary parts of the S-matrix element, $\left\{ \begin{matrix} abc \\ def \end{matrix} \right\}$ represent 6–j symbols, and we adopt the notation $[x, y] = (2x + 1)(2y + 1)$. Both $S_I$ and $S_F$ are calculated for the same incident electron energy $\varepsilon = mv^2/2$. Equation (1) takes into account the fine structure effects and relativistic corrections resulting from the breakdown of the $LS$ coupling approximation for the target.

The main goal is the evaluation of the real ($\mathrm{Re}\,\mathbf{S}$) and the imaginary parts ($\mathrm{Im}\,\mathbf{S}$) of the scattering matrix $\mathbf{S}$ in the initial $I$ and the final $F$ level. The calculation starts with the study of the atomic structure. The structure problem has been treated using the SUPERSTRUCTURE (SST) code described in Eissner et al. [14], taking into account configuration interaction, where each individual configuration is an expansion in terms of Slater states built from orthonormal orbitals. The radial functions were calculated assuming a scaled Thomas–Fermi–Dirac–Amaldi (TFDA) potential. The potential depends upon parameters $\lambda_l$ which are determined variationally by optimizing the weighted sum of the term energies. Relativistic corrections (spin-orbit, mass, Darwin, and one-body) are introduced according to the Breit–Pauli approach [15] as a perturbation to the non-relativistic Hamiltonian. The SST program also produces the term coupling coefficients (TCCs), which are used to transform the scattering $\mathbf{S}$ or reactance $\mathbf{R}$-matrices to intermediate coupling [7].

The second step is the treatment of the scattering problem. The calculation is carried out in the non-relativistic distorted wave approximation using the UCL distorted wave (DW) program [16]. The reactance matrices are calculated in LS coupling. The program JAJOM [17] uses these reactance matrices and the TCC to calculate collision strengths in intermediate coupling. In the present work, we have transformed JAJOM into JAJPOLARI (Elabidi and Dubau, unpublished results) to produce the collision strengths and the reactance matrices $\mathbf{R}$ in intermediate coupling, which will be used by the program RtoS (Dubau, unpublished results) to evaluate the real and the imaginary parts of the scattering matrix according to

$$\mathrm{Re}\,\mathbf{S} = \left(1 - \mathbf{R}^2\right)\left(1 + \mathbf{R}^2\right)^{-1} \tag{3}$$

and

$$\mathrm{Im}\,\mathbf{S} = 2\mathbf{R}\left(1 + \mathbf{R}^2\right)^{-1}. \tag{4}$$

The two expressions (3) and (4) have been deduced from the relation $\mathbf{S} = (1 + i\mathbf{R})(1 - i\mathbf{R})^{-1}$, and such expressions guarantee the unitarity of the $\mathbf{S}$-matrix.

Finally, in the code JAJPOLARI, the reactance matrices $R_I$ in intermediate coupling corresponding to the initial $I$ level are evaluated for each channel and at a total energy $E_I = E_i + \varepsilon$. The same procedure is done for $R_F$ but at a total energy $E_F = E_f + \varepsilon$. $E_i$ ($E_f$) are the energies of the initial (final) atomic levels. The program RtoS receives $\mathbf{R}$-matrices and transforms them into real and imaginary parts of $\mathbf{S}$-matrices according to Equations (3) and (4) at total energies $E_I$ and $E_F$, and combines a given matrix element $S_I$ for an initial level $I$ with a number of matrix element $S_F$ for the final level $F$. The obtained matrix elements $\mathrm{Re}\,\mathbf{S}$ and $\mathrm{Im}\,\mathbf{S}$ enter into Equation (2).

The integral over the Maxwell distribution (Equation (1)) is evaluated numerically using a trapezoid integration with a variable step to provide the line width $W$. The energy step is chosen to be as small as possible around the threshold region where the variation of $\Gamma_w$ in (1) is fast. For large energies and far from the threshold region, the variation of $\Gamma_w$ becomes slow and then the step is gradually increased.

*2.2. Semiclassical Perturbation Method*

We give here a detailed description of the semiclassical perturbation formalism for line broadening calculations. The profile $F(\omega)$ is Lorentzian for isolated lines:

$$F(\omega) = \frac{w/\pi}{(\omega - \omega_{if} - d)^2 + w^2} \tag{5}$$

where

$$\omega_{if} = \frac{E_i - E_f}{\hbar}$$

$i$ and $f$ denote the initial and final atomic states and $E_i$ and $E_f$ their corresponding energies.

The total width at half maximum ($W = 2w$) in angular frequency units of a spectral line can be expressed as

$$W = N \int v f(v) dv \left( \sum_{i' \neq i} \sigma_{ii'}(v) + \sum_{f' \neq f} \sigma_{ff'}(v) + \sigma_{el} \right) \tag{6}$$

where $N$ is the electron density, $f(v)$ the Maxwellian velocity distribution function for electrons, $i'$ (resp. $f'$) denotes the perturbing levels of the initial state $i$ (resp. final state $f$). The inelastic cross section $\sigma_{ii'}(v)$ (resp. $\sigma_{ff'}(v)$) can be expressed by an integral over the impact parameter $\rho$ of the transition probability $P_{ii'}(\rho, v)$ (resp. $P_{ff'}(\rho, v)$ ) as

$$\sum_{i' \neq i} \sigma_{ii'}(v) = \frac{1}{2} \pi R_1^2 + \int_{R_1}^{R_D} 2\pi \rho d\rho \sum_{i' \neq i} P_{ii'}(\rho, v) \tag{7}$$

where $\rho$ denotes the impact parameter of the incoming electron. The elastic cross section is given by

$$\sigma_{el} = 2\pi R_2^2 + \int_{R_2}^{R_D} 2\pi \rho d\rho \sin^2 \delta + \sigma_r \tag{8}$$

$$\delta = (\varphi_p^2 + \varphi_q^2)^{\frac{1}{2}}.$$

Strong collisions are evaluated for $\rho < R_1, R_2$. The phase shifts $\varphi_p$ and $\varphi_q$, due respectively to the polarization potential ($r^{-4}$) and to the quadrupolar potential ($r^{-3}$), are given in Section 3 of Chapter 2 in Sahal-Bréchot [18], and $R_D$ is the Debye radius. The cut-offs $R_1$ and $R_2$ are described in Section 1 of Chapter 3 in Sahal-Bréchot [19]. Detailed calculations of the interference term $\sigma_{el}$ can be found in Formulas 18 and 24–30 on pages 109–110 of Sahal-Bréchot [18]. $\sigma_r$ is the contribution of the Feshbach resonances [20], which concerns only ionized radiating atoms colliding with electrons. It is an extrapolation of the excitation collision strengths (and not the cross-sections) under the threshold by means of the semiclassical limit of the Gailitis approximation (see page 601 of [20] for details of the calculations). A review of the theory, all approximations and the details of applications are given in Sahal-Bréchot et al. [21].

## 3. Results and Discussions

### 3.1. Atomic Structure and Electron Scattering Data

We have used the following nine configurations in our calculation: $1s^2 2s^2 2p^6$: $3s^2$, $3s3p$, $3p^2$, $3s3d$, $3p3d$, $3s4s$, $3s4p$, $3s4d$, and $3s5s$, which give rise to 38 levels, which are listed in Table 1 with their energies in cm$^{-1}$. These values have been compared with the observed ones taken from the tables of the National Institute of Standards and Technology database: NIST [22] which are originally from Saloman [23]. We compare also with the energies computed using the multiconfiguration Hartree–Fock method (MCHF) [24] and with those obtained using the AUTOSTRUCTURE code [25]. The averaged disagreement between these three results is less than 1%. We detect an inversion between the two levels 10/13 and 25/26 regarding those of NIST and MCHF. This inversion does not affect the calculations since the agreement is still acceptable (about 5%).

**Table 1.** Our present fine-structure energy levels E (in cm$^{-1}$) for Ar VII compared with those of NIST [22], with those obtained from the multiconfiguration Hartree–Fock method (MCHF) [24], and with those from the R-matrix calculation (AS2014) [25]. Levels denoted by asterisks (*) are inverted compared to the NIST values.

| $i$ | Conf. | Level | E | NIST | MCHF | AS2014 | $\frac{|E-NIST|}{NIST}$ (%) |
|---|---|---|---|---|---|---|---|
| 1 | $3s^2$ | $^1S_0$ | 0.0 | 0.0 | 0.0 | 0.0 | — |
| 2 | $3s3p$ | $^3P^o_0$ | 110,717 | 113,101 | 112,817.66 | 112,070 | 2.1 |
| 3 | $3s3p$ | $^3P^o_1$ | 111,488 | 113,906 | 113,632.14 | 112,889 | 2.1 |
| 4 | $3s3p$ | $^3P^o_2$ | 113,088 | 115,590 | 115,324.84 | 114,593 | 2.2 |
| 5 | $3s3p$ | $^1P_1$ | 172,878 | 170,722 | 170,598.08 | 173,751 | 1.3 |
| 6 | $3p^2$ | $^1D_2$ | 263,439 | 264,749 | 264,797.88 | 264,530 | 0.5 |
| 7 | $3p^2$ | $^3P^o_0$ | 271,494 | 269,836 | 269,688.15 | 270,704 | 0.6 |
| 8 | $3p^2$ | $^3P^o_1$ | 272,341 | 270,777 | 270,667.14 | 271,641 | 0.6 |
| 9 | $3p^2$ | $^3P^o_2$ | 273,971 | 272,562 | 272,474.76 | 273,432 | 0.5 |
| 10 | $3s3d$ | $^3D_1$ | 325,254 | 324,104 | 324,950.35 | 326,054 | 0.4 |
| 11 | $3s3d$ | $^3D_2$ | 325,335 | 324,141 | 324,966.00 | 326,141 | 0.4 |
| 12 | $3s3d$ | $^3D_3$ | 325,456 | 324,205 | 325,056.68 | 326,273 | 0.4 |
| 13 | $3p^2$ | $^1S_0$ | 333,116 | 316,717 | 317,014.73 | 320,974 * | 5.2 |
| 14 | $3s3d$ | $^1D_2$ | 384,031 | 370,294 | 371,275.29 | 377,167 | 3.7 |
| 15 | $3p3d$ | $^3F^o_2$ | 443,952 | 443,362 | 444,508.36 | 444,677 | 0.1 |
| 16 | $3p3d$ | $^3F_{3o}$ | 444,892 | 444,780 | 445,556.29 | 445,701 | 0.0 |
| 17 | $3p3d$ | $^3F^o_4$ | 446,051 | 446,011 | 446,849.87 | 446,969 | 0.0 |
| 18 | $3p3d$ | $^1D_2$ | 450,025 | 450,477 | 450,808.06 | 451352 | 0.1 |
| 19 | $3p3d$ | $^3P^o_2$ | 474,314 | 472,282 | 473,009.27 | 475,022 | 0.4 |
| 20 | $3p3d$ | $^3P^o_1$ | 474,956 | 472,875 | 473,782.67 | 475,699 | 0.4 |
| 21 | $3p3d$ | $^3P^o_0$ | 475,497 | 473,810 | 474,466.36 | 476,301 | 0.4 |
| 22 | $3p3d$ | $^3D_1$ | 477,133 | 475,217 | 475,932.22 | 477,901 | 0.4 |
| 23 | $3p3d$ | $^3D_2$ | 477,515 | 475,585 | 476,306.50 | 478,313 | 0.4 |
| 24 | $3p3d$ | $^3D_3$ | 477,753 | 475,762 | 476,474.91 | 478,560 | 0.4 |
| 25 | $3s4s$ | $^3S_1$ | 513,685 | 514,076 | 508,971.69 | 511,372 | 0.1 |
| 26 | $3p3d$ | $^1F^o_3$ | 521,897 | 510,268 | 514,890.47 | 515,169 * | 2.3 |
| 27 | $3p3d$ | $^1P^o_1$ | 527,518 | 517,105 | 517,788.24 | 524,282 | 2.0 |
| 28 | $3s4s$ | $^1S_0$ | 529,866 | 528,910 | 526,205.45 | 523,618 | 0.2 |
| 29 | $3s4p$ | $^3P^o_0$ | 567,050 | 563,880 | 568,040.66 | 565,087 | 0.6 |
| 30 | $3s4p$ | $^3P^o_1$ | 567,287 | 564,418 | 568,275.74 | 565,295 | 0.5 |
| 31 | $3s4p$ | $^3P^o_2$ | 567,811 | 564,728 | 568,944.94 | 565,840 | 0.5 |
| 32 | $3s4p$ | $^1P^o_1$ | 576,576 | 569,797 | 570,403.78 | 568,205 | 0.2 |
| 33 | $3s4d$ | $^3D_1$ | 635,209 | 634,605 | 635,580.25 | 632,497 | 0.1 |
| 34 | $3s4d$ | $^3D_2$ | 635,241 | 634,639 | 635,659.10 | 632,562 | 0.1 |
| 35 | $3s4d$ | $^3D_3$ | 635,290 | 634,701 | 635,749.02 | 632,659 | 0.1 |
| 36 | $3s4d$ | $^1D_2$ | 639,087 | 635,295 | 636,353.38 | 633,443 | 0.6 |
| 37 | $3s5s$ | $^3S_1$ | 713,912 | 715,747 | — | 717,638 | 0.3 |
| 38 | $3s5s$ | $^1S_0$ | 719,473 | 714,794 | — | 717,997 | 0.7 |

We present also in Table 2 radiative decay rates $A_{ij}$, weighted oscillator strengths $gf$, and line strengths $S$ for some Ar VII lines up to the level 14 (3s3d $^1D_2$). Our $A_{ij}$ values have been compared with those obtained from the AUTOSTRUCTURE code [25], and with those from the SUPERSTRUCTURE code [26] using five configurations ($1s^22s^22p^6$: $3s^2$, 3s3p, $3p^2$, 3s3d, and 3s4s). The averaged difference is about 20% with the results of [25] and about 24% with those of Christensen et al. [26]. Some transitions present a high difference, especially those for which $A_{i-j}$ are relatively small (about $10^6$ s$^{-1}$ and below). The $gf$ values have been compared only with Christensen et al. [26] and the difference is about 24 %. The $gf$ values are calculated in [26], but we took them from the database CHIANTI version 8.0 [27].

With the code JAJOM, fine structure collision strengths are calculated for low partial waves $l$ of the incoming electron up to 29. For large partial waves $l$, this method becomes cumbersome and inaccurate, but their contributions to collision strengths cannot be neglected. For $30 \leq l \leq 50$, two different procedures have been used: for dipole transitions, the contribution has been calculated using the JAJOM-CBe code (Dubau, unpublished results) based upon the Coulomb–Bethe formulation of Burgess and Sheorey [28] and adapted to JAJOM approximation. For non-dipole transitions, the contribution has been estimated by the SERIE-GEOM code assuming a geometric series behavior for high partial wave collision strengths [29,30].

**Table 2.** Present radiative decay rates $A_{ij}$ (in s$^{-1}$) and weighted oscillator strengths $gf$ compared to those from Christensen et al. [26] (SST86) and to those from [25] (AS2014) for some Ar VII allowed transitions. Line strengths $S$ are also presented. $i$ and $j$ label the levels as in Table 1.

| $i-j$ | $A_{i-j}$ | $A_{i-j}$(AS2014) | $A_{i-j}$(SST86) | $gf$ | $gf$(SST86) | $S$ |
|---|---|---|---|---|---|---|
| $3-1$ | $5.968 \times 10^5$ | $7.13 \times 10^5$ | $1.65 \times 10^5$ | $2.160 \times 10^{-4}$ | $5.820 \times 10^{-5}$ | 0.000638 |
| $5-1$ | $8.114 \times 10^9$ | $8.30 \times 10^9$ | $8.21 \times 10^9$ | $1.221 \times 10^0$ | $1.270 \times 10^0$ | 2.325423 |
| $6-3$ | $1.854 \times 10^7$ | $2.97 \times 10^7$ | $2.39 \times 10^7$ | $6.018 \times 10^{-3}$ | $7.780 \times 10^{-3}$ | 0.013038 |
| $6-4$ | $3.748 \times 10^7$ | $6.18 \times 10^7$ | $6.07 \times 10^7$ | $1.243 \times 10^{-2}$ | $2.020 \times 10^{-2}$ | 0.027216 |
| $6-5$ | $3.719 \times 10^8$ | $4.00 \times 10^8$ | $3.98 \times 10^8$ | $3.400 \times 10^{-1}$ | $3.380 \times 10^{-1}$ | 1.235825 |
| $7-3$ | $7.249 \times 10^9$ | $6.94 \times 10^9$ | $6.93 \times 10^9$ | $4.245 \times 10^{-1}$ | $4.280 \times 10^{-1}$ | 0.873472 |
| $7-5$ | $7.818 \times 10^5$ | $1.72 \times 10^6$ | $1.02 \times 10^6$ | $1.205 \times 10^{-4}$ | $1.590 \times 10^{-4}$ | 0.000402 |
| $8-2$ | $2.492 \times 10^9$ | $2.40 \times 10^9$ | $2.35 \times 10^9$ | $4.291 \times 10^{-1}$ | $4.250 \times 10^{-1}$ | 0.874104 |
| $8-3$ | $1.842 \times 10^9$ | $1.77 \times 10^9$ | $1.74 \times 10^9$ | $3.202 \times 10^{-1}$ | $3.180 \times 10^{-1}$ | 0.655396 |
| $8-4$ | $2.976 \times 10^9$ | $2.85 \times 10^9$ | $2.84 \times 10^9$ | $5.278 \times 10^{-1}$ | $5.310 \times 10^{-1}$ | 1.090986 |
| $8-5$ | $1.274 \times 10^5$ | $1.33 \times 10^5$ | $3.65 \times 10^4$ | $5.793 \times 10^{-5}$ | $1.670 \times 10^{-5}$ | 0.000192 |
| $9-3$ | $1.877 \times 10^9$ | $1.80 \times 10^9$ | $1.77 \times 10^9$ | $5.329 \times 10^{-1}$ | $5.270 \times 10^{-1}$ | 1.079810 |
| $9-4$ | $5.481 \times 10^9$ | $5.24 \times 10^9$ | $5.22 \times 10^9$ | $1.587 \times 10^0$ | $1.590 \times 10^0$ | 3.248175 |
| $9-5$ | $4.965 \times 10^6$ | $8.35 \times 10^6$ | $7.11 \times 10^6$ | $3.642 \times 10^{-3}$ | $5.240 \times 10^{-3}$ | 0.011859 |
| $10-2$ | $5.993 \times 10^9$ | $5.86 \times 10^9$ | $5.92 \times 10^9$ | $5.857 \times 10^{-1}$ | $5.980 \times 10^{-1}$ | 0.898754 |
| $10-3$ | $4.449 \times 10^9$ | $4.35 \times 10^9$ | $4.40 \times 10^9$ | $4.379 \times 10^{-1}$ | $4.480 \times 10^{-1}$ | 0.674448 |
| $10-4$ | $2.906 \times 10^8$ | $2.84 \times 10^8$ | $2.88 \times 10^8$ | $2.904 \times 10^{-2}$ | $2.980 \times 10^{-2}$ | 0.045058 |
| $10-5$ | $5.117 \times 10^5$ | $5.88 \times 10^5$ | $1.52 \times 10^5$ | $9.913 \times 10^{-5}$ | $2.950 \times 10^{-5}$ | 0.000214 |
| $11-3$ | $8.015 \times 10^9$ | $7.84 \times 10^9$ | $7.92 \times 10^9$ | $1.314 \times 10^0$ | $1.340 \times 10^0$ | 2.022700 |
| $11-4$ | $2.618 \times 10^9$ | $2.56 \times 10^9$ | $2.60 \times 10^9$ | $4.356 \times 10^{-1}$ | $4.480 \times 10^{-1}$ | 0.675718 |
| $11-5$ | $7.418 \times 10^5$ | $7.91 \times 10^5$ | $1.15 \times 10^5$ | $2.392 \times 10^{-4}$ | $3.710 \times 10^{-5}$ | 0.000517 |
| $12-4$ | $1.048 \times 10^{10}$ | $1.02 \times 10^{10}$ | $1.04 \times 10^{10}$ | $2.439 \times 10^0$ | $2.510 \times 10^0$ | 3.781547 |
| $13-3$ | $4.347 \times 10^6$ | $9.07 \times 10^6$ | $5.79 \times 10^6$ | $1.327 \times 10^{-4}$ | $2.030 \times 10^{-4}$ | 0.000197 |
| $13-5$ | $8.643 \times 10^9$ | $6.98 \times 10^9$ | $6.97 \times 10^9$ | $5.047 \times 10^{-1}$ | $4.840 \times 10^{-1}$ | 1.036879 |
| $14-3$ | $9.459 \times 10^6$ | $9.90 \times 10^6$ | $1.40 \times 10^6$ | $9.546 \times 10^{-4}$ | $1.590 \times 10^{-4}$ | 0.001153 |
| $14-4$ | $4.427 \times 10^5$ | $4.77 \times 10^5$ | $4.47 \times 10^5$ | $4.521 \times 10^{-5}$ | $5.130 \times 10^{-5}$ | 0.000055 |
| $14-5$ | $2.085 \times 10^{10}$ | $1.90 \times 10^{10}$ | $1.88 \times 10^{10}$ | $3.506 \times 10^0$ | $3.540 \times 10^0$ | 5.466647 |

We present our collision strengths from the lowest five levels to the first 14 levels in Table 3 at electron energy values 7.779, 13.674, and 23.336 Ry. We compared them with the 5-configurations collision strengths of Christensen et al. [26]. Some important discrepancies exist for transitions involving levels arising from the $3p^3$ configuration (levels 7, 8, and 9). Except for these transitions, the agreement (averaged over the three energies and all the other transitions) is about 20%. The agreement between our results and those of [26] is the worse for the electron energy 7.779 Ry. This energy is close to the excitation energy of the last calculated level (here the energy 6.80 Ry of the level 38). In this situation, the contribution of elastic collisions (which are mostly due to close/strong collisions) is important. We remark also that the agreement is better for transitions from higher levels: for example, $\Delta\Omega_{ij}$ is about 39% for transitions from the level $i = 1$, and it is about 15% for transitions from the levels $i = 4, 5$. We note that, in [26], calculations have been carried out for partial waves $l \le 11$. This may be the origin of the above disagreement for some transitions (we have taken into account partial waves up to 50 in the present work). The difference in the configurations number may also affect the collision strength values.

**Table 3.** Our collision strengths $\Omega_{ij}$ (Present) and those from [26] (DW86) where $1 \le i \le 5$ and $i+1 \le j \le 14$. $i$ and $j$ label the levels as in Table 1.

| $i - j$ | 7.779 Ry | | 13.674 Ry | | 23.336 Ry | |
|---|---|---|---|---|---|---|
| | Present | DW86 | Present | DW86 | Present | DW86 |
| $1-2$ | $8.901 \times 10^{-3}$ | $9.29 \times 10^{-3}$ | $4.523 \times 10^{-3}$ | $4.54 \times 10^{-3}$ | $2.067 \times 10^{-3}$ | $2.01 \times 10^{-3}$ |
| $1-3$ | $2.936 \times 10^{-2}$ | $2.87 \times 10^{-2}$ | $1.652 \times 10^{-2}$ | $1.46 \times 10^{-2}$ | $9.202 \times 10^{-3}$ | $7.08 \times 10^{-3}$ |
| $1-4$ | $4.430 \times 10^{-2}$ | $4.64 \times 10^{-2}$ | $2.251 \times 10^{-2}$ | $2.26 \times 10^{-2}$ | $1.029 \times 10^{-2}$ | $1.00 \times 10^{-2}$ |
| $1-5$ | $8.841 \times 10^{0}$ | $8.60 \times 10^{0}$ | $9.848 \times 10^{0}$ | $1.03 \times 10^{1}$ | $1.006 \times 10^{1}$ | $1.21 \times 10^{1}$ |
| $1-6$ | $4.208 \times 10^{-1}$ | $3.57 \times 10^{-1}$ | $4.712 \times 10^{-1}$ | $3.48 \times 10^{-1}$ | $5.067 \times 10^{-1}$ | $3.17 \times 10^{-1}$ |
| $1-7$ | $5.400 \times 10^{-5}$ | $1.06 \times 10^{-4}$ | $2.900 \times 10^{-5}$ | $5.40 \times 10^{-5}$ | $1.200 \times 10^{-5}$ | $2.33 \times 10^{-5}$ |
| $1-8$ | $1.560 \times 10^{-4}$ | $2.81 \times 10^{-4}$ | $8.400 \times 10^{-5}$ | $1.30 \times 10^{-4}$ | $3.300 \times 10^{-5}$ | $4.64 \times 10^{-5}$ |
| $1-9$ | $4.102 \times 10^{-3}$ | $5.74 \times 10^{-3}$ | $4.449 \times 10^{-3}$ | $5.38 \times 10^{-3}$ | $4.785 \times 10^{-3}$ | $4.80 \times 10^{-3}$ |
| $1-10$ | $2.239 \times 10^{-2}$ | $2.39 \times 10^{-2}$ | $1.051 \times 10^{-2}$ | $1.10 \times 10^{-2}$ | $4.603 \times 10^{-3}$ | $4.73 \times 10^{-3}$ |
| $1-11$ | $3.731 \times 10^{-2}$ | $3.99 \times 10^{-2}$ | $1.752 \times 10^{-2}$ | $1.84 \times 10^{-2}$ | $7.671 \times 10^{-3}$ | $7.87 \times 10^{-3}$ |
| $1-12$ | $5.222 \times 10^{-2}$ | $5.58 \times 10^{-2}$ | $2.452 \times 10^{-2}$ | $2.57 \times 10^{-2}$ | $1.074 \times 10^{-2}$ | $1.10 \times 10^{-2}$ |
| $1-13$ | $3.460 \times 10^{-4}$ | $1.28 \times 10^{-2}$ | $1.030 \times 10^{-4}$ | $1.10 \times 10^{-2}$ | $5.800 \times 10^{-5}$ | $8.06 \times 10^{-3}$ |
| $1-14$ | $5.431 \times 10^{-1}$ | $7.40 \times 10^{-1}$ | $6.408 \times 10^{-1}$ | $8.03 \times 10^{-1}$ | $7.164 \times 10^{-1}$ | $8.46 \times 10^{-1}$ |
| $2-3$ | $6.257 \times 10^{-2}$ | $9.55 \times 10^{-2}$ | $3.113 \times 10^{-2}$ | $5.28 \times 10^{-2}$ | $1.475 \times 10^{-2}$ | $3.03 \times 10^{-2}$ |
| $2-4$ | $3.621 \times 10^{-1}$ | $2.98 \times 10^{-1}$ | $3.562 \times 10^{-1}$ | $2.79 \times 10^{-1}$ | $3.553 \times 10^{-1}$ | $2.50 \times 10^{-1}$ |
| $2-5$ | $1.103 \times 10^{-2}$ | $1.19 \times 10^{-2}$ | $5.152 \times 10^{-3}$ | $5.09 \times 10^{-3}$ | $2.222 \times 10^{-3}$ | $2.13 \times 10^{-3}$ |
| $2-6$ | $2.860 \times 10^{-2}$ | $2.77 \times 10^{-2}$ | $1.381 \times 10^{-2}$ | $1.40 \times 10^{-2}$ | $5.991 \times 10^{-3}$ | $6.05 \times 10^{-3}$ |
| $2-7$ | $5.657 \times 10^{-3}$ | $2.09 \times 10^{-2}$ | $2.784 \times 10^{-3}$ | $2.95 \times 10^{-3}$ | $1.210 \times 10^{-3}$ | $1.27 \times 10^{-3}$ |
| $2-8$ | $3.379 \times 10^{0}$ | $3.32 \times 10^{0}$ | $3.796 \times 10^{0}$ | $4.05 \times 10^{0}$ | $3.867 \times 10^{0}$ | $4.70 \times 10^{0}$ |
| $2-9$ | $4.773 \times 10^{-3}$ | $6.85 \times 10^{-2}$ | $2.338 \times 10^{-3}$ | $6.45 \times 10^{-3}$ | $1.012 \times 10^{-3}$ | $3.57 \times 10^{-3}$ |
| $2-10$ | $2.744 \times 10^{0}$ | $2.75 \times 10^{0}$ | $3.272 \times 10^{0}$ | $3.43 \times 10^{0}$ | $3.563 \times 10^{0}$ | $4.10 \times 10^{0}$ |
| $2-11$ | $2.286 \times 10^{-2}$ | $4.89 \times 10^{-2}$ | $1.027 \times 10^{-2}$ | $1.23 \times 10^{-2}$ | $4.178 \times 10^{-3}$ | $5.19 \times 10^{-3}$ |
| $2-12$ | $5.517 \times 10^{-2}$ | $6.10 \times 10^{-2}$ | $5.424 \times 10^{-2}$ | $5.58 \times 10^{-2}$ | $5.713 \times 10^{-2}$ | $5.14 \times 10^{-2}$ |
| $2-13$ | $1.746 \times 10^{-3}$ | $1.68 \times 10^{-3}$ | $8.200 \times 10^{-4}$ | $9.19 \times 10^{-4}$ | $3.490 \times 10^{-4}$ | $3.75 \times 10^{-4}$ |
| $2-14$ | $9.861 \times 10^{-3}$ | $1.25 \times 10^{-2}$ | $4.171 \times 10^{-3}$ | $4.77 \times 10^{-3}$ | $1.639 \times 10^{-3}$ | $1.85 \times 10^{-3}$ |
| $3-4$ | $8.916 \times 10^{-1}$ | $7.87 \times 10^{-1}$ | $8.398 \times 10^{-1}$ | $6.89 \times 10^{-1}$ | $8.177 \times 10^{-1}$ | $5.96 \times 10^{-1}$ |
| $3-5$ | $3.383 \times 10^{-2}$ | $3.61 \times 10^{-2}$ | $1.603 \times 10^{-2}$ | $1.55 \times 10^{-2}$ | $7.172 \times 10^{-3}$ | $6.55 \times 10^{-3}$ |
| $3-6$ | $1.349 \times 10^{-1}$ | $1.43 \times 10^{-1}$ | $9.775 \times 10^{-2}$ | $1.19 \times 10^{-1}$ | $7.539 \times 10^{-2}$ | $1.10 \times 10^{-1}$ |
| $3-7$ | $3.379 \times 10^{0}$ | $3.51 \times 10^{0}$ | $3.926 \times 10^{0}$ | $4.27 \times 10^{0}$ | $4.410 \times 10^{0}$ | $4.89 \times 10^{0}$ |
| $3-8$ | $2.556 \times 10^{0}$ | $2.67 \times 10^{0}$ | $2.857 \times 10^{0}$ | $3.12 \times 10^{0}$ | $2.904 \times 10^{0}$ | $3.60 \times 10^{0}$ |
| $3-9$ | $4.186 \times 10^{0}$ | $4.06 \times 10^{0}$ | $4.695 \times 10^{0}$ | $4.81 \times 10^{0}$ | $4.780 \times 10^{0}$ | $5.63 \times 10^{0}$ |
| $3-10$ | $2.088 \times 10^{0}$ | $2.17 \times 10^{0}$ | $2.467 \times 10^{0}$ | $2.66 \times 10^{0}$ | $2.677 \times 10^{0}$ | $3.15 \times 10^{0}$ |
| $3-11$ | $6.232 \times 10^{0}$ | $6.04 \times 10^{0}$ | $7.416 \times 10^{0}$ | $7.47 \times 10^{0}$ | $8.072 \times 10^{0}$ | $9.00 \times 10^{0}$ |
| $3-12$ | $1.428 \times 10^{-1}$ | $1.90 \times 10^{-1}$ | $1.231 \times 10^{-1}$ | $1.28 \times 10^{-1}$ | $1.202 \times 10^{-1}$ | $1.09 \times 10^{-1}$ |
| $3-13$ | $6.369 \times 10^{-3}$ | $6.50 \times 10^{-3}$ | $3.504 \times 10^{-3}$ | $4.24 \times 10^{-3}$ | $2.005 \times 10^{-3}$ | $2.75 \times 10^{-3}$ |
| $3-14$ | $3.305 \times 10^{-2}$ | $3.80 \times 10^{-2}$ | $1.675 \times 10^{-2}$ | $1.49 \times 10^{-2}$ | $9.606 \times 10^{-3}$ | $6.27 \times 10^{-3}$ |
| $4-5$ | $5.693 \times 10^{-2}$ | $6.08 \times 10^{-2}$ | $2.671 \times 10^{-2}$ | $2.61 \times 10^{-2}$ | $1.169 \times 10^{-2}$ | $1.10 \times 10^{-2}$ |
| $4-6$ | $2.318 \times 10^{-1}$ | $2.89 \times 10^{-1}$ | $1.793 \times 10^{-1}$ | $2.75 \times 10^{-1}$ | $1.471 \times 10^{-1}$ | $2.74 \times 10^{-1}$ |
| $4-7$ | $6.345 \times 10^{-3}$ | $7.20 \times 10^{-2}$ | $3.119 \times 10^{-3}$ | $7.70 \times 10^{-3}$ | $1.355 \times 10^{-3}$ | $4.21 \times 10^{-3}$ |
| $4-8$ | $4.238 \times 10^{0}$ | $4.63 \times 10^{0}$ | $4.751 \times 10^{0}$ | $5.46 \times 10^{0}$ | $4.835 \times 10^{0}$ | $6.23 \times 10^{0}$ |
| $4-9$ | $1.261 \times 10^{1}$ | $1.27 \times 10^{1}$ | $1.414 \times 10^{1}$ | $1.52 \times 10^{1}$ | $1.439 \times 10^{1}$ | $1.76 \times 10^{1}$ |
| $4-10$ | $2.468 \times 10^{-1}$ | $2.79 \times 10^{-1}$ | $2.679 \times 10^{-1}$ | $2.93 \times 10^{-1}$ | $2.979 \times 10^{-1}$ | $3.16 \times 10^{-1}$ |
| $4-11$ | $2.208 \times 10^{0}$ | $2.29 \times 10^{0}$ | $2.581 \times 10^{0}$ | $2.75 \times 10^{0}$ | $2.795 \times 10^{0}$ | $3.23 \times 10^{0}$ |
| $4-12$ | $1.165 \times 10^{1}$ | $1.07 \times 10^{1}$ | $1.382 \times 10^{1}$ | $1.32 \times 10^{1}$ | $1.501 \times 10^{1}$ | $1.62 \times 10^{1}$ |
| $4-13$ | $1.042 \times 10^{-2}$ | $1.01 \times 10^{-2}$ | $4.917 \times 10^{-3}$ | $5.27 \times 10^{-3}$ | $2.094 \times 10^{-3}$ | $2.16 \times 10^{-3}$ |
| $4-14$ | $4.929 \times 10^{-2}$ | $6.27 \times 10^{-2}$ | $2.100 \times 10^{-2}$ | $2.39 \times 10^{-2}$ | $8.410 \times 10^{-3}$ | $9.45 \times 10^{-3}$ |
| $5-6$ | $6.396 \times 10^{0}$ | $6.54 \times 10^{0}$ | $6.502 \times 10^{0}$ | $7.69 \times 10^{0}$ | $6.069 \times 10^{0}$ | $8.82 \times 10^{0}$ |
| $5-7$ | $1.214 \times 10^{-2}$ | $1.21 \times 10^{-2}$ | $6.841 \times 10^{-3}$ | $8.61 \times 10^{-3}$ | $3.973 \times 10^{-3}$ | $5.89 \times 10^{-3}$ |
| $5-8$ | $3.337 \times 10^{-2}$ | $2.99 \times 10^{-2}$ | $1.671 \times 10^{-2}$ | $1.68 \times 10^{-2}$ | $7.776 \times 10^{-3}$ | $7.31 \times 10^{-3}$ |
| $5-9$ | $1.169 \times 10^{-1}$ | $1.34 \times 10^{-1}$ | $8.954 \times 10^{-2}$ | $1.27 \times 10^{-1}$ | $7.004 \times 10^{-2}$ | $1.27 \times 10^{-1}$ |
| $5-10$ | $4.295 \times 10^{-2}$ | $4.18 \times 10^{-2}$ | $1.955 \times 10^{-2}$ | $1.84 \times 10^{-2}$ | $8.514 \times 10^{-3}$ | $7.74 \times 10^{-3}$ |
| $5-11$ | $7.151 \times 10^{-2}$ | $6.94 \times 10^{-2}$ | $3.279 \times 10^{-2}$ | $3.04 \times 10^{-2}$ | $1.459 \times 10^{-2}$ | $1.25 \times 10^{-2}$ |
| $5-12$ | $9.742 \times 10^{-2}$ | $9.63 \times 10^{-2}$ | $4.322 \times 10^{-2}$ | $4.18 \times 10^{-2}$ | $1.770 \times 10^{-2}$ | $1.69 \times 10^{-2}$ |
| $5-13$ | $3.939 \times 10^{0}$ | $4.46 \times 10^{0}$ | $4.478 \times 10^{0}$ | $5.32 \times 10^{0}$ | $4.587 \times 10^{0}$ | $6.12 \times 10^{0}$ |
| $5-14$ | $1.653 \times 10^{1}$ | $1.64 \times 10^{1}$ | $1.990 \times 10^{1}$ | $2.06 \times 10^{1}$ | $2.181 \times 10^{1}$ | $2.54 \times 10^{1}$ |

*3.2. Line Broadening Results*

Two methods for line broadening calculations have been used in our work. The first is the quantum mechanical approach ($Q$), and the second is the semiclassical perturbation method $SCP$. To evaluate the line broadening through the second method, we need atomic parameters such as energy levels and oscillator strengths. In our SCP calculations ($SCP_{SST}$), we have taken atomic data of the code SST [14]. We compare our results ($Q$ and $SCP_{SST}$) to the SCP calculations ($SCP_{BD}$) performed in [12], where atomic data have been taken from the method of Bates and Damgaard [13]. This method has been used many times with different ions, and it has been shown that the corresponding results (using the Bates and Damgaard or the SST data) are in good agreement with experimental and other theoretical results [31–33]. Many of these SCP results have been stored in the database STARK-B [34].

We have performed quantum ($Q$) and semiclassical perturbation ($SCP$) Stark broadening for 12 lines of Ar VII for electron temperature range $(2-50) \times 10^4$ K and at electron density $N_e = 10^{18}$ cm$^{-3}$. We present our results in Table 4 for transitions between singlets, in Table 5 for the resonance line 3s$^2$ $^1$S$_0$–3s3p $^1$P$^o_1$, and in Table 6 for transitions between triplets. A comparison was made between our quantum and our semiclassical perturbation results $SCP_{SST}$ in Tables 4 and 5. We also included the semiclassical results $SCP_{BD}$ [12] in Table 6 in our comparison. Tables 4 and 6 show that the quantum line widths are always higher than the two semiclassical ones ($SCP_{SST}$ and $SCP_{BD}$). We also found that, except for the resonance line, the ratio $\frac{Q}{SCP}$ increases and decreases with temperature. The decreasing part starts in general at $T \simeq 10^5$ K. For the resonance line 3s$^2$ $^1$S$_0$–3s3p $^1$P$^o_1$, the ratio $\frac{Q}{SCP}$ increases with $T$. As per Table 5, this ratio has the same behavior as that of the other lines (increasing and after decreasing) but starts to decrease for higher temperatures ($T \simeq 10^6$ K). Table 6 shows that, in all studied cases, the $SCP_{SST}$ widths are closer to the quantum results than the $SCP_{BD}$ ones. The disagreement between $SCP_{SST}$ and $SCP_{BD}$ results is due to the difference in the source of the used atomic data.

To understand the difference between SCP and quantum calculations, we present also, in Tables 4 and 6, the contributions of elastic ($\frac{Elastic}{Total}$) and strong ($\frac{Strong}{Total}$) collisions to the $SCP_{SST}$ line broadening. Firstly, we remark that, for $T > 10^5$ K and except the resonance line, the ratios $\frac{Elastic}{Total}$ and $\frac{Strong}{Total}$ decrease with the temperature. Secondly, we see that, for each line, as the elastic and strong collisions contributions decrease, the two results ($Q$ and SCP) become close to each other. For electron temperature $T \leq 5 \times 10^4$ K, we can detect in some cases an opposite behavior between $\frac{Elastic}{Total}$ and $\frac{Strong}{Total}$ on the one hand and the ratio $\frac{Q}{SCP}$ on the other hand. This may be due to the contributions of resonances that are dominant at low temperatures. These contributions are taken into account differently in the quantum and the semiclassical perturbative methods. Figure 1 shows the behavior of the ratios $\frac{Q}{SCP}$ and $\frac{Strong}{Total}$ with the electron temperature for the 3s3d $^3$D$_2$– 3s4p $^3$P$^o_1$, 3s3p $^3$P$^o_2$–3s4d $^3$D$_3$, 3s4p $^3$P$^o_2$–3s4d $^3$D$_3$, and 3s3p $^1$P$^o_1$–3s4s $^1$S$_0$ transitions. In fact, the Ar VII perturbing levels $i'$ and $f'$ are so far from the initial ($i$) and final ($f$) levels of the considered transition ($\Delta E_{ii'}$ and $\Delta E_{ff'}$ are high) and, due to this fact, for collisions by electrons, the close collisions are important. Furthermore, with the used temperature values, the ratio $\Delta E/k_B T$ is high and consequently, the inelastic cross sections are small compared to the elastic ones that become dominant (mostly due to the close collisions). The perturbative treatment in the semiclassical approach does not correctly estimate this contribution. In that situation, it is necessary to perform more sophisticated calculations such as the quantum ones. We have shown in Elabidi et al. [11], through extensive comparisons between quantum and semiclassical Stark broadening of Ar XV lines, that the disagreement between the two results increases with the increase in strong collision contributions. Figure 2 displays the Stark widths as a function of the electron temperature at a constant electron density for two selected lines between singlets : 3s$^2$ $^1$S$_0$–3s4p $^1$P$^o_1$ and 3s4p $^1$P$^o_1$–3s5s $^1$S$_0$ and two lines between triplets: 3s3p $^3$P$^o_2$–3s4d $^3$D$_3$, and 3s3d $^3$D$_2$– 3s4p $^3$P$^o_1$.

The obtained Stark broadening parameters will be useful for the investigation and modeling of the plasma of stellar atmospheres. They will be also important for the investigation of laser-produced and inertial fusion plasmas.

**Table 4.** Our Stark line widths (FWHM) $Q$ for Ar VII at electron density $N_e = 10^{18}$ cm$^{-3}$ compared to the semiclassical results $SCP_{SST}$ obtained using the atomic data of the code SST. $\frac{Elastic}{Total}$ and $\frac{Strong}{Total}$ are respectively the contributions of elastic and strong collisions to $SCP$ line broadening. $T$ is expressed in $10^4$ K.

| Transition | $T$ | $Q$ (Å) | $SCP_{SST}$ | $\frac{Elastic}{Total}$ | $\frac{Strong}{Total}$ | $\frac{Q}{SCP}$ |
|---|---|---|---|---|---|---|
| $3s^2\ ^1S_0 - 3s3p\ ^1P_1^o$ $\lambda = 585.75$ Å | 2 | $2.684 \times 10^{-2}$ | $2.380 \times 10^{-2}$ | 0.966 | 0.288 | 1.13 |
| | 5 | $1.872 \times 10^{-2}$ | $1.500 \times 10^{-2}$ | 0.937 | 0.287 | 1.25 |
| | 10 | $1.453 \times 10^{-2}$ | $1.070 \times 10^{-2}$ | 0.891 | 0.286 | 1.36 |
| | 20 | $1.135 \times 10^{-2}$ | $7.640 \times 10^{-3}$ | 0.789 | 0.282 | 1.49 |
| | 30 | $9.875 \times 10^{-3}$ | $6.360 \times 10^{-3}$ | 0.706 | 0.276 | 1.55 |
| | 50 | $8.322 \times 10^{-3}$ | $5.160 \times 10^{-3}$ | 0.605 | 0.267 | 1.61 |
| $3s^2\ ^1S_0 - 3s4p\ ^1P_1^o$ $\lambda = 175.5$ Å | 2 | $9.941 \times 10^{-3}$ | $6.070 \times 10^{-3}$ | 0.897 | 0.575 | 1.64 |
| | 5 | $6.242 \times 10^{-3}$ | $2.930 \times 10^{-3}$ | 0.827 | 0.361 | 2.13 |
| | 10 | $4.370 \times 10^{-3}$ | $2.110 \times 10^{-3}$ | 0.766 | 0.357 | 2.07 |
| | 20 | $3.038 \times 10^{-3}$ | $1.550 \times 10^{-3}$ | 0.689 | 0.345 | 1.96 |
| | 30 | $2.445 \times 10^{-3}$ | $1.320 \times 10^{-3}$ | 0.641 | 0.334 | 1.85 |
| | 50 | $1.848 \times 10^{-3}$ | $1.090 \times 10^{-3}$ | 0.588 | 0.319 | 1.70 |
| $3s3p\ ^1P_1^o - 3s4s\ ^1S_0$ $\lambda = 279.2$ Å | 2 | $6.322 \times 10^{-2}$ | $9.750 \times 10^{-3}$ | 0.944 | 0.155 | 6.48 |
| | 5 | $3.923 \times 10^{-2}$ | $5.400 \times 10^{-3}$ | 0.767 | 0.175 | 7.26 |
| | 10 | $2.676 \times 10^{-2}$ | $3.190 \times 10^{-3}$ | 0.616 | 0.172 | 6.84 |
| | 20 | $1.759 \times 10^{-2}$ | $2.920 \times 10^{-3}$ | 0.472 | 0.162 | 6.02 |
| | 30 | $1.344 \times 10^{-2}$ | $2.500 \times 10^{-3}$ | 0.401 | 0.155 | 5.38 |
| | 50 | $9.332 \times 10^{-3}$ | $2.080 \times 10^{-3}$ | 0.328 | 0.145 | 4.49 |
| $3s4p\ ^1P_1^o - 3s5s\ ^1S_0$ $\lambda = 662.6$ Å | 2 | $4.517 \times 10^{-1}$ | $1.360 \times 10^{-1}$ | 0.792 | 0.193 | 3.32 |
| | 5 | $2.942 \times 10^{-1}$ | $8.840 \times 10^{-2}$ | 0.595 | 0.189 | 3.33 |
| | 10 | $1.975 \times 10^{-1}$ | $6.570 \times 10^{-2}$ | 0.458 | 0.179 | 3.01 |
| | 20 | $1.266 \times 10^{-1}$ | $5.050 \times 10^{-2}$ | 0.359 | 0.166 | 2.51 |
| | 30 | $9.671 \times 10^{-2}$ | $4.400 \times 10^{-2}$ | 0.315 | 0.156 | 2.20 |
| | 50 | $6.843 \times 10^{-2}$ | $3.740 \times 10^{-2}$ | 0.274 | 0.144 | 1.83 |
| $3s3d\ ^1D_2 - 3s4p\ ^1P_1^o$ $\lambda = 489.6$ Å | 2 | $1.910 \times 10^{-1}$ | $4.150 \times 10^{-2}$ | 0.859 | 0.361 | 4.60 |
| | 5 | $1.355 \times 10^{-1}$ | $2.650 \times 10^{-2}$ | 0.809 | 0.356 | 5.11 |
| | 10 | $1.011 \times 10^{-1}$ | $1.930 \times 10^{-2}$ | 0.737 | 0.346 | 5.24 |
| | 20 | $6.854 \times 10^{-2}$ | $1.430 \times 10^{-2}$ | 0.661 | 0.335 | 4.79 |
| | 30 | $5.206 \times 10^{-2}$ | $1.210 \times 10^{-2}$ | 0.611 | 0.323 | 4.30 |
| | 50 | $3.602 \times 10^{-2}$ | $1.010 \times 10^{-2}$ | 0.561 | 0.307 | 3.57 |
| $3s3p\ ^1P_1^o - 3s5s\ ^1S_0$ $\lambda = 176.5$ Å | 2 | $1.779 \times 10^{-2}$ | $8.070 \times 10^{-3}$ | 0.825 | 0.085 | 2.20 |
| | 5 | $1.119 \times 10^{-2}$ | $4.750 \times 10^{-3}$ | 0.559 | 0.092 | 2.36 |
| | 10 | $7.845 \times 10^{-3}$ | $3.520 \times 10^{-3}$ | 0.400 | 0.087 | 2.23 |
| | 20 | $5.460 \times 10^{-3}$ | $2.710 \times 10^{-3}$ | 0.282 | 0.081 | 2.01 |
| | 30 | $4.390 \times 10^{-3}$ | $2.360 \times 10^{-3}$ | 0.229 | 0.076 | 1.86 |
| | 50 | $3.303 \times 10^{-3}$ | $2.000 \times 10^{-3}$ | 0.177 | 0.069 | 1.65 |

**Table 5.** Our Stark widths (FWHM) $Q$ for the Ar VII $3s^2\ ^1S_0 - 3s3p\ ^1P_1^o$ resonance line at electron density $N_e = 10^{18}$ cm$^{-3}$ compared to the semiclassical results $SCP_{SST}$ obtained using the atomic data of the code SST. $\frac{Elastic}{Total}$ and $\frac{Strong}{Total}$ are respectively the contributions of elastic and strong collisions to $SCP$ line broadening. $T$ is expressed in $10^5$ K.

| Transition | $T$ | $Q$ (Å) | $SCP_{SST}$ | $\frac{Elastic}{Total}$ | $\frac{Strong}{Total}$ | $\frac{Q}{SCP}$ |
|---|---|---|---|---|---|---|
| | 0.2 | $2.684 \times 10^{-2}$ | $2.380 \times 10^{-2}$ | 0.966 | 0.288 | 1.13 |
| | 0.5 | $1.872 \times 10^{-2}$ | $1.500 \times 10^{-2}$ | 0.937 | 0.287 | 1.25 |
| | 1 | $1.453 \times 10^{-2}$ | $1.070 \times 10^{-2}$ | 0.891 | 0.286 | 1.36 |
| | 2 | $1.135 \times 10^{-2}$ | $7.640 \times 10^{-3}$ | 0.789 | 0.282 | 1.49 |
| $3s^2\ ^1S_0 - 3s3p\ ^1P_1^o$ | 3 | $9.875 \times 10^{-3}$ | $6.360 \times 10^{-3}$ | 0.706 | 0.276 | 1.55 |
| $\lambda = 585.75$ Å | 5 | $8.322 \times 10^{-3}$ | $5.160 \times 10^{-3}$ | 0.605 | 0.267 | 1.61 |
| | 7.5 | $7.271 \times 10^{-3}$ | $4.430 \times 10^{-3}$ | 0.537 | 0.255 | 1.64 |
| | 10 | $6.593 \times 10^{-3}$ | $4.020 \times 10^{-3}$ | 0.494 | 0.245 | 1.64 |
| | 15 | $5.699 \times 10^{-3}$ | $3.530 \times 10^{-3}$ | 0.444 | 0.232 | 1.61 |
| | 30 | $4.284 \times 10^{-3}$ | $2.890 \times 10^{-3}$ | 0.389 | 0.211 | 1.48 |
| | 50 | $3.327 \times 10^{-3}$ | $2.530 \times 10^{-3}$ | 0.366 | 0.199 | 1.32 |

**Table 6.** Same as in Table 4 but we add the semiclassical results $SCP_{BD}$ obtained in Dimitrijević et al. [12] using the atomic data from Bates and Damgaard [13]. Electron density is $N_e = 10^{18}$ cm$^{-3}$ and $T$ is expressed in $10^4$ K.

| Transition | $T$ | $Q$ | $SCP_{SST}$ | $SCP_{BD}$ | $\frac{Q}{SCP_{SST}}$ | $\left(\frac{Elastic}{Total}\right)_{SST}$ | $\left(\frac{Strong}{Total}\right)_{SST}$ | $\frac{Q}{SCP_{BD}}$ |
|---|---|---|---|---|---|---|---|---|
| | 2 | $7.510 \times 10^{-3}$ | $4.83 \times 10^{-3}$ | $4.21 \times 10^{-3}$ | 1.55 | 0.889 | 0.444 | 1.77 |
| | 5 | $4.748 \times 10^{-3}$ | $2.86 \times 10^{-3}$ | $2.49 \times 10^{-3}$ | 1.66 | 0.877 | 0.475 | 1.90 |
| $3s3p\ ^3P_2^o - 3s4d\ ^3D_3$ | 10 | $3.369 \times 10^{-3}$ | $2.04 \times 10^{-3}$ | $1.81 \times 10^{-3}$ | 1.65 | 0.861 | 0.472 | 1.85 |
| $\lambda = 192.3$ Å | 20 | $2.389 \times 10^{-3}$ | $1.48 \times 10^{-3}$ | $1.35 \times 10^{-3}$ | 1.62 | 0.823 | 0.461 | 1.77 |
| | 30 | $1.967 \times 10^{-3}$ | $1.25 \times 10^{-3}$ | $1.15 \times 10^{-3}$ | 1.47 | 0.798 | 0.453 | 1.70 |
| | 50 | $1.532 \times 10^{-3}$ | $1.02 \times 10^{-3}$ | $9.51 \times 10^{-4}$ | 1.50 | 0.764 | 0.437 | 1.61 |
| | 2 | $1.613 \times 10^{0}$ | $3.76 \times 10^{-1}$ | $3.09 \times 10^{-1}$ | 4.29 | 0.901 | 0.457 | 5.22 |
| | 5 | $1.061 \times 10^{0}$ | $2.38 \times 10^{-1}$ | $1.99 \times 10^{-1}$ | 4.46 | 0.873 | 0.455 | 5.33 |
| $3s4p\ ^3P_2^o - 3s4d\ ^3D_3$ | 10 | $7.141 \times 10^{-1}$ | $1.71 \times 10^{-1}$ | $1.45 \times 10^{-1}$ | 4.18 | 0.824 | 0.448 | 4.92 |
| $\lambda = 1425.9$ Å | 20 | $4.285 \times 10^{-1}$ | $1.26 \times 10^{-1}$ | $1.08 \times 10^{-1}$ | 3.40 | 0.767 | 0.435 | 3.97 |
| | 30 | $3.073 \times 10^{-1}$ | $1.07 \times 10^{-1}$ | $9.27 \times 10^{-2}$ | 2.87 | 0.740 | 0.424 | 3.31 |
| | 50 | $2.013 \times 10^{-1}$ | $8.87 \times 10^{-2}$ | $7.77 \times 10^{-2}$ | 2.27 | 0.705 | 0.406 | 2.59 |
| | 2 | $7.585 \times 10^{-2}$ | $2.33 \times 10^{-2}$ | $1.84 \times 10^{-2}$ | 3.26 | 0.928 | 0.406 | 4.12 |
| | 5 | $4.562 \times 10^{-2}$ | $1.52 \times 10^{-2}$ | $1.16 \times 10^{-2}$ | 3.00 | 0.878 | 0.397 | 3.93 |
| $3s3d\ ^3D_2 - 3s4p\ ^3P_1^o$ | 10 | $2.896 \times 10^{-2}$ | $1.10 \times 10^{-2}$ | $8.35 \times 10^{-3}$ | 2.64 | 0.809 | 0.389 | 3.47 |
| $\lambda = 416.0$Å | 20 | $1.806 \times 10^{-2}$ | $8.14 \times 10^{-3}$ | $6.12 \times 10^{-3}$ | 2.22 | 0.723 | 0.375 | 2.95 |
| | 30 | $1.389 \times 10^{-2}$ | $6.91 \times 10^{-3}$ | $5.18 \times 10^{-3}$ | 2.01 | 0.675 | 0.364 | 2.68 |
| | 50 | $1.040 \times 10^{-2}$ | $5.73 \times 10^{-3}$ | $4.27 \times 10^{-3}$ | 1.82 | 0.625 | 0.346 | 2.43 |
| | 2 | $7.605 \times 10^{-1}$ | $5.36 \times 10^{-1}$ | $5.34 \times 10^{-1}$ | 1.42 | 0.942 | 0.337 | 1.42 |
| | 5 | $5.453 \times 10^{-1}$ | $3.44 \times 10^{-1}$ | $3.33 \times 10^{-1}$ | 1.59 | 0.863 | 0.331 | 1.64 |
| $3s4s\ ^3S_1 - 3s4p\ ^3P_1^o$ | 10 | $4.272 \times 10^{-1}$ | $2.49 \times 10^{-1}$ | $2.40 \times 10^{-1}$ | 1.72 | 0.757 | 0.326 | 1.78 |
| $\lambda = 1982.0$ Å | 20 | $3.336 \times 10^{-1}$ | $1.85 \times 10^{-1}$ | $1.79 \times 10^{-1}$ | 1.80 | 0.632 | 0.309 | 1.86 |
| | 30 | $2.871 \times 10^{-1}$ | $1.59 \times 10^{-1}$ | $1.53 \times 10^{-1}$ | 1.81 | 0.577 | 0.298 | 1.88 |
| | 50 | $2.350 \times 10^{-1}$ | $1.33 \times 10^{-1}$ | $1.28 \times 10^{-1}$ | 1.77 | 0.518 | 0.281 | 1.84 |
| | 2 | $1.396 \times 10^{-2}$ | $8.72 \times 10^{-3}$ | $5.06 \times 10^{-3}$ | 1.56 | 0.996 | 0.358 | 2.75 |
| | 5 | $8.893 \times 10^{-3}$ | $5.35 \times 10^{-3}$ | $2.68 \times 10^{-3}$ | 1.66 | 0.947 | 0.368 | 3.30 |
| $3s3p\ ^3P_2^o - 3s4s\ ^3S_1$ | 10 | $6.351 \times 10^{-3}$ | $3.80 \times 10^{-3}$ | $1.91 \times 10^{-3}$ | 1.67 | 0.860 | 0.368 | 3.31 |
| $\lambda = 250.4$ Å | 20 | $4.549 \times 10^{-3}$ | $2.78 \times 10^{-3}$ | $1.41 \times 10^{-3}$ | 1.64 | 0.750 | 0.358 | 3.21 |
| | 30 | $3.742 \times 10^{-3}$ | $2.35 \times 10^{-3}$ | $1.21 \times 10^{-3}$ | 1.59 | 0.690 | 0.349 | 3.08 |
| | 50 | $2.915 \times 10^{-3}$ | $1.94 \times 10^{-3}$ | $1.01 \times 10^{-3}$ | 1.50 | 0.626 | 0.333 | 2.87 |
| | 2 | $5.727 \times 10^{-2}$ | $1.69 \times 10^{-2}$ | $9.11 \times 10^{-3}$ | 3.39 | 0.986 | 0.334 | 6.24 |
| | 5 | $2.896 \times 10^{-2}$ | $9.92 \times 10^{-3}$ | $5.85 \times 10^{-3}$ | 2.92 | 0.961 | 0.363 | 4.91 |
| $3s3p\ ^3P_2^o - 3s3d\ ^3D_3$ | 10 | $1.708 \times 10^{-2}$ | $7.02 \times 10^{-3}$ | $4.17 \times 10^{-3}$ | 2.43 | 0.921 | 0.362 | 4.06 |
| $\lambda = 477.5$ Å | 20 | $1.022 \times 10^{-2}$ | $5.02 \times 10^{-3}$ | $2.96 \times 10^{-3}$ | 2.04 | 0.830 | 0.357 | 3.43 |
| | 30 | $7.755 \times 10^{-3}$ | $4.18 \times 10^{-3}$ | $2.45 \times 10^{-3}$ | 1.86 | 0.764 | 0.353 | 3.14 |
| | 50 | $5.734 \times 10^{-3}$ | $3.37 \times 10^{-3}$ | $1.95 \times 10^{-3}$ | 1.70 | 0.684 | 0.343 | 2.92 |

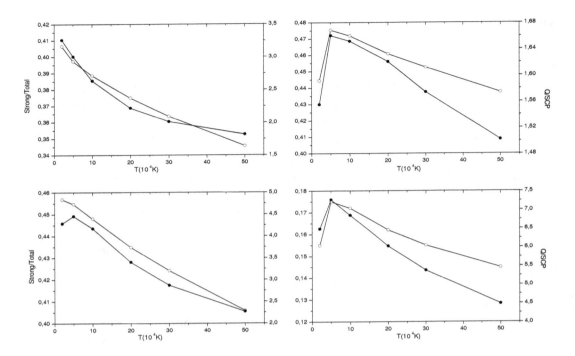

**Figure 1.** Ratios $\frac{Q}{SCP_{SST}}$ (•) and $\frac{Strong}{Total}$ (○) as a function of the electron temperature for the transitions: $3s3d\ ^3D_2 - 3s4p\ ^3P^o_1$ (**left up**), $3s3p\ ^3P^o_2 - 3s4d\ ^3D_3$ (**right up**), $3s4p\ ^3P^o_2 - 3s4d\ ^3D_3$ (**left down**), and $3s3p\ ^1P^o_1 - 3s4s\ ^1S_0$ (**right down**).

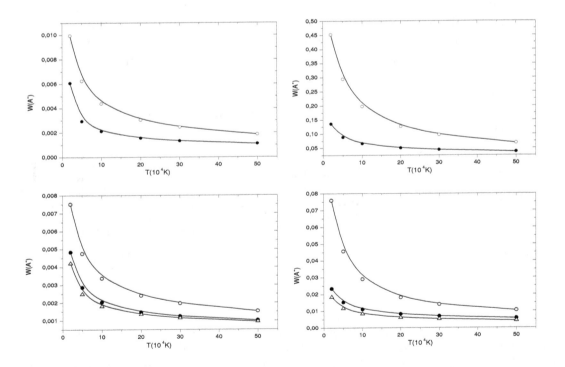

**Figure 2.** Stark width (FWHM) $W$ as a function of the electron temperature for transitions $3s^2\ ^1S_0 - 3s4p\ ^1P^o_1$ (**left up**) and $3s4p\ ^1P^o_1 - 3s5s\ ^1S_0$ (**right up**) at electron density $N_e = 10^{17}$ cm$^{-3}$, and for transitions $3s3p\ ^3P^o_2 - 3s4d\ ^3D_3$ (**left down**) and $3s3d\ ^3D_2 - 3s4p\ ^3P^o_1$ (**right down**) at electron density $N_e = 10^{18}$ cm$^{-3}$. ○: Present quantum results. •: Present SCP results. △: SCP results from [12].

## 4. Conclusions

We have calculated in the present work quantum and semiclassical perturbation Stark broadening parameters for 12 Ar VII lines at electron temperatures from $2 \times 10^4$ to $5 \times 10^5$ K and at electron density $N_e = 10^{18}$ cm$^{-3}$. The structure and collision problem has also been treated for this ion. We have used nine configurations ($1s^2 2s^2 2p^6$: $3s^2$, 3s3p, $3p^2$, 3s3d, 3p3d, 3s4s, 3s4p, 3s4d, and 3s5s). The structure and collisional parameters have been used in our quantum mechanical line broadening calculations. Since it is important to check their accuracy, we compared our energies to those of [22], to those obtained by the multiconfiguration Hartree-Fock method [24], and to those obtained from the AUTOSTRUCTURE code [25]. An acceptable agreement was found with the NIST results (better than 1%). We also compared our $A_{ij}$ values with those obtained from the AUTOSTRUCTURE code [25] and with those from the SUPERSTRUCTURE code [26] using five configurations ($1s^2 2s^2 2p^6$: $3s^2$, 3s3p, $3p^2$, 3s3d, and 3s4s). The averaged difference is about 20% with the results of [25] and about 24% with those of Christensen et al. [26]. The oscillator strengths have been compared only with the results of Christensen et al. [26], and we found an averaged agreement of about 24%. The electron-ion collision process was also studied, and collision strengths from the lowest five levels to the first 14 levels are presented at three electron energies 7.779 Ry, 13.674 Ry, and 23.336 Ry. The comparison with the collision strengths of Christensen et al. [26] indicates an agreement (averaged over the considered transitions and energies) of about 20%. The reason for the disagreement between the two results could be the difference in the number of the configurations and the difference in the partial waves taken into account in the two calculations. Stark line widths for 12 lines have been calculated using our quantum formalism. We perform also a semiclassical perturbation calculations using the structure data of the SST code. We present other semiclassical widths [12] obtained using the atomic data from Bates and Damgaard [13]. Firstly, the disagreement between the two semiclassical calculations is due to the difference in the source of atomic data. Secondly, we have shown that the disagreement between the quantum and the semiclassical widths increases with the increase in the contributions to line broadening of elastic collisions (which are mostly due to strong collisions). This is because the perturbative treatment in the semiclassical approach does not estimate very well the strong collisions. We hope that the present results can fill the lack of line broadening parameters or improve the available results for the Ar VII ion, which are of interest in the investigation and modeling of astrophysical and laboratory plasmas.

**Acknowledgments:** This work has been supported by the Tunisian Research Unit UR11ES03 and the French one UMR 8112. It has also been supported by the Paris Observatory and the CNRS. We also acknowledge financial support from the "Programme National de Physique Stellaire" (PNPS) of CNRS/INSU, CEA, and CNES, France. This work has also been supported by the Ministry of Education, Science and Technological Development of Serbia (project 176002). Some results have been taken from the database CHIANTI, which is a collaborative project involving George Mason University, the University of Michigan (USA), and the University of Cambridge (UK).

**Author Contributions:** R.A. and H.E. performed the calculations; H.E., S.S.B., and M.S.D. analyzed the data; H.E. wrote the paper.

## References

1.    Dimitrijević, M.S. Stark broadening in astrophysics (applications of Belgrade school results and collaboration with former Soviet republics. *Astron. Astrophys. Trans.* **2003**, *22*, 389–412.

2.    Popović, L.Č.; Simić, S.; Milovanović N.; Dimitrijević, M.S. Stark Broadening Effect in Stellar Atmospheres: Nd II Lines. *Astrophys. J. Suppl. Ser.* **2001**, *135*, 109–114.

3.    Dimitrijević, M.S.; Ryabchikova, T.; Simić, Z.; Popović, L.Č.; Dačić, M. The influence of Stark broadening on Cr II spectral line shapes in stellar atmospheres. *Astron. Astrophys.* **2007**, *469*, 681–686.

4.    Rauch, T.; Ziegler, M.; Werner, K.; Kruk, J.W.; Oliveira, C.M.; Putte, D.V.; Mignani, R.P.; Kerber, F. High-resolution FUSE and HST ultraviolet spectroscopy of the white dwarf central star of Sh 2-216. *Astron. Astrophys.* **2007**, *470*, 317–329.

5.    Werner, K.; Rauch, T.; Kruk, J.W. Discovery of photospheric argon in very hot central stars of planetary nebulae and white dwarfs. *Astron. Astrophys.* **2007** *466*, 317–322.

6.  Djurović, S.; Mar, S.; Peláez, R.J.; Aparicio, J.A. Stark broadening of ultraviolet Ar III spectral lines. *Mon. Not. R. Astron. Soc.* **2011**, *414*, 1389–1396.

7.  Elabidi, H.; Ben Nessib, N.; Sahal-Bréchot, S. Quantum mechanical calculations of the electron-impact broadening of spectral lines for intermediate coupling. *J. Phys. B* **2004**, *37*, 63–71.

8.  Elabidi, H.; Ben Nessib, N.; Cornille, M. Dubau, J.; Sahal-Bréchot, S. Electron impact broadening of spectral lines in Be-like ions: Quantum calculations. *J. Phys. B* **2008**, *41*, 025702.

9.  Elabidi, H.; Sahal-Bréchot, S.; Ben Nessib, N. Quantum Stark broadening of 3s-3p spectral lines in Li-like ions; Z-scaling and comparison with semi-classical perturbation theory. *Eur. Phys. J. D* **2009**, *54*, 51–64.

10. Elabidi, H.; Sahal-Bréchot, S. Checking the dependence on the upper level ionization potential of electron impact widths using quantum calculations. *Eur. Phys. J. D* **2011**, *61*, 285–290.

11. Elabidi, H.; Sahal-Bréchot, S.; Dimitrijević, M.S. Quantum Stark broadening of Ar XV lines. Strong collision and quadrupolar potential contributions. *Adv. Res. Space* **2014**, *54*, 1184–1189.

12. Dimitrijević, M.S.; Valjarević, A.; Sahal-Bréchot, S. Semiclassical Stark broadening parameters of Ar VII spectral lines. *Atoms* **2017**, *5*, 27.

13. Bates, D.R.; Damgaard, A. The calculation of the absolute strengths of spectral lines. *Phil. Trans. R. Soc. Lond. A* **1949**, *242*, 101–122.

14. Eissner, W.; Jones, M.; Nussbaumer, H. Techniques for the calculation of atomic structures and radiative data including relativistic corrections. *Comput. Phys. Commun.* **1974**, *8*, 270–306.

15. Bethe, H.A.; Salpeter, E.E. *Quantum Mechanics of One- and Two-Electron Atoms*; Springer: Berlin/Göttingen, Germany, 1957.

16. Eissner, W. The UCL distorted wave code. *Comput. Phys. Commun.* **1998**, *114*, 295–341.

17. Saraph, H.E. Fine structure cross sections from reactance matrices. *Comput. Phys. Commun.* **1972**, *4*, 256–268.

18. Sahal-Bréchot, S. Impact Theory of the Broadening and Shift of Spectral Lines due to Electrons and Ions in a Plasma. *Astron. Astrophys.* **1969**, *1*, 91–123.

19. Sahal-Bréchot, S. Impact Theory of the Broadening and Shift of Spectral Lines due to Electrons and Ions in a Plasma (Continued). *Astron. Astrophys.* **1969**, *2*, 322–354.

20. Fleurier, C.; Sahal-Bréchot, S.; Chapelle, J. Stark profiles of some ion lines of alkaline earth elements. *J. Quant. Spectrosc. Radiat. Transfer* **1977**, *17*, 595–603.

21. Sahal-Bréchot, S.; Dimitrijević, M.S.; Ben Nessib, N. Widths and Shifts of Isolated Lines of Neutral and Ionized Atoms Perturbed by Collisions with Electrons and Ions: An Outline of the Semiclassical Perturbation Method and of the Approximations Used for the Calculations. *Atoms* **2014**, *2*, 225–252.

22. Kramida, A.; Ralchenko, Y.; Reader, J.; NIST ASD Team. *NIST Atomic Spectra Database (Version 5.5.1)*; National Institute of Standards and Technology: Gaithersburg, MD, USA, 2015. Available online: http://physics.nist.gov/asd (accessed on 8 November 2017).

23. Saloman, E.B. Energy Levels and Observed Spectral Lines of Ionized Argon, Ar II through Ar XVIII. *J. Phys. Chem. Ref. Data* **2010**, *39*, 033101.

24. Froese Fischer, C.; Tachiev, G.; Irimia, A. Relativistic energy levels, lifetimes, and transition probabilities for the sodium-like to argon-like sequences. *At. Data Nucl. Data Tables* **2006**, *92*, 607–812.

25. Fernández-Menchero, L.; Del Zanna, G.; Badnell, N.R. R-matrix electron-impact excitation data for the Mg-like iso-electronic sequence. *Astron. Astrophys.* **2014**, *572*, A115.

26. Christensen, R.B.; Norcross, D.W.; Pradhan, A.K. Electron-impact excitation of ions in the magnesium sequence. II. SV, ArVII, CaIX, CrXIII, and NiXVII. *Phys. Rev. A* **1986**, *34*, 4704–4715.

27. Del Zanna, G.; Dere, K.P.; Young, P.R.; Landi, E.; Mason, H.E. CHIANTI—An atomic database for emission lines. *Astron. Astrophys.* **2015**, *582*, A56.

28. Burgess, A.; Sheorey, V.B. Electron impact excitation of the resonance lines of alkali-like positive ions. *J. Phys. B* **1974**, *7*, 2403–2416.

29. Chidichimo, M.C.; Haig, S.P. Electron-impact excitation of quadrupole-allowed transitions in positive ions. *Phys. Rev. A* **1989**, *39*, 4991–4997.

30. Chidichimo, M.C. Electron-impact excitation of electric octupole transitions in positive ions: Asymptotic behavior of the sum over partial-collision strengths. *Phys. Rev. A* **1992**, *45*, 1690–1700.

31. Hamdi, R.; Ben Nessib, N.; Dimitrijević, M.S.; Sahal-Bréchot, S. Stark broadening of Pb IV lines. *Mon. Not. R. Astron. Soc.* **2013**, *431*, 1039–1047.

32. Hamdi, R.; Ben Nessib, N.; Sahal-Bréchot, S.; Dimitrijević, M.S. Stark widths of Ar III spectral lines in the atmospeheres of subdwarfs B stars. *Adv. Res. Space* **2014**, *54*, 1223–1230.
33. Hamdi, R.; Ben Nessib, N.; Sahal-Bréchot, S.; Dimitrijević, M.S. Stark widths of Ar II spectral lines in the atmospeheres of subdwarfs B stars. *Atoms* **2017**, *5*, 26.
34. Sahal-Bréchot, S.; Dimitrijević, M.S.; Moreau, N. STARK-B Database. Observatory of Paris, LERMA and Astronomical Observatory of Belgrade, 2017. Available online: http://stark-b.obspm.fr (accessed on 8 November 2017).

# 23

# A New Analysis of Stark and Zeeman Effects on Hydrogen Lines in Magnetized DA White Dwarfs

Ny Kieu [1], Joël Rosato [1,*], Roland Stamm [1], Jelena Kovačević-Dojcinović [2],
Milan S. Dimitrijević [2], Luka Č. Popović [2] and Zoran Simić [2]

[1]    Laboratoire PIIM, Aix-Marseille Université and CNRS, 13397 Marseille CEDEX 20, France;
missny0909@gmail.com (N.K.); roland.stamm@univ-amu.fr (R.S.)

[2]    Astronomical Observatory, Volgina 7, 11060 Belgrade 38, Serbia; jkovacevic@aob.bg.ac.rs (J.K.-D.);
mdimitrijevic@aob.rs (M.S.D.); lpopovic@aob.rs (L.C.P.); zsimic@aob.rs (Z.S.)

*    Correspondence: joel.rosato@univ-amu.fr

**Abstract:** White dwarfs with magnetic field strengths larger than 10 T are understood to represent more than 10% of the total population of white dwarfs. The presence of such strong magnetic fields is clearly indicated by the Zeeman triplet structure visible on absorption lines. In this work, we discuss the line broadening mechanisms and focus on the sensitivity of hydrogen lines on the magnetic field. We perform new calculations in conditions relevant to magnetized DA stellar atmospheres using models inspired from magnetic fusion plasma spectroscopy. A white dwarf spectrum from the Sloan Digital Sky Survey (SDSS) database is analyzed. An effective temperature is provided by an adjustment of the background radiation with a Planck function, and the magnetic field is inferred from absorption lines presenting a Zeeman triplet structure. An order-of-magnitude estimate for the electron density is also performed from Stark broadening analysis.

**Keywords:** line shapes; Stark broadening; Zeeman effect; white dwarfs

## 1. Introduction

White dwarfs—the end products of stellar evolution—occupy a key position in astrophysical theory. Their properties provide clues to the physical processes that take place during the evolutionary stages near the end of stellar lifetimes. With an understanding of white dwarf magnetic fields, one can indicate the formation or the existence of centered or non-centered dipoles of these magnetic fields [1]. However, only 10% of white dwarfs are thought to carry strong magnetic fields. Most of them have so far only been analyzed by visual comparison of the observation with relatively simple models of the radiation transport in a magnetized stellar atmosphere [2,3]. Some of them have been analyzed by using the least squares minimization scheme to find the best fit for magnetic field geometry [1,4]. Studies of white dwarf atmospheres have shown that the majority of white dwarfs have an atmosphere of pure hydrogen as a result of gravitational setting, which removes helium and heavier elements from the atmosphere and moves them towards inner layers [5,6]. These atmospheres can be considered as hydrogen plasmas, which are similar to some created in laboratory. Such white dwarfs are classified as of DA type due to the strong hydrogen absorption lines they present. The electron density in a white dwarf atmosphere is high enough (up to $10^{17}$ cm$^{-3}$, and higher) so that the line shapes are dominated by Stark broadening, and hence can serve as a probe for the electron density $N_e$. A noticeable feature of magnetized DA white dwarfs is that their absorption lines exhibit a Zeeman triplet structure, which stems from the perturbation of the atomic energy levels due to the magnetic field. The design of a line shape model accounting for both Stark broadening and Zeeman splitting is not straightforward, because these two effects do not act as additive perturbations. Previous investigations in low-density (tokamak edge) magnetized plasma conditions have indicated an alteration of the broadening of each

Zeeman component due to the change of the atomic energy level structure (degeneracy removal) in response to an external magnetic field [7,8]. This issue also concerns magnetized white dwarfs with spectra exhibiting Zeeman splitting. In this work, we report on the current status of the design of a line shape model accounting for Stark and Zeeman effects simultaneously, for applications to white dwarf atmosphere analysis. We perform new Stark–Zeeman line shape calculations. The applicability of spectroscopy as a diagnostic means is illustrated through the analysis of a white dwarf spectrum obtained from the Sloan Digital Sky Survey (SDSS) database.

## 2. Stark Line Shape Modeling

In a plasma, the energy levels of atomic species (either neutral atoms or multicharged ions) emitting line radiation is perturbed by the microscopic electric field, which results in a broadening of spectral lines. The spectral profile of an atomic line is proportional to the Fourier transform of the dipole autocorrelation function $C(t)$ [9]

$$L(\omega) = \frac{1}{\pi}\mathrm{Re}\int_0^\infty dt C(t)e^{i\omega t}, \tag{1}$$

$$C(t) = \sum_{\vec{e}}\langle\langle\vec{d}\cdot\vec{e}|\{\hat{U}(t)\}|\rho\vec{d}\cdot\vec{e}\rangle\rangle, \tag{2}$$

where $\{...\}$ is the average over plasma particles, $\rho$ is the atomic density operator, and $\vec{d}\cdot\vec{e}$ is the projection of the atomic dipole operator into the polarization plane. The double bracket notation $\langle\langle...|...\rangle\rangle$ is used for the scalar product in the Liouville space which is formed by the tensor product $\mathcal{H}\otimes\mathcal{H}^*$ between the atomic state space $\mathcal{H}$ and its dual $\mathcal{H}^*$. The evolution operator $\hat{U}(t)$ obeys the time-dependent Schrödinger equation in the Liouville space

$$i\frac{d\hat{U}(t)}{dt}(t) = [\hat{L}_0 + \hat{V}(t)]\hat{U}(t), \tag{3}$$

which is also called the Liouville equation. $\hat{L}_0$ is the energy superoperator and $\hat{V} = -\hat{\vec{d}}\cdot\vec{E}$ is the interaction superoperator corresponding to the Stark perturbation. The action of a superoperator $\hat{A}$ on an operator $X$ is defined by $\hat{A}X = [A, X]/\hbar$, where $A$ is the related operator and $[,]$ is the commutator. In a plasma, the electric field is caused by the electrons and the ions and the corresponding Stark perturbation decomposes into two parts

$$\hat{V} = \hat{\vec{d}}\cdot\vec{E} = -\hat{\vec{d}}\cdot\vec{E}_i - \hat{\vec{d}}\cdot\vec{E}_e, \tag{4}$$

where the $e$ and $i$ subscripts correspond to the electrons and ions, respectively. In the framework of the so-called "standard model", the ions are assumed motionless and a constant electric field is used for their contribution, while the electron contribution is described within a collisional picture by a relaxation ("collision") operator $\hat{K}_e$, that is through the formal substitution $-\hat{\vec{d}}\cdot\vec{E}_e \to -i\hat{K}_e$. In this way, the Liouville equation has the exponential solution $\hat{U}(t) = \exp[-i(\hat{L}_0 - \hat{\vec{d}}.\vec{E}_i - i\hat{K}_e)]t$ and the line shape function is obtained from a matrix inversion

$$L(\omega) = -\frac{1}{\pi}\mathrm{Im}\sum_{\vec{e}}\int d^3E_i W(\vec{E}_i)\langle\langle\vec{d}\cdot\vec{e}|(\omega - \hat{L}_0 + \hat{\vec{d}}.\vec{E}_i + i\hat{K}_e)^{-1}|\rho\vec{d}\cdot\vec{e}\rangle\rangle \tag{5}$$

The integral is performed over the ionic electric field $\vec{E}_i$ and involves its probability density function $W(\vec{E}_i)$. This integration can be performed numerically by using a discretization scheme or by Monte Carlo method. The standard model is convenient for numerical applications due to the structure of Equation (5), which can be interpreted as a sum of complex Lorentzian functions after decomposition of the Liouville double brackets onto a basis is done. The static ion approximation holds at conditions such that the collision time $r_0/v_i$ ($r_0$ and $v_i$ being the mean interparticle distance and the thermal ion velocity, respectively) is much larger than the characteristic time for dipole decorrelation (also called "time of interest"). Extensions of the standard model accounting for ion dynamics effects can be set up through suitable modifications; e.g., by formally adding a non-Hermitian contribution to $L_0$ (Model Microfield Method [10,11], Frequency Fluctuation Model [12,13]). The ion dynamics leads to an additional broadening, which can be important especially at low densities, high temperatures, or for lines with a low upper principal quantum number. Simulations involving a numerical integration of the Liouville Equation (3) and not referring to Equation (5) can also be performed, but they become time-consuming if the atomic system is complex [7,14,15].

For hydrogen lines, the electron collision operator can be evaluated using the Griem–Kolb–Shen model [16], which provides an analytical expression for the contribution of weak collisions based on the time-dependent perturbation series. Such weak collisions involve an impact parameter larger than the Weisskopf radius $b_W = \hbar n^2 / m_e v$ ($n$ being the principal quantum number and $v$ being the velocity), and they give a dominant contribution to the Stark broadening due to electrons. The strong collision contribution is usually estimated using a Lorentz model, assuming complete interruption of the wave train. Extensions to the Griem–Kolb–Shen model, accounting for "static" perturbations due to the electrons (especially present in the line wings), can be devised through the use of kinetic theory techniques (such as the "unified theory" [17–19]) and they lead to a frequency-dependent collision operator. In white dwarf atmospheres, the plasma is partially ionized and there can be a significant additional broadening caused by collisions with neutrals, either due to resonant interactions or to van der Waals interactions. This broadening is also described with a collision operator $\hat{K}_0$, and it is added to $\hat{K}_e$. Simple models can be found in [9].

## 3. Zeeman Effect in Magnetized White Dwarfs

In the presence of a magnetic field, the atomic Hamiltonian and the corresponding Liouvillian $\hat{L}_0$ contain an additional term due to the coupling with the field. Observations of white dwarf spectra have indicated that the magnetic field can attain values larger than 100 T, which renders the Zeeman effect visible on spectra. A description of this effect in line shape modeling requires the following substitution be done in the Liouville Equation (3):

$$\hat{L}_0 \to \hat{L}_0 - \hat{\vec{\mu}} \cdot \vec{B}, \tag{6}$$

where $\hat{\vec{\mu}}$ is the magnetic moment expressed in the Liouville space. We do not consider the weak ("anomalous") Zeeman effect that perturbs fine structure lines, because it is negligible in white dwarf spectra due to high density and strong magnetic fields. If the standard model is used, the same substitution as Equation (6) must be done in the line shape formula (5). An illustration of the Zeeman effect in white dwarf atmosphere conditions is shown in Figure 1. The H$\alpha$ line shape has been calculated assuming an electron density of $10^{17}$ cm$^{-3}$, a temperature of 1 eV, and a magnetic field of 200 T. An observation direction perpendicular to the magnetic field has been assumed. The line shape exhibits a clean Zeeman triplet structure. The splitting is proportional to the magnetic field, and hence can be used to estimate the magnetic field in white dwarfs. In some white dwarfs, the magnetic field can be sufficiently large so that the quadratic Zeeman effect becomes visible in spectra. This effect can be retained through an additional term in the Hamiltonian which is proportional to $B^2$. This term stems from the expansion of the quantity $(\vec{p} + e\vec{A})^2$ present in the kinetic energy to the second order in the vector potential $\vec{A}$. Algebraic manipulations lead to the following expression for the quadratic Zeeman effect Hamiltonian:

$$V_{Z2} = \frac{e^2 B^2 r_\perp^2}{8m_e}, \tag{7}$$

where $\vec{r}_\perp$ stands for the atomic electron position operator projected onto the plane perpendicular to $\vec{B}$. A similar relation holds for the corresponding superoperator in the Liouville space. The quadratic

Zeeman effect is illustrated in Figure 2. A calculation of Hα at the same plasma conditions as above, assuming a magnetic field of 2000 T, is shown. Note the asymmetry of the spectrum and the presence of more components.

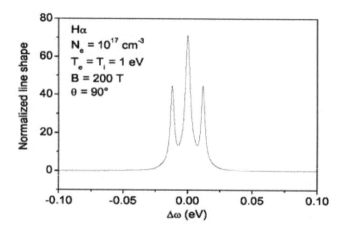

**Figure 1.** Plot of Hα calculated at $N_e = 10^{17}$ cm$^{-3}$, $T_e = T_i = 1$ eV, with a magnetic field of 200 T. An observation direction perpendicular to the magnetic field has been assumed. The spectrum exhibits a clean Zeeman triplet structure.

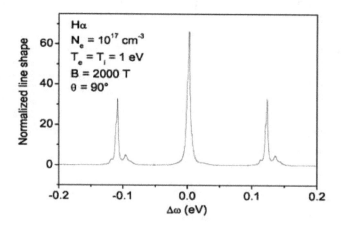

**Figure 2.** Some white dwarfs can have a magnetic field strong enough so that the quadratic Zeeman effect becomes significant. The spectrum becomes asymmetric and presents more components. Here, a calculation has been done assuming $B = 2000$ T.

## 4. White Dwarf Spectrum Analysis

In this section, we illustrate the applicability of spectroscopy as a diagnostic means for the characterization of white dwarf atmosphere. A large amount of spectra presenting Zeeman splitting are available on the SDSS database (www.sdss.org) [1,4]. Figure 3 shows an example of such a spectrum. We have performed a fitting of the Hα and Hβ absorption lines using an exponential attenuation model (Beer-Lambert formula) assuming a homogeneous atmosphere slab, leaving the product $N_2 \times L$ between absorber density $N_2$ and thickness $L$ as an adjustable parameter. The temperature that enters the formula (through the electron collision operator and the microfield probability density function) is estimated from adjustment of the background radiation with a Planck function, assuming local thermodynamic equilibrium. Figure 4 shows the fitting result. The effective temperature is around 7000 K. In order to get a reasonable adjustment on each of these lines, we had to assume different values for the electronic density and the $N_2L$ parameter: $N_e = 5 \times 10^{17}$ cm$^{-3}$ and $N_2L = 5 \times 10^{17}$ m$^{-2}$ for Hα, $N_e = 2 \times 10^{16}$ cm$^{-3}$ and $N_2L = 2 \times 10^{18}$ m$^{-2}$ for Hβ. A value of 130 T for the magnetic was found in both cases. These results can be used as a first order-of-magnitude estimate. The necessity of using different values for densities stems from inaccuracies intrinsic to the model (e.g., due to the neglect of self-emission), and this indicates that a further refinement is required.

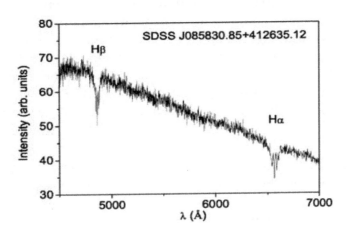

**Figure 3.** Plot of the Sloan Digital Sky Survey (SDSS) + J085830 + 412635.12 spectrum in the 4500–7000 Å wavelength range. This spectrum corresponds to a magnetized DA white dwarf. The Hα and Hβ lines present a triplet structure, which is characteristic of Zeeman effect.

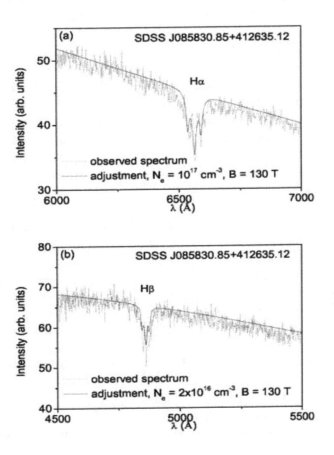

**Figure 4.** An analysis of the shape of absorption lines provides information on the plasma parameters. Here, an adjustment of **(a)** Hα and **(b)** Hβ using the Stark-Zeeman line shape model is presented. These lines have been analyzed independently assuming a homogeneous slab geometry. The obtained values for $N_e$ and $B$ provide a first order-of-magnitude estimate.

## 5. Conclusions

According to observations, a significant proportion of white dwarfs have a strong magnetic field, sufficiently so that the absorption lines present a Zeeman triplet structure. In this work, we have done new calculations of Stark–Zeeman hydrogen line shapes in preparation for the development of a model. Specific issues such as the quadratic Zeeman effect need to be investigated further. The applicability of spectroscopy as a diagnostic means for magnetized white dwarf atmospheres has been illustrated through adjustments performed on a spectrum from the Sloan Digital Sky Survey (SDSS) database. It is found that line width adjustments provide an order of magnitude for the electron density and the magnetic field in the atmosphere. An extension of the work to the accounting for plasma non-homogeneity is presently ongoing. Comparisons to other white dwarf spectra analyses, such as in [1,4], will be done.

**Acknowledgments:** This work is supported by the funding agency Campus France (Pavle Savic PHC project 36237PE); L. Č. Popović is supported by the Ministry of Science of Serbia (the project 146002).

**Author Contributions:** All authors contributed equally to this work.

## References

1. Külebi, B.; Jordan, S.; Euchner, F.; Gänsicke, B.T.; Hirsch, H. Analysis of hydrogen-rich magnetic white dwarfs detected in the Sloan Digital Sky Survey. *Astron. Astrophys.* **2009**, *506*, 1341–1350.
2. Schmidt, G.D.; Harris, H.C.; Liebert, J.; Eisenstein, D.J.; Anderson, S.F.; Brinkmann, J.; Hall, P.B.; Harvanek, M.; Hawley, S.; Kleinman, S.J.; et al. Magnetic White Dwarfs from the Sloan Digital Sky Survey: The First Data Release. *Astrophys. J.* **2003**, *595*, 1101.
3. Vanlandingham, K.M.; Schmidt, G.D.; Eisenstein, D.J.; Harris, H.C.; Anderson, S.F.; Hall, P.B.; Liebert, J.; Schneider, D.P.; Silvestri, N.M.; Stinson, G.S.; et al. Magnetic White Dwarfs from the SDSS II. The Second and Third Data Releases. *Astron. J.* **2005**, *130*, 734–741.
4. Kepler, S.O.; Pelisoli, I.; Jordan, S.; Kleinman, S.J.; Kulebi, B.; Koester, D.; Peçanha, V.; Castanheira, B.G.; Nitta, A.; da Silveira Costa, J.E.; et al. Magnetic white dwarf stars in the Sloan Digital Sky Survey. *Mon. Not. R. Astron. Soc.* **2013**, *429*, 2934–2944.
5. Fontaine, G.; Michaud, G. Diffusion time scales in white dwarfs. *Astrophys. J.* **1979**, *231*, 826–840.
6. Rohrmann, R.D. Hydrogen-model atmospheres for white dwarf stars. *Mon. Not. R. Astron. Soc.* **2001**, *323*, 699–712.
7. Rosato, J.; Kieu, N.; Hannachi, I.; Koubiti, M.; Marandet, Y.; Stamm, R.; Dimitrijević, M.S.; Simić, Z. Stark-Zeeman Line Shape Modeling for Magnetic White Dwarf and Tokamak Edge Plasmas: Common Challenges. *Atoms* **2017**, *5*, 36.
8. Rosato, J.; Marandet, Y.; Capes, H.; Ferri, S.; Mossé, C.; Godbert-Mouret, L.; Koubiti, M.; Stamm, R. Stark broadening of hydrogen lines in low-density magnetized plasmas. *Phys. Rev. E* **2009**, *79*, 46408.
9. Griem, H.R. *Principles of Plasma Spectroscopy*; Cambridge University Press: Cambridge, UK, 1997.
10. Brissaud, A.; Frisch, U. Theory of Stark broadening—II exact line profile with model microfield. *J. Quant. Spectrosc. Radiat. Transf.* **1971**, *11*, 1767–1783.
11. Stehlé, C.; Hutcheon, R. Extensive tabulations of Stark broadened hydrogen line profiles. *Astron. Astrophys. Suppl. Ser.* **1999**, *140*, 93–97.
12. Calisti, A.; Mossé, C.; Ferri, S.; Talin, B.; Rosmej, F.; Bureyeva, L.A.; Lisitsa, V.S. Dynamic Stark broadening as the Dicke narrowing effect. *Phys. Rev. E* **2010**, *81*, 16406.
13. Stambulchik, E.; Maron, Y. Quasicontiguous frequency-fluctuation model for calculation of hydroge. *Phys. Rev. E* **2013**, *87*, 53108.
14. Stamm, R.; Smith, E.W.; Talin, B. Study of hydrogen Stark profiles by means of computer simulation. *Phys. Rev. A* **1984**, *30*, 2039.
15. Stambulchik, E.; Maron, Y. Plasma line broadening and computer simulations: A mini-review. *High Energy Density Phys.* **2010**, *6*, 9–14.
16. Griem, H.R.; Kolb, A.C.; Shen, K.Y. Stark Broadening of Hydrogen Lines in a Plasma. *Phys. Rev.* **1959**, *116*, 4.

17. Voslamber, D. Unified Model for Stark Broadening. *Z. Naturforsch* **1969**, *24*, 1458–1472.

18. Smith, E.W.; Cooper, J.; Vidal, C.R. Unified Classical-Path Treatment of Stark Broadening in Plasmas. *Phys. Rev.* **1969**, *185*, 140.

19. Rosato, J.; Capes, H.; Stamm, R. Influence of correlated collisions on Stark-broadened lines in plasmas. *Phys. Rev. E* **2012**, *86*, 46407.

# 24

# Doppler Broadening of Spectral Line Shapes in Relativistic Plasmas

**Mohammed Tayeb Meftah** [1,2,*], **Hadda Gossa** [1], **Kamel Ahmed Touati** [1,3], **Keltoum Chenini** [1,4] **and Amel Naam** [1,2]

[1]   Laboratoire de Recherche de Physique des Plasmas et Surfaces (LRPPS), UKMO Ouargla 30000, Algerie; hadda.gossa@gmail.com (H.G.); ktouati@yahoo.com (K.A.T.); k1_chenini@yahoo.fr (K.C.); naamnaam10@gmail.com (A.N.)

[2]   Département de Physique, Faculté de Mathématiques et Sciences de la matière, Université Kasdi-Merbah, Ouargla 30000, Algerie

[3]   Lycée professionnel les Alpilles, Rue des Lauriers, 13140 Miramas, France

[4]   Département des Sciences et Technologies, Faculté des Sciences et Technologies, Université de Ghardaia, Ghardaia 47000, Algerie

[*]   Correspondence: mewalid@yahoo.com or meftah.tayeb@univ-ouargla.dz

**Abstract:** In this work, we report some relativistic effects on the spectral line broadening. In particular, we give a new Doppler broadening in extra hot plasmas that takes into account the possible high velocity of the emitters. This suggests the use of an appropriate distribution of the velocities for the emitters. Indeed, the Juttner-Maxwell distribution of the velocities is more adequate for relativistic velocities of the emitters when the latter are in plasma with an extra high temperature. We find an asymmetry in the Doppler line shapes unlike the case of the traditional Doppler effect.

**Keywords:** plasmas; Maxwell; Juttner-Maxwell; relativistic; Doppler effect; asymmetry

## 1. Introduction

The Doppler effect, discovered by physicist and mathematician Christian Doppler in the nineteenth century, is the modification of the frequency of a wave when the emitting source and the receiver are in relative motion. The frequency change also implies that of the period and the wavelength. This effect concerns both mechanical waves and electromagnetic waves. In plasmas, the neutral atoms, molecules or ions moving inside the plasma are similar to the moving antennae. Atoms or ions subjected to the Doppler effect, exhibit the well-known phenomenon: the Doppler broadening of the line profile. The investigation in recent decades of the derivation and illustration of the Doppler effect, especially the generalized relativistic Doppler effect, is still being actively pursued today [1–5]. In the following we will illustrate this method to formulate the classical Doppler effect first, and then the same method is formulated to get the relativistic Doppler effect on the Doppler broadening of the line profile. In our work, we will present a better derivation allowing quick and exact expressions of the classic and relativistic Doppler effect on the broadening of the spectral lines observed in the plasmas. In the formulation of the classical Doppler effect, we used the Maxwell velocity distribution for the emitters, while in the formulation of the relativistic Doppler effect we used the relativistic Juttner-Maxwell distribution. The latter is justified for the case of very high temperatures (in the range $10^5$–$10^8$ K) such as that encountered in fusion plasmas, in astrophysics, in cosmology (primordial Universe) and in unstable Z Pinch experiments [5]. Indeed, the Juttner-Maxwell distribution remains valid for all temperatures since it is more general than the Maxwell distribution.

## 2. Doppler Broadening

### 2.1. Classical Doppler Broadening: Non Relativistic Case

Often the emission (or absorption) of radiation by a particle (atom, ion, etc.) occurs during the movement. By the Doppler effect, the observed frequency in the observer (at rest) frame (see Figure 1) is different from the frequency emitted in the atom frame. The mean particle velocity at thermodynamic equilibrium is related to the temperature of the medium. Hence the broadening of the statistical Doppler effect is related to the distribution of the velocities of the emitter at the temperature T of the medium and the mass $m$ of the emitter.

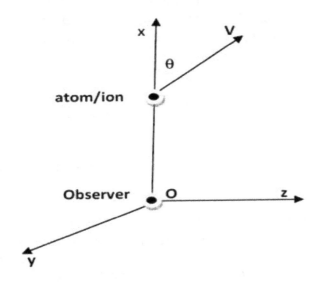

**Figure 1.** The fixed frame where the emitter moves with a velocity $V$ forming an angle $\theta$ with the observation direction Ox.

One can assume a motionless observer, looking at an emitting atom moving with a velocity $V$ in a direction forming an angle $\theta$ with the direction of observation (Ox) (see Figure 1), records a shifted angular frequency $\omega$ with respect to the angular eigenfrequency $\omega_0$ of the emitter assumed to be stationary. This angular frequency is given by

$$\omega(V_x) = \omega_0(1 - \frac{V}{c}\cos\theta) = \omega_0(1 - \frac{V_x}{c}) \tag{1}$$

where $c$ is the velocity of the light in vacuum. The normalized intensity (normalized to one) of the line at the angular frequency $\omega$ is given by the average over the normalized Maxwell distribution (normalized to one).

$$f_{Maxwell}(V_x) = (m/(2\pi k_B T))^{1/2} \exp(-mV_x^2/(2k_B T)) \tag{2}$$

of the Dirac delta distribution as the following

$$I(\omega) = < \delta(\omega - \omega(V_x)) >_{Maxwell} = \sqrt{\frac{m}{2\pi k_B T}} \int\limits_{-\infty}^{+\infty} \exp(-\frac{m}{2k_B T}V_x^2)\delta(\omega - \omega(V_x))dV_x \tag{3}$$

Using the integral representation of the Dirac delta distribution ($u$ is the integration variable whose unit is the second)

$$\delta(\omega - \omega(V_x)) = \frac{1}{2\pi} \int_{-\infty}^{+\infty} \exp(iu(\omega - \omega(V_x)))du \tag{4}$$

We find the normalized intensity (normalized to one) as

$$I(\omega) = \frac{1}{2\pi}\sqrt{\frac{m}{2\pi k_B T}} \int_{-\infty}^{+\infty} \exp(iu\omega)du \int_{-\infty}^{+\infty} \exp(-\frac{m}{2k_B T}V_x^2 - iu\omega(V_x))dV_x \tag{5}$$

$$= \frac{1}{2\pi}\sqrt{\frac{2\pi mc^2}{k_B T\omega_0^2}} \exp\left(-\frac{mc^2}{2k_B T}(\hat{\omega}-1)^2\right) \tag{6}$$

where $\hat{\omega} = \frac{\omega}{\omega_0}$, $k_B$ is the Boltzmann constant and $m$ is the emitter mass. This is the formula of the intensity of the line in the non relativistic case. It is symmetric (Gaussian) around the central angular frequency $\omega_0$. We note that the integrals in Formula (6) are strongly convergent because we deal with purely Gaussian integrals. The full width at the half maximum (FWHM) is given by the well known formula (in angular frequency unit)

$$\Delta\omega_{Doppler} = \omega_0\sqrt{\left(\frac{8k_B T\ln(2)}{mc^2}\right)} = 7.1574\times 10^{-7}\times\omega_0\sqrt{\frac{T}{M}} \tag{7}$$

where $M$ is the mass of the emitter in atomic mass unit whereas $T$ is the temperature in Kelvin.

*2.2. Relativistic Doppler Broadening*

When an observer at a rest, recording the emitted radiation from a moving atom (or ion) with relativistic velocity $V$, they find that the angular frequency of this radiation is equal to [6]:

$$\omega(\beta) = \omega_0\gamma(1+\beta\cos\theta) \tag{8}$$

where

$$\beta = V/c, \tag{9}$$

$$\gamma = 1/\sqrt{(1-\beta^2)} \tag{10}$$

and $\omega_0$ is the angular eigenfrequency and $\theta$ is the angle between the velocity of the emitter and the observation direction $(Ox)$ (see Figure 1). By using the normalized Juttner-Maxwell distribution (normalized to one) [7]

$$W_{J-M}(\beta)d\beta = \lambda\frac{\gamma^5\beta^2 d\beta}{K_2(\lambda)}\exp(-\lambda\gamma) \tag{11}$$

where

$$\lambda = mc^2/(k_B T) \tag{12}$$

and $K_2(X)$ is the modified Bessel function of order two, we obtained the normalized relativistic intensity (normalized to one) of the line profile

$$I(\omega) = <\delta(\omega-\omega(\beta))>_{Juttner-Maxwell} = \int W_{J-M}(\beta)d\beta\cdot\delta(\omega-\omega(\beta)) \tag{13}$$

$$= \frac{1}{4\pi}\int_{-\infty}^{+\infty}du\int\int\int W_{J-M}(\beta)d\beta\exp(iu(\omega-\omega(\beta)))\sin\theta d\theta d\phi \tag{14}$$

Here, we have introduced the integral over the spherical angles that makes the emitter velocity with the fixed frame axis (see Figure 1). We have replaced the Dirac delta distribution by its integral representation by integrating over the variable $u$. Replacing the Juttner-Maxwell distribution $W_{J-M}(\beta)$

given by (11) and $\omega(\beta)$ given by (8) in Formula (14), we reach a more suitable expression of the relativistic intensity of the line profile

$$I(\omega) = \frac{\lambda}{4\pi \cdot K_2(\lambda)} \int\limits_{-\infty}^{+\infty} \exp(iu\omega)du \int_0^1 \gamma^5\beta^2 \exp(-\lambda\gamma)\exp(-iu\omega_0\gamma)d\beta \int_0^\pi \exp(iu\omega_0\gamma\beta\cos\theta)\sin\theta d\theta \tag{15}$$

or after integration on $\theta$ between zero and $\pi$;

$$I(\omega) = \frac{\lambda}{2\cdot K_2(\lambda)} \int_0^1 \gamma^4\beta \exp(-\lambda\gamma)d\beta \int\limits_{-\infty}^{+\infty} \frac{\exp(iu\omega-iu\omega_0\gamma+iu\omega_0\gamma\beta)}{iu\omega_0}du$$
$$- \frac{\lambda}{2\cdot K_2(\lambda)} \int_0^1 \gamma^4\beta \exp(-\lambda\gamma)d\beta \int\limits_{-\infty}^{+\infty} \frac{\exp(iu\omega-iu\omega_0\gamma-iu\omega_0\gamma\beta)}{iu\omega_0}du \tag{16}$$

Finally, the integration over $u$, allows us to get the relativistic intensity of the line profile

$$I(\widehat\omega) = \frac{\lambda}{2\cdot K_2(\lambda)} \int_1^\infty \gamma d\gamma \exp(-\lambda\gamma)\left(S(\widehat\omega-\gamma+\sqrt{\gamma^2-1})-S(\widehat\omega-\gamma-\sqrt{\gamma^2-1})\right) \tag{17}$$

where $S(t) = +1$ if $t > 0$ and $S(t) = -1$ if $t < 0$ and $\widehat\omega = \frac{\omega}{\omega_0}$ is the reduced angular frequency. We can manage the formula to be more suitable for the numerical treatment:

$$I(\widehat\omega) = \frac{\lambda}{2\cdot K_2(\lambda)} \exp(-\lambda) \int_1^\infty \gamma d\gamma \exp(-\lambda(\gamma-1))\left(S(\widehat\omega-\gamma+\sqrt{\gamma^2-1})-S(\widehat\omega-\gamma-\sqrt{\gamma^2-1})\right) \tag{18}$$

if we put $\lambda(\gamma-1) = y$, ($\lambda$ is given by formula (12)) we find

$$I(\widehat\omega) = \frac{\exp(-\lambda)}{2\cdot K_2(\lambda)} \int_0^\infty (\tfrac{y}{\lambda}+1)dy \exp(-y)*$$
$$\left(S(\widehat\omega-\tfrac{y}{\lambda}-1+\sqrt{(\tfrac{y}{\lambda})^2+2\tfrac{y}{\lambda}})-S(\widehat\omega-\tfrac{y}{\lambda}-1-\sqrt{(\tfrac{y}{\lambda})^2+2\tfrac{y}{\lambda}})\right) \tag{19}$$

We note that the integral in the last formula is convergent because we deal with the integral in distribution sense [8].

Unlike the classical Doppler effect, the relativistic one has a property: an asymmetric broadening as it is shown clearly in the following table corresponding to the temperatures in the range $10^5$–$10^9$ K. We remark also that in the relativistic case, as in the classical case, the central frequency is unchanged (see Figure 2). The maximum value of the intensity is at $\widehat\omega = 1$ both for classical and relativistic case but the maximum value of the relativistic case is smaller than the maximum of the classical case (the asymmetry at $\widehat\omega = 1$) is negative, see Figure 3.

We mention that in the Table 1, we have denoted by $\widehat\omega_{L,R}$ the value of the reduced angular frequency at the left and the right of the peak of the line (centred at $\widehat\omega = 1$). $\widehat\omega_L$ and $\widehat\omega_R$ are chosen to be symmetrical with respect the centre of the line at $\widehat\omega = 1$ and giving intensities very close to the half of the maximum of the intensity. Strictly speaking, we have considered $\widehat\omega_R = 1 + g$ and $\widehat\omega_R = 1 - g$ with g = 0.00045 for $Fe^{+25}$, 0.00075 for $W^{+73}$, 0.00085 for $Fm^{+99}$ (produced in nuclear reactions) and 0.00095 for $Cn^{+111}$ (it is synthesized in laboratories for use in nuclear reactions). As we see in this table, the value of the intensity at the right $I(\widehat\omega_R)$ is greater than the intensity at the left $I(\widehat\omega_L)$. This remark shows clearly that the line profile has an asymmetry as defined by [9] (see the definition at the last line in the above table). Another feature in this study is that we have not specified the line profile, because we have used the reduced angular frequency $\widehat\omega = \frac{\omega}{\omega_0}$: for each specific line (specific transition), we must multiply $\widehat\omega$ (the x-axis) by the corresponding angular eigenfrequency $\omega_0$ to obtain the intensity $I(\omega)$. If we define the asymmetry as [10]

$$Asym = I(\omega, relativistic) - I(\omega, classical) \tag{20}$$

which is the difference between the normalized relativistic intensity given by (18) and the normalized classical intensity given by (6), we obtain the following figure representing the asymmetry for $Cn^{+111}$ at $1.9 \times 10^9$ K.

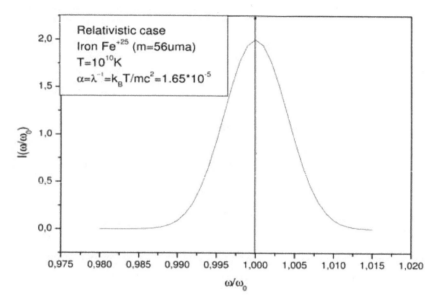

**Figure 2.** Relativistic intensity as defined by Formula (19) for Iron at $T = 10^{10}$ K.

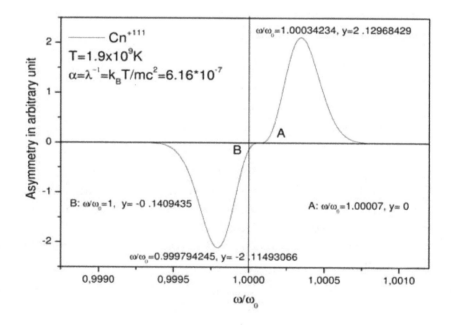

**Figure 3.** Asymmetry as defined by Formula (20) for $Cn^{+111}$ at $T = 1.9 \times 10^9$ K.

**Table 1.** Asymmetry percentages for different hydrogen-like ions.

|  | $T = 10^8$ K, $Fe^{+25}$ | $T = 8.5 \times 10^8$ K, $W^{+73}$ | $T = 1.5 \times 10^9$ K, $Fm^{+99}$ | $T = 1.9 \times 10^9$ K, $Cn^{+111}$ |
|---|---|---|---|---|
| $\hat{\omega}_L$ | 0.99955 | 0.99925 | 0.99915 | 0.99905 |
| $\hat{\omega}_R$ | 1.00045 | 1.00075 | 1.00085 | 1.00095 |
| $I(\hat{\omega}_L)$ | 1.0802 | 1.0303 | 1.0195 | 0.95715 |
| $I(\hat{\omega}_R)$ | 1.0807 | 1.0313 | 1.0206 | 0.95849 |
| $\frac{I(\hat{\omega}_R)-I(\hat{\omega}_L)}{I(\hat{\omega}_R)+I(\hat{\omega}_L)} * 100$ | 0.030 | 0.051 | 0.056 | 0.060 |

In Figure 3, we see clearly that, in the left of $\hat{\omega} = \omega/\omega_0 = 1.00007$, the intensity of the relativistic profile is lower than of the classical profile, whereas it is higher in the right of $\hat{\omega} = \omega/\omega_0 = 1.00007$. It can be seen clearly in this figure that the asymmetry is a function of ($\hat{\omega} = \omega/\omega_0$) and that means that for any line, the asymmetry is as indicated in this figure. To obtain the asymmetry, for a specific line centred at $w_0$, we must multiply $\hat{\omega}$ by $w_0$. The same remark holds for Figure 4 for the hydrogen-like Iron, but with a more pronounced asymmetry since the maximum of the asymmetry is equal to 2.40 for the iron ($Fe^{+25}$) whereas for the Copernicium ($Cn^{+111}$) is equal to 2.12.

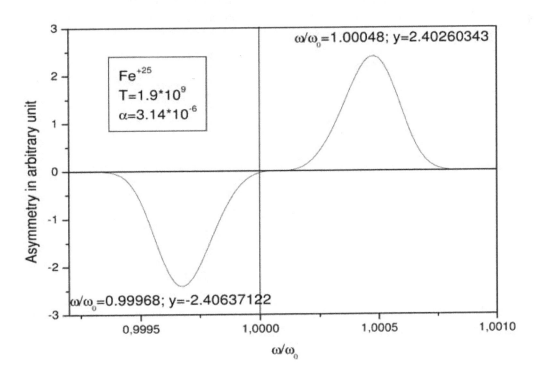

**Figure 4.** Asymmetry as defined by Formula (20) for $Fe^{+25}$ at $T = 1.9 \times 10^9$ K.

## 3. Conclusions

In this work, we report some relativistic effects on the spectral line broadening. In particular, we obtained a new expression for the Doppler broadening that takes into account the possible high velocity of the emitters. This suggests the use of an appropriate distribution of the velocities for emitters. We find, an asymmetry in the Doppler broadening unlike the well known classical Gaussian Doppler broadening.

**Acknowledgments:** We wish to acknowledge the support of LRPPS laboratory and its director Pr: Fethi Khelfaoui, by offering us the encouragement, and some technical materials for developing this work.

**Author Contributions:** All authors M.T. Meftah, H. Gossa, K.A. Touati, K. Chenini and A. Naam were participated equivalently to this theoretical work.

## References

1. Huang, Y.-S. Formulation of relativistic Doppler-broadened absorption line profile. *Europhys. Lett.* **2012**, *97*, 23001.
2. Huang, Y.-S. Formulation of the classical and the relativistic Doppler effect by a systematic method. *Can. J. Phys.* **2004**, *82*, 957–964.
3. Kichenassamy, S.; Krikorian, R.; Nikogosian, A. The relativistic Doppler broadening of the line absorption profile. *J. Quant. Spectrosc. Radiat. Transf.* **1982**, *27*, 653–655.
4. McKinley, J.M. Relativistic transformations of light power. *Am. J. Phys.* **1979**, *47*, 602–605.

5.   Haines, M.G.; LePell, P.D.; Coverdale, C.A.; Jones, B.; Deeney, C.; Apruzese, J.P. Ion Viscous Heating in a Magnetohydrodynamically Unstable Z Pinch at Over $2 \times 10^9$ Kelvin. *Phys. Rev. Lett.* **2006**, *96*, 075003.

6.   Jackson, J.D. Special Theory of Relativity. In *Classical Electrodynamics*, 3rd ed.; John Wiley: New York, NY, USA, 1962; Chapter 11, pp. 360–364.

7.   Zenitani, S. Loading relativistic Maxwell distributions in particle simulations. *Phys. Plamas* **2015**, *22*, 042116.

8.   Stehlé, C.; Gilles, D.; Demura, A.V. Asymmetry of Stark profiles: The microfield point of view. *Eur. Phys. J. D* **2000**, *12*, 355–367.

9.   Schwartz, L. *Théorie des Distributions*; Editions Hernmann: Paris, France, 1967.

10.  Huang, Y.-S.; Chiue, J.-H.; Huang, Y.-C.; Hsiung, T.-C. Relativistic formulation for the Doppler-broadened line profile. *Phys. Rev. A* **2010**, *82*, 010102(R).

# 25

# Stark Widths of Na IV Spectral Lines

**Milan S. Dimitrijević** [1,2,†,‡], **Zoran Simić** [1,‡], **Aleksandar Valjarević** [3,‡] and **Cristina Yubero** [4,*,‡]

1    Astronomical Observatory, Volgina 7, 11060 Belgrade 38, Serbia; mdimitrijevic@aob.rs (M.S.D.);
     zsimic@aob.rs (Z.S.)

2    Laboratoire d'Etudes du Rayonnement et de la Matière en Astrophysique et Atmosphères—LERMA,
     Observatoire de Paris, 5 Place Jules Janssen, 92195 Meudon CEDEX, France

3    Department of Geography, Faculty of Natural Sciences and Mathematics, University of Kosovska Mitrovica,
     Ive Lole Ribara 29, 38220 Kosovska Mitrovica, Serbia; aleksandar.valjarevic@pr.ac.rs

4    Edificio A. Einstein (C-2), Universidad de Córdoba, Campus de Rabanales, 14071 Córdoba, Spain

*    Correspondence: mdimitrijevic@aob.rs

†    Current address: Astronomical Observatory, Volgina 7, 11060 Belgrade 38, Serbia.

‡    These authors contributed equally to this work.

Academic Editor: Robert C. Forrey

**Abstract:** Sodium is a very important element for the research and analysis of astrophysical, laboratory, and technological plasmas, but neither theoretical nor experimental data on Stark broadening of Na IV spectral lines are present in the literature. Using the modified semiempirical method of Dimitrijević and Konjević, here Stark widths have been calculated for nine Na IV transitions. Na IV belongs to the oxygen isoelectronic sequence, and we have calculated Stark widths belonging to singlets, triplets, and quintuplets, as well as with different parent terms. This is used to discuss similarities within one transition array with different multiplicities and parent terms.

**Keywords:** stark broadening; line profiles; atomic data; Na IV

## 1. Introduction

In spite of the fact that sodium is important for the research and analysis of various astrophysical (e.g., [1]), laboratory (e.g., [2]), and technological (e.g., [3]) plasmas, Stark broadening data for its different ionization stages are very scarce or even missing. Experimental results exist only for Na I [4]. There are also a number of theoretical results for neutral sodium. For example, semiclassical perturbation (SCP) Stark broadening parameters from [5] have been used for non-LTE calculations for neutral Na in late-type stars [6]. For Na II, there are the semiclassical results of Jones et al. [7] published also by Griem [8], while for Na VI, Na IX, and Na X, there are theoretical SCP results in the STARK-B database [9,10]. There are also quantum mechanical calculations of Stark width for one Na VII, and one Na VIII spectral line [11], and that is all. There are no results for Na IV and Na V.

The creation of a set of Stark broadening parameters for as large as possible number of spectral lines is useful for a number of problems such as stellar spectra analysis and synthesis, opacity calculations, and the modelling of stellar atmospheres. In order to contribute to this aim , we will calculate full widths at half intensity maximum (FWHM), due to collisions with surrounding electrons, for nine Na IV spectral lines using the modified semiempirical method (MSE) [12–14], since a set of atomic data needed for an adequate application of the more sophisticated semiclassical perturbation method [15–17] does not exist. The obtained results will be used for a consideration of the regular behavior of Stark widths within a transition array.

## 2. The Modified Semiempirical Method

For calculation of Na IV Stark widths, we will use the modified semiempirical method (MSE) [12–14]. Since it is analysed in more details in Dimitrijević et al. [18], only basic information will be summarized in this work. The expression for the electron impact full width (FHWM) of an isolated ion line is [12]:

$$w_{MSE} = N \frac{4\pi}{3c} \frac{\hbar^2}{m^2} \left(\frac{2m}{\pi kT}\right)^{1/2} \frac{\lambda^2}{\sqrt{3}} \times \left\{ \sum_{\ell_i \pm 1} \sum_{L_{i'} J_{i'}} \vec{\mathfrak{R}}^2_{\ell_i,\ell_i \pm 1} \widetilde{g}(x_{\ell_i,\ell_i \pm 1}) + \right.$$

$$\sum_{\ell_f \pm 1} \sum_{L_{f'} J_{f'}} \vec{\mathfrak{R}}^2_{\ell_f,\ell_f \pm 1} \widetilde{g}(x_{\ell_f,\ell_f \pm 1}) + \left(\frac{3n_i^*}{2Z}\right)^2 \frac{1}{9}(n_i^{*2} + 3\ell_i^2 + 3\ell_i + 11) g(x_{n_i,n_i+1}) + \qquad (1)$$

$$\left. \left(\frac{3n_f^*}{2Z}\right)^2 \frac{1}{9}(n_f^{*2} + 3\ell_f^2 + 3\ell_f + 11) g(x_{n_f,n_f+1}) \right\},$$

where $i$ and $f$ are for initial and final levels, $\vec{\mathfrak{R}}^2_{\ell_k,\ell_{k'}}$, $k = i, f$ is the square of the matrix element, $x_{l_k,l_{k'}}$ and $x_{n_k,n_k+1}$ are the ratios of electron kinetic energy and the energy difference between the corresponding energy levels (see [18]), $N$ and $T$ are electron density and temperature, and $g(x)$ [19] and $\widetilde{g}(x)$ [12] are the corresponding Gaunt factors.

## 3. Results and Discussion

The required atomic energy levels for Na IV have been taken from Sansonetti [20], and the corresponding matrix elements have been calculated within the Coulomb approximation [21]. The results for the Stark widths of nine transitions from the Na IV spectrum—obtained by using the modified semiempirical method [12] (see also the review of innovations and applications in [14])—are given in Table 1 for perturber density of $10^{17}$ cm$^{-3}$ and temperatures from 10,000 K up to 160,000 K. These data are the first for the Stark broadening of Na IV, so there are no other experimental or theoretical data to compare with the present values.

**Table 1.** FWHM (full width at half intensity maximum; Å) for Na IV spectral lines, for a perturber density of $10^{17}$ cm$^{-3}$ and temperatures from 10,000 to 160,000 K. Calculated wavelength ($\lambda$) of the transitions (in Å) is also given.

| Element | Transition | $\lambda$ (Å) | T (K) = 10,000 | 20,000 | 40,000 | 80,000 | 160,000 |
|---|---|---|---|---|---|---|---|
| | | | **FWHM (Å)** | | | | |
| Na IV | $(^2$D$)3$s$^1$D$^o$– $(^2$D$)3$p$^1$D | 1534.5 | 0.259E−01 | 0.183E−01 | 0.130E−01 | 0.916E−02 | 0.705E−02 |
| Na IV | $(^2$D$)3$s$^1$D$^o$–$(^2$D$)3$p$^1$F | 2156.4 | 0.466E−01 | 0.329E−01 | 0.233E−01 | 0.165E−01 | 0.126E−01 |
| Na IV | $(^2$P$)3$s$^1$P$^o$–$(^2$P$)3$p$^1$P | 1998.6 | 0.408E−01 | 0.289E−01 | 0.204E−01 | 0.144E−01 | 0.109E−01 |
| Na IV | $(^2$P$)3$s$^1$P$^o$–$(^2$P$)3$p$^1$D | 1791.6 | 0.337E−01 | 0.238E−01 | 0.169E−01 | 0.119E−01 | 0.903E−02 |
| Na IV | $(^4$S$^o)3$s$^3$S$^o$–$(^4$S$^o)3$p$^3$P | 2019.3 | 0.425E−01 | 0.300E−01 | 0.212E−01 | 0.150E−01 | 0.114E−01 |
| Na IV | $(^2$D$)3$s$^3$D$^o$–$(^2$D$)3$p$^3$D | 2111.7 | 0.425E−01 | 0.300E−01 | 0.212E−01 | 0.150E−01 | 0.114E−01 |
| Na IV | $(^2$D$)3$s$^3$D$^o$–$(^2$D$)3$p$^3$F | 1971.2 | 0.376E−01 | 0.266E−01 | 0.188E−01 | 0.133E−01 | 0.101E−01 |
| Na IV | $(^2$P$)3$s$^3$P$^o$–$(^2$P$)3$p$^3$D | 1985.9 | 0.382E−01 | 0.270E−01 | 0.191E−01 | 0.135E−01 | 0.102E−01 |
| Na IV | $(^4$S$^o)3$s$^5$S$^o$–$(^4$S$^o)3$p$^5$P | 1963.6 | 0.367E−01 | 0.259E−01 | 0.183E−01 | 0.130E−01 | 0.976E−02 |

All calculated data belong to one transition array: 3s–3p, but they have not only different multiplicities, but also different parent terms. This is an interesting set of data which can be used to test how similar Stark broadening parameters are. Wiese and Konjević [22] concluded in their article where regularities and similarities in plasma broadened spectral line widths were considered that "line widths within transition arrays normally stay within a range of about ±40%", while for supermultiplets, Stark line widths are within about 30%. In order to see how it is in the case of the

Na IV 3s–3p transition array considered in this work, first we should convert results from Å units to angular frequency units. For this purpose, the following formula can be used:

$$W(\text{Å}) = \frac{\lambda^2}{2\pi c} W(\text{s}^{-1})$$ 
(2)

where $c$ is the speed of light. The results in angular frequency units—convenient for the consideration of regularities—are shown in Table 2. We can see that the lowest value is 13.5% smaller than the largest one at 10,000 K and for 15.4% at 160,000 K. Within the data for the considered transition array, there are three cases when two multiplets belong to the same supermultiplet. Namely, two singlets with the parent term ($^2$D), two singlets with the parent term ($^2$P), and two triplets with the parent term ($^2$D). For them, the lower value is 8.7%, 3%, and 1.6% smaller from the larger one at 10,000 K and 9.4%, 2.6%, and 1.2% at 160,000 K. This is in excellent accordance with conclusions of Wiese and Konjević [22], and we can conclude that missing values for other lines from the considered transition array are similar, and if needed, averaged data from Table 2 (in angular frequency units) can be used for the estimation of their Stark widths.

**Table 2.** Same as in Table 1, but FWHM is in angular frequency units.

| Element | Transition | $\lambda$ (Å) | T (K) = 10,000 | 20,000 | 40,000 | 80,000 | 160,000 |
|---------|------------|---------------|----------------|--------|--------|--------|---------|
| | | | FWHM (s$^{-1}$) $\times$ 10$^{-11}$ | | | | |
| Na IV | ($^2$D)3s$^1$D$^\circ$– ($^2$D)3p$^1$D | 1534.5 | 2.07 | 1.47 | 1.04 | 0.733 | 0.564 |
| Na IV | ($^2$D)3s$^1$D$^\circ$–($^2$D)3p$^1$F | 2156.4 | 1.89 | 1.33 | 0.943 | 0.667 | 0.511 |
| Na IV | ($^2$P)3s$^1$P$^\circ$–($^2$P)3p$^1$P | 1998.6 | 1.92 | 1.36 | 0.962 | 0.681 | 0.516 |
| Na IV | ($^2$P)3s$^1$P$^\circ$–($^2$P)3p$^1$D | 1791.6 | 1.98 | 1.40 | 0.989 | 0.700 | 0.530 |
| Na IV | ($^4$S$^\circ$)3s$^3$S$^\circ$–($^4$S$^\circ$)3p$^3$P | 2019.3 | 1.96 | 1.39 | 0.981 | 0.694 | 0.526 |
| Na IV | ($^2$D)3s$^3$D$^\circ$–($^2$D)3p$^3$D | 2111.7 | 1.79 | 1.27 | 0.897 | 0.634 | 0.483 |
| Na IV | ($^2$D)3s$^3$D$^\circ$–($^2$D)3p$^3$F | 1971.2 | 1.82 | 1.29 | 0.911 | 0.644 | 0.489 |
| Na IV | ($^2$P)3s$^3$P$^\circ$–($^2$P)3p$^3$D | 1985.9 | 1.83 | 1.29 | 0.913 | 0.645 | 0.487 |
| Na IV | ($^4$S$^\circ$)3s$^5$S$^\circ$–($^4$S$^\circ$)3p$^5$P | 1963.6 | 1.79 | 1.27 | 0.895 | 0.633 | 0.477 |

Our values of Stark widths for Na IV spectral lines will be added to the STARK-B database [9,10], which is also a part of the set of databases included in the Virtual Atomic and Molecular Data Center (VAMDC) [23,24], enabling a much better and easier search and mining of atomic and molecular data.

We hope that the new Stark broadening data obtained in this work will be of interest for a number of problems in astrophysics and for the diagnostics of laboratory plasmas, as well as for the investigation of laser-produced inertial fusion plasma and for plasma in industry (e.g., laser welding or plasma-based light sources).

**Acknowledgments:** The support of the Ministry of Education, Science and Technological Development of the Republic of Serbia through project 176002 is gratefully acknowledged.

**Author Contributions:** These authors contributed equally to this work.

## References

1. Černiauskas, A.; Kučinskas, A.; Klevas, J.; Prakapavičius, D.; Korotin, S.; Bonifacio, P.; Ludwig, H.-G.; Caffau, E.; Steffen, M. Abundances of Na, Mg, and K in the atmospheres of red giant branch stars of Galactic globular cluster 47 Tucanae. *Astron. Astrophys.* **2017**, *604*, 35.

2. Lesage, A. Experimental Stark widths and shifts for spectral lines of neutral and ionized atoms A critical review of selected data for the period 2001–2007. *New Astron.* **2009**, *52*, 471–535.

3.    Vilela, J.A.; Perin, A.J. Pulsed Voltage-Mode Supply for High-Pressure Sodium Lamps. *IEEE Trans. Plasma Sci.* **2015**, *43*, 3242–3248.

4.    Konjević, N.; Lesage, A.; Fuhr, J.R.; Wiese, W.L. Experimental Stark widths and shifts for spectral lines of neutral and ionized atoms. *J. Phys. Chem. Ref. Data* **2002**, *31*, 819–927.

5.    Dimitrijević, M.S.; Sahal-Bréchot, S. Stark broadening of Na (I) lines with principal quantum number of the upper state between 6 and 10. *J. Quant. Spectrosc. Radiat. Transf.* **1990**, *44*, 421–431.

6.    Lind, K.; Asplund, M.; Barklem, P.S.; Belyaev, A.K. Non-LTE calculations for neutral Na in late- type stars using improved atomic data. *Astron. Astrophys.* **2011**, *528*, A103.

7.    Jones, W.W.; Benett, S.M.; Griem, H.R. *Calculated Electron Impact Broadening Parameters for Isolated Spectral Lines from the Singly Charged Ions: Lithium through Calcium*; University of Maryland Technical Report; University of Maryland: College Park, MD, USA, 1971; Volumes 71–128, pp. 1–53.

8.    Griem, H.R. *Spectral Line Broadening by Plasmas*; Academic Press, Inc.: New York, NY, USA, 1974.

9.    Sahal-Bréchot, S.; Dimitrijević, M.S.; Moreau, N. STARK-B Database, Observatory of Paris, LERMA and Astronomical Observatory of Belgrade, 2017. Available online: http://stark-b.obspm.fr (accessed on 1 August 2017).

10.   Sahal-Bréchot, S.; Dimitrijević, M.S.; Moreau, N.; Ben Nessib, N. The STARK-B database VAMDC node: A repository for spectral line broadening and shifts due to collisions with charged particles. *Phys. Scr.* **2015**, *50*, 054008.

11.   Elabidi, H.; Sahal-Bréchot, S. Checking the dependence on the upper level ionization potential of electron impact widths using quantum calculations. *Eur. Phys. J. D* **2011**, *61*, 285–290.

12.   Dimitrijević, M.S.; Konjević, N. Stark widths of doubly- and triply-ionized atom lines. *J. Quant. Spectrosc. Radiat. Transf.* **1980**, *24*, 451–459.

13.   Dimitrijević, M.S.; Kršljanin, V. Electron-impact shifts of ion lines—Modified semiempirical approach. *Astron. Astrophys.* **1986**, *165*, 269–274.

14.   Dimitrijević, M.S.; Popović, L.Č. Modified Semiempirical Method. *J. Appl. Spectrosc.* **2001**, *68*, 893–901.

15.   Sahal-Bréchot, S. Impact theory of the broadening and shift of spectral lines due to electrons and ions in a plasma. *Astron. Astrophys.* **1969**, *1*, 91–123.

16.   Sahal-Bréchot, S. Impact theory of the broadening and shift of spectral lines due to electrons and ions in a plasma (continued). *Astron. Astrophys.* **1969**, *2*, 322–354.

17.   Sahal-Bréchot, S.; Dimitrijević, M.S.; Ben Nessib, N. Widths and Shifts of Isolated Lines of Neutral and Ionized Atoms Perturbed by Collisions With Electrons and Ions: An Outline of the Semiclassical Perturbation (SCP) Method and of the Approximations Used for the Calculations. *Atoms* **2014**, *2*, 225–252.

18.   Dimitrijević, M.S.; Simić, Z.; Stamm, R.; Rosato, J.; Milovanović, N.; Yubero, C. Stark Broadening of Se IV, Sn IV, Sb IV and Te IV Spectral Lines. *Atoms* **2017**, submitted.

19.   Griem, H.R. Semiempirical Formulas for the Electron-Impact Widths and Shifts of Isolated Ion Lines in Plasmas. *Phys. Rev.* **1968**, *165*, 258–266.

20.   Sansonetti, J.E. Wavelengths, Transition Probabilities, and Energy Levels for the Spectra of Sodium (Na I–Na XI). *J. Phys. Chem. Ref. Data* **2008**, *37*, 1659–1763.

21.   Bates, D.R.; Damgaard, A. The Calculation of the Absolute Strengths of Spectral Lines. In *Philosophical Transactions of the Royal Society of London. Series A. Mathematical and Physical Sciences*; The Royal Society Publishing: London, UK, 1949; Volume 242, pp. 101–122.

22.   Wiese, W.L.; Konjević, N. Regularities and similarities in plasma broadened spectral line widths (Stark widths). *J. Quant. Spectrosc. Radiat. Transf.* **1982**, *28*, 185–198.

23.   Dubernet, M.L.; Boudon, V.; Culhane, J.L.; Dimitrijevic, M.S.; Fazliev, A.Z.; Joblin, C.; Kupka, F.; Leto, G.; le Sidaner, P.; Loboda, P.A.; et al. Virtual atomic and molecular data centre. *J. Quant. Spectrosc. Radiat. Transf.* **2010**, *111*, 2151–2159.

24.   Dubernet, M.L.; Antony, B.K.; Ba, Y.A.; Babikov, Y.L.; Bartschat, K.; Boudon, V.; Braams, B.J.; Chunf, H.-K.; Danial, F.; Delahaye, F.; et al. The virtual atomic and molecular data centre (VAMDC) consortium. *J. Phys. B* **2016**, *49*, 074003.

# Permissions

The contributors of this book come from diverse backgrounds, making this book a truly international effort. This book will bring forth new frontiers with its revolutionizing research information and detailed analysis of the nascent developments around the world.

We would like to thank all the contributing authors for lending their expertise to make the book truly unique. They have played a crucial role in the development of this book. Without their invaluable contributions this book wouldn't have been possible. They have made vital efforts to compile up to date information on the varied aspects of this subject to make this book a valuable addition to the collection of many professionals and students.

This book was conceptualized with the vision of imparting up-to-date information and advanced data in this field. To ensure the same, a matchless editorial board was set up. Every individual on the board went through rigorous rounds of assessment to prove their worth. After which they invested a large part of their time researching and compiling the most relevant data for our readers.

The editorial board has been involved in producing this book since its inception. They have spent rigorous hours researching and exploring the diverse topics which have resulted in the successful publishing of this book. They have passed on their knowledge of decades through this book. To expedite this challenging task, the publisher supported the team at every step. A small team of assistant editors was also appointed to further simplify the editing procedure and attain best results for the readers.

Apart from the editorial board, the designing team has also invested a significant amount of their time in understanding the subject and creating the most relevant covers. They scrutinized every image to scout for the most suitable representation of the subject and create an appropriate cover for the book.

The publishing team has been an ardent support to the editorial, designing and production team. Their endless efforts to recruit the best for this project, has resulted in the accomplishment of this book. They are a veteran in the field of academics and their pool of knowledge is as vast as their experience in printing. Their expertise and guidance has proved useful at every step. Their uncompromising quality standards have made this book an exceptional effort. Their encouragement from time to time has been an inspiration for everyone.

The publisher and the editorial board hope that this book will prove to be a valuable piece of knowledge for researchers, students, practitioners and scholars across the globe.

# List of Contributors

**Zlatko Majlinger**
Astronomical Observatory, Volgina 7, 11060 Belgrade, Serbia

**Milan S. Dimitrijević**
Astronomical Observatory, Volgina 7, 11060 Belgrade 38, Serbia
LERMA, Observatoire de Paris, PSL Research University, CNRS, Sorbonne Universités, UPMC (Univ. Pierre & Marie Curie) Paris 06, 5 Place Jules Janssen, 92190 Meudon, France
Laboratoire d'Etudes du Rayonnement et de la Matière en Astrophysique et Atmosphères – LERMA, Observatoire de Paris, 5 Place Jules Janssen, 92195 Meudon CEDEX, France
Astronomical Observatory, Volgina 7, 11060 Belgrad, Serbia
IHIS-Technoexperts, Bezanijska 23, 11080 Zemun, Serbia

**Vladimir A. Srećković**
Institute of Physics, University of Belgrade, Pregrevica 118, Zemun, 11080 Belgrade, Serbia

**Cristóbal Colón, María Isabel de Andrés-García and Andrés Moya**
Department of Applied Physics, E.T.S.I.D. Industrial, Universidad Politécnica de Madrid, Calle Ronda de Valencia 3, 28012 Madrid, Spain

**Lucía Isidoro-García**
Department of Industrial Chemistry and Polymers, E.T.S.I.D. Industrial, Universidad Politécnica de Madrid, Calle Ronda de Valencia 3, 28012 Madrid, Spain

**Sylvie Sahal-Bréchot**
LERMA, Observatoire de Paris, PSL Research University, CNRS, Sorbonne Universités, UPMC (Univ. Pierre & Marie Curie) Paris 06, 5 Place Jules Janssen, 92190 Meudon, France

**Magdalena Christova**
Department of Applied Physics, Technical University–Sofia, 1000 Sofia, Bulgaria

**Eugene Oks**
Physics Department, 380 Duncan Drive, Auburn University, Auburn, AL 36849, USA

**Joël Rosato, Ny Kieu, Mohammed Koubiti, Yannick Marandet and Roland Stamm**
Département de physique, Aix-Marseille Université, CNRS, PIIM UMR 7345, 13397 Marseille CEDEX 20, France

**Zoran Simić**
Astronomical Observatory, Volgina 7, 11060 Belgrade 38, Serbia

**Antonio Rodero, Antonio Gamero and Maria del Carmen García**
Grupo de Física de Plasmas: Diagnosis, Modelos y Aplicaciones (FQM-136) Edificio A. Einstein (C-2), Campus de Rabanales, Universidad de Córdoba, 14071 Córdoba, Spain

**Milan S. Dimitrijevic**
Astronomical Observatory, Volgina 7, 11060 Belgrade, Serbia

**Paola Marziani**
Osservatorio Astronomico di Padova, Istituto Nazionale di Astrofisica (INAF), IT 35122 Padova, Italy

**Ascensión del Olmo and Mary Loli Martínez-Aldama**
Instituto de Astrofisíca de Andalucía (IAA-CSIC), E-18008 Granada, Spain

**Deborah Dultzin and Alenka Negrete**
Instituto de Astronomía, Universidad Nacional Autónoma de México (UNAM), México D.F. 04510, Mexico

**Edi Bon and Natasa Bon**
Astronomical Observatory, Volgina 7, 11060 Belgrade 38, Serbia

**Mauro D'Onofrio**
Dipartimento di Fisica & Astronomia "Galileo Galilei", Università di Padova, IT35122 Padova, Italy

**Mutia Meireni and Laurence Godbert-Mouret**
Département de Physique, Aix-Marseille Université, CNRS, PIIM UMR 7345, 13397 Marseille CEDEX 20, France

**Ibtissem Hannachi**
Département de Physique, Aix-Marseille Université, CNRS, PIIM UMR 7345, 13397 Marseille CEDEX 20, France
PRIMALAB, Faculty of Sciences, University of Batna 1, Batna 05000, Algeria

**Ljubinko M. Ignjatović**
Institute of Physics, University of Belgrade, Pregrevica 118, Zemun, 11080 Belgrade, Serbia

**Darko Jevremović**
Astronomical Observatory, Volgina 7, 11060 Belgrade, Serbia

**Veljko Vujčić**
Astronomical Observatory, Volgina 7, 11060 Belgrade, Serbia
Faculty of Organizational Sciences, University of Belgrade, Jove Ilica 154, 11000 Belgrade, Serbia

**Nenad M. Sakan**
Institute of Physics, Belgrade University, Pregrevica 118, 11080 Zemun, Belgrade, Serbia

**Zoran J. Simić**
Astronomical Observatory, Volgina 7, 11060 Belgrade, Serbia

**Aleksandar Valjarević**
Department of Geography, Faculty of Natural Sciences and Mathematics, University of Kosovska Mitrovica, Ive Lole Ribara 29, 38220 Kosovska Mitrovica, Serbia

**Khadra Arif**
Laboratoire de Recherche de Physique des Plasmas et Surfaces, Ouargla 30000, Algérie

**Mohammed Tayeb Meftah**
Département de Physique, Faculté de Mathématiques et Sciences de la matière, Université Kasdi-Merbah, Ouargla 30000, Algérie
Laboratoire de Recherche de Physique des Plasmas et Surfaces (LRPPS), UKMO Ouargla 30000, Algerie

**Said Douis**
Laboratoire de Recherche de Physique des Plasmas et Surfaces, Ouargla 30000, Algérie
Département de Physique, Faculté de Mathématiques et Sciences de la matière, Université Kasdi-Merbah, Ouargla 30000, Algérie

**Giovanni La Mura, Sina Chen, Abhishek Chougule, Stefano Ciroi, Valentina Cracco, Michele Frezzato, Sabrina Mordini and Piero Rafanelli**
Department of Physics and Astronomy, University of Padua, Vicolo dell'Osservatorio 3, 35122 Padova, Italy

**Marco Berton and Enrico Congiu**
Department of Physics and Astronomy, University of Padua, Vicolo dell'Osservatorio 3, 35122 Padova, Italy
Astronomical Observatory of Brera, National Institute of Astrophysics (INAF), Via Bianchi 46, 23807 Merate, Italy

**Nenad Milovanović**
Astronomical Observatory, Volgina 7, 11060 Belgrade 38, Serbia

**Bratislav P. Marinković**
Institute of Physics Belgrade, University of Belgrade, Pregrevica 118, Belgrade 11080, Serbia

**Jan Hendrik Bredehöft**
Institute for Applied and Physical Chemistry, Fachbereich 2 (Biologie/Chemie), Universität Bremen, Leobener Straße 5, Bremen 28359, Germany

**Nigel J. Mason**
Department of Physical Sciences, The Open University, Milton Keynes MK7 6AA, UK

**Nelly Bonifaci**
Laboratoire G2Elab, CNRS and Grenoble University, 25 rue des Martyrs, 38042 Grenoble, France

**Zhiling Li**
Guizhou Institute of Technology, Caiguan Road 1, Guiyang 550003, China

**Paul Génésio**
Département de physique, Aix-Marseille Université, CNRS, PIIM UMR 7345, 13397 Marseille CEDEX 20, France

**Nikolai N. Bezuglov and Andrey N. Klyucharev**
Saint Petersburg State University, St. Petersburg State University, 7/9 Universitetskaya nab., St. Petersburg 199034, Russia

**Dmitry K. E imov**
Saint Petersburg State University, St. Petersburg State University, 7/9 Universitetskaya nab., St. Petersburg 199034, Russia
Instytut Fizyki im. Mariana Smoluchowskiego, Uniwersytet Jagiellónski, 30-348 Kraków, Poland

**Martins Bruvelis**
Laser Centre, University of Latvia, LV-1002 Riga, Latvia

**Vladimir A. Srećković**
Institute of Physics, University of Belgrad, 11001 Belgrad, Serbia

**Yurij N. Gnedin**
Pulkovo Observatory, Russian Academy of Sciences, St. Petersburg 196140, Russia

**Francesco Fuso**
Dipartimento di Fisica Enrico Fermi and CNISM, Università di Pisa, I-56127 Pisa, Italy

**Rihab Aloui**
Laboratoire Dynamique Moléculaire et Matériaux Photoniques, GRePAA, École Nationale Supérieure d'ingénieurs de Tunis, University of Tunis, 1008 Tunis, Tunisia

**Haykel Elabidi**
LDMMP, GRePAA, Faculté des Sciences de Bizerte, University of Carthage, 7021 Bizerte, Tunisia

**Jelena Kovačević-Dojcinović and Luka Č. Popović**
Astronomical Observatory, Volgina 7, 11060 Belgrade 38, Serbia

**Hadda Gossa**
Laboratoire de Recherche de Physique des Plasmas et Surfaces (LRPPS), UKMO Ouargla 30000, Algerie

**Amel Naam**
Laboratoire de Recherche de Physique des Plasmas et Surfaces (LRPPS), UKMO Ouargla 30000, Algerie

Département de Physique, Faculté de Mathématiques et Sciences de la matière, Université Kasdi-Merbah, Ouargla 30000, Algerie

**Kamel Ahmed Touati**
Laboratoire de Recherche de Physique des Plasmas et Surfaces (LRPPS), UKMO Ouargla 30000, Algerie
Lycée professionnel les Alpilles, Rue des Lauriers, 13140 Miramas, France

**Keltoum Chenini**
Laboratoire de Recherche de Physique des Plasmas et Surfaces (LRPPS), UKMO Ouargla 30000, Algerie
Département des Sciences et Technologies, Faculté des Sciences et Technologies, Université de Ghardaia, Ghardaia 47000, Algerie

**Cristina Yubero**
Grupo de Física de Plasmas: Diagnosis, Modelos y Aplicaciones (FQM-136) Edificio A. Einstein (C-2), Campus de Rabanales, Universidad de Córdoba, 14071 Córdoba, Spain

# Index